Yurij A. Drozd   Vladimir V. Kirichenko

Finite Dimensional Algebras

With an Appendix by Vlastimil Dlab

Translated from the Russian
by Vlastimil Dlab

Springer-Verlag
Berlin Heidelberg New York
London Paris Tokyo
Hong Kong Barcelona
Budapest

Yurij A. Drozd
Vladimir V. Kirichenko
Faculty of Mechanics and Mathematics
Kiev Taras Shevchenko University
252 617 Kiev, Ukraina

Translator:
Vlastimil Dlab
Department of Mathematics and Statistics
Carleton University
Ottawa K1S 5B6, Canada

Title of the Russian edition: *Konechnomernye algebry.*
Publisher of Kiev State University, Vishcha Shkola, Kiev 1980

Mathematics Subject Classification (1991): 16P10, 16G10, 16K20

ISBN-13:978-3-642-76246-8 e-ISBN-13:978-3-642-76244-4
DOI: 10.1007/978-3-642-76244-4

Cataloging-in-Publication Data applied for

Softcover reprint of the hardcover 1st edition 1994

Typesetting: Camera-ready copy from the translator
SPIN 10028296 41/3140 - 5 4 3 2 1 0 - Printed on acid-free paper

Preface to the English Edition

This English edition has an additional chapter "Elements of Homological Algebra". Homological methods appear to be effective in many problems in the theory of algebras; we hope their inclusion makes this book more complete and self-contained as a textbook. We have also taken this occasion to correct several inaccuracies and errors in the original Russian edition.

We should like to express our gratitude to V. Dlab who has not only meticulously translated the text, but has also contributed by writing an Appendix devoted to a new important class of algebras, viz. quasi-hereditary algebras. Finally, we are indebted to the publishers, Springer-Verlag, for enabling this book to reach such a wide audience in the world of mathematical community.

Kiev, February 1993

Yu.A. Drozd
V.V. Kirichenko

Preface

The theory of finite dimensional algebras is one of the oldest branches of modern algebra. Its origin is linked to the work of Hamilton who discovered the famous algebra of quaternions, and Cayley who developed matrix theory. Later finite dimensional algebras were studied by a large number of mathematicians including B. Peirce, C.S. Peirce, Clifford, Weierstrass, Dedekind, Jordan and Frobenius. At the end of the last century T. Molien and E. Cartan described the semisimple algebras over the complex and real fields and paved the first steps towards the study of non-semisimple algebras.

A new period in the development of the theory of finite dimensional algebras opened with the work of Wedderburn; the fundamental results of this theory belong to him: a description of the structure of semisimple algebras over an arbitrary field; the theorem on lifting the quotient algebra by its radical; the theorem on commutativity of finite division rings, etc. Most of his results were extended to rings with minimal condition (artinian rings) and the theory of semisimple algebras found its present form in the work of algebraists from the German school headed by E. Noether, E. Artin and R. Brauer. Moreover, their work exposed the fundamental role of the concept of a module (or representation).

Further development of the theory advanced basically in two directions. The first led to establishing the theory of infinite dimensional algebras (and rings without chain conditions); those results are reflected in the monograph "Structure of rings" by N. Jacobson. The second direction – the study of the structure of non-semisimple algebras – met considerable difficulties, most of which were not overcome till now. Therefore papers which single out and describe "natural" classes of algebras occupy here an important place. This direction originates in the investigations of Köthe, Asano and Nakayama of principal ideal algebras and their generalizations.

Throughout its development, the theory of finite dimensional algebras was closely related to various branches of mathematics, acquiring from them new ideas and methods, and in turn exerting influence on their development. In the initial period, the most profound connections were to linear algebra, the theory of groups and their representations and Galois theory. Recently, in particular in connection with the study of non-semisimple algebras, an important role is being played by methods of homological algebra, category theory and algebraic geometry.

The present book is intended to be a modern textbook on the theory of finite dimensional algebras. The basic tools of investigation are methods of the theory of modules (representations), which, in our opinion, allow a very simple and clear approach both to classical and new results. Naturally, we cannot pursue all directions equally. The principal goal of this book is structure theory, i. e. investigation of the structure of algebras. In particular, the general theory of representations of algebras, which is presently undergoing a remarkable resurgence, is almost not touched upon. Undoubtedly, specialists will notice the absence of some other areas recently developed. Nevertheless, we hope that the present book will enable the reader to learn the basic results of the classical theory of algebras and to acquire sufficient background to follow and be able to get familiar with contemporary investigations.

A large portion of the book is based on the standard university course in abstract and linear algebra and is fully accessible to students of the second and third year. In particular, we do not assume knowledge of any preliminary information on the theory of rings and modules (moreover, the word "ring" is almost absent in the book). The chapters devoted to group representations and Galois theory require, of course, familiarity with elements of group theory (for instance, to the extent of A.I. Kostrikin's textbook "An introduction to algebra"). At the end of each chapter, we provide exercises of varied complexity which contain examples instrumental for understanding the material as well as fragments of theories which are not reflected in the main text. We strongly recommend that readers (and in particular, beginners) work through most of the exercises on their first reading. The most difficult ones are accompanied by rather explicit hints.

The content of the book is divided into three parts. The first part consists of Chapters 1–3; here the basic concepts of the theory of algebras are discussed, and the classical theory of semisimple algebras and radicals is explained. The second part, Chapters 4–6, can be called the "subtle theory of semisimple algebras". Here, using the technique of tensor products and bimodules, the theory of central simple algebras, elements of Galois field theory, the concept of the Brauer group and the theory of separable algebras are presented. Finally, the third part, Chapters 8–10, is devoted to more recent results: to the Morita theorem on equivalence of module categories, to the theory of quasi-Frobenius, uniserial, hereditary and serial algebras. Some of the results of these latter chapters until now have been available only in journal articles. A somewhat special place is occupied by Chapter 7; in it are developed, based on the results from semisimple algebras, the theory of group representations up to the integral theorems and the Burnside theorem on solvability of a group of order $p^a q^b$.

Naturally, we have not tried to formulate and prove theorems in their utmost generality. Besides, we have used the fact that we deal only with the finite dimensional case whenever, in our view, it simplified our presentation. An experienced reader will certainly note that many results hold, for example, for arbitrary artinian rings. Very often such generalizations follow almost automatically, and when this is not the case, it would be necessary to introduce sufficiently complex new concepts, which, in our opinion, could make reading the book substantially more difficult.

We are not presenting a complete list of references on finite dimensional algebras because, even when restricted to topics covered in the book, it would probably be comparable in length to the entire work. We point out only several textbooks and monographs in which the reader can get acquainted with other aspects of the theory of rings and algebras [1,2,4–9]. The questions of arithmetic of semisimple algebras are dealt with in the book [10], or in the classical textbook of Deuring [3].

We follow generally used notation. In particular, the symbols $\mathbb{Q}$, $\mathbb{R}$, $\mathbb{C}$ denote, respectively, the fields of rational, real and complex numbers. Numbering of statements is done in the book by sections. For instance, "Theorem 4.6.5" denotes the fifth theorem in Section 6 of Chapter 4.

Contents

1. Introduction

1.1 Basic Concepts. Examples

An *algebra* over a field K, or a K-algebra, is a vector space A over the field K together with a bilinear associative multiplication. In other words, to any two elements a and b from the space A, taken in a definite order, there corresponds a uniquely defined element from A which is usually called their product and denoted by ab, whereby the following axioms are satisfied:

$$1)\ a(b+c) = ab + ac\,;$$
$$2)\ (b+c)a = ba + ca\,;$$
$$3)\ (\alpha a)b = a(\alpha b) = \alpha(ab)\,;$$
$$4)\ (ab)c = a(bc)\,,$$

where a, b, c are arbitrary elements from A and α an arbitrary element (scalar) of the field K.

An algebra A is said to be *finite dimensional* or *infinite dimensional* according to whether the space A is finite dimensional or infinite dimensional. We shall consider mainly finite dimensional algebras, although in some chapters we shall deal with infinite dimensional ones.

The dimension of the vector space A is called the *dimension of the algebra* A and is denoted by $[A : K]$.

It follows from the bilinearity of the multiplication that, given a basis $\{a_1, a_2, \dots, a_n\}$ of the space A, the multiplication is uniquely determined by the products of the basis vectors $b_{ij} = a_i a_j$. Indeed, if $a = \sum_{i=1}^{n} \alpha_i a_i$ and $b = \sum_{j=1}^{n} \beta_j a_j$, then

$$ab = \Big(\sum_{i=1}^{n} \alpha_i a_i\Big)\Big(\sum_{j=1}^{n} \beta_j a_j\Big) = \sum_{i,j=1}^{n} \alpha_i \beta_j (a_i a_j) = \sum_{i,j=1}^{n} \alpha_i \beta_j b_{ij}\,.$$

Now, decompose the vectors b_{ij} with respect to the basis: $b_{ij} = \sum_{k=1}^{n} \gamma_{ij}^{k} a_k$. We see that the structure of the algebra over the space A with a fixed basis is uniquely given by a choice of n^3 elements γ_{ij}^{k} $(i, j, k = 1, 2, \dots, n)$ of the field K. These elements are called the *structure constants* of the algebra A.

Of course, the vectors b_{ij} (and thus the structure constants γ_{ij}^k) cannot be chosen arbitrarily; although the bilinearity of the multiplication (i. e. the validity of the axioms 1)–3)) is guaranteed by the defining formula

$$\left(\sum_{i=1}^{n} \alpha_i a_i\right)\left(\sum_{j=1}^{n} \beta_j a_j\right) = \sum_{i,j,k=1}^{n} \alpha_i \beta_j \gamma_{ij}^k a_k ,$$

the associativity is, in general, not satisfied. Indeed,

$$(a_i a_j)a_k = \sum_{\ell=1}^{n} \gamma_{ij}^{\ell} a_\ell a_k = \sum_{\ell,m=1}^{n} \gamma_{ij}^{\ell} \gamma_{\ell k}^{m} a_m ,$$

$$a_i(a_j a_k) = a_i \sum_{\ell=1}^{n} \gamma_{jk}^{\ell} a_\ell = \sum_{\ell,m=1}^{n} \gamma_{jk}^{\ell} \gamma_{i\ell}^{m} a_m ,$$

and thus it follows that, for arbitrary i, j, k, m,

$$\sum_{\ell=1}^{n} \gamma_{ij}^{\ell} \gamma_{\ell k}^{m} = \sum_{\ell=1}^{n} \gamma_{jk}^{\ell} \gamma_{il}^{m} . \tag{1.1.1}$$

Conversely, the relations (1.1.1) imply that multiplication is associative for the basis vectors, and one can therefore easily verify that multiplication is, indeed, associative.

Assume that $\{\tilde{a}_1, \tilde{a}_2, \ldots, \tilde{a}_n\}$ is another basis of the space A, related with the original basis by the transformation matrix $S = (s_{ij})$. Then

$$\tilde{a}_i \tilde{a}_j = \left(\sum_{\ell=1}^{n} s_{i\ell} a_\ell\right)\left(\sum_{r=1}^{n} s_{jr} a_r\right) = \sum_{\ell,r=1}^{n} s_{i\ell} s_{jr} (a_\ell a_r) =$$
$$= \sum_{\ell,r,m=1}^{n} s_{i\ell} s_{jr} \gamma_{\ell r}^{m} a_m = \sum_{k,\ell,r,m=1}^{n} s_{i\ell} s_{jr} \gamma_{\ell r}^{m} s'_{mk} \tilde{a}_k ,$$

where s'_{mk} are the entries of the inverse matrix S^{-1}. Consequently the structure constants $\tilde{\gamma}_{ij}^k$ corresponding to the new basis have the form

$$\tilde{\gamma}_{ij}^k = \sum_{\ell,r,m=1}^{n} s_{i\ell} s_{jk} s'_{mk} \gamma_{\ell r}^{m} ,$$

i. e. the elements γ_{ij}^k can be considered as the coordinates of a three valent tensor (twice covariant and once contravariant).

An element e of an algebra A is called the *identity* of the algebra if

$$ae = ea = a \text{ for an arbitrary element } a \in A.$$

In what follows we shall always assume that A has the identity. Observe that the identity e is unique: if e' is another identity, then $e = ee' = e'$.

The existence of the identity is a usual and non-essential restriction. If A is an algebra without the identity, then it is always possible to "adjoin" it by considering the algebra $\tilde{A}$ consisting of the pairs (a, α), where $a \in A$, $\alpha \in K$ with the componentwise addition and scalar multiplication, and the multiplication defined by

$$(a, \alpha)(b, \beta) = (ab + \alpha b + a\beta, \alpha\beta) .$$

It is easy to verify that $\tilde{A}$ is an algebra and that the element $(0, 1)$ is its identity. All the properties of the algebras A and $\tilde{A}$ are essentially the same; we shall illustrate this fact in the exercises.

Examples. 1. The set of all square matrices of order n with entries from a field K forms an algebra with respect to the ordinary operations on the matrices. It is a finite dimensional algebra of dimension n^2 which will be denoted by $M_n(K)$.

2. The polynomials in one variable over a field K form an infinite dimensional algebra $K[x]$.

3. If V is a vector space over the field K, then the linear transformations of the space V form also an algebra $E(V)$. This algebra is finite dimensional if and only if V is finite dimensional.

4. Consider the four-dimensional vector space over the field $\mathbb{R}$ of the real numbers, with the basis $\{e, i, j, k\}$. Define the multiplication by means of the following table:

	e	i	j	k
e	e	i	j	k
i	i	$-e$	k	$-j$
j	j	$-k$	$-e$	i
k	k	j	$-i$	$-e$

(The product ab is written in the row denoted by a and in the column denoted by b.)

It is easy to verify that one obtains in this way an algebra with identity e over the field $\mathbb{R}$. This algebra is called the *quaternion algebra* $\mathbb{H}$. Historically, it is one of the first examples of an algebra.

5. Every extension L of a field K, i. e. a field containing K as a subfield, can be considered as an algebra over K. If this algebra is finite dimensional then the extension is called finite; otherwise, it is called infinite.

6. Let G be a group. Consider the elements of this group as basis elements of a vector space, i. e. consider the set KG of all formal sums of the form $\sum_{g \in G} \alpha_g g$, where α_g are elements of the field K which are, except for a finite number, all equal to zero. The group multiplication (products of the basis elements) defines the algebra structure over the space KG. This algebra is called

the *group algebra* of the group G over the field K and plays a fundamental role in the theory of representations of groups.

7. Consider the n-dimensional vector space of all n-tuples $(\alpha_1, \alpha_2, \ldots, \alpha_n)$, $\alpha_i \in K$, with coordinatewise addition and scalar multiplication. By defining the multiplication coordinatewise

$$(\alpha_1, \alpha_2, \ldots, \alpha_n)(\beta_1, \beta_2, \ldots, \beta_n) = (\alpha_1\beta_1, \alpha_2\beta_2, \ldots, \alpha_n\beta_n),$$

we obtain an algebra over the field K which will be denoted by K^n.

8. Let $A_1, A_2, \ldots, A_n$ be algebras over the field K. Consider their Cartesian product A, i. e. the set of all sequences $(a_1, a_2, \ldots, a_n)$, $a_i \in A_i$, and define the operations coordinatewise:

$$\begin{aligned}(a_1, a_2, \ldots, a_n) + (b_1, b_2, \ldots, b_n) &= (a_1 + b_1, a_2 + b_2, \ldots, a_n + b_n),\\ \alpha(a_1, a_2, \ldots, a_n) &= (\alpha a_1, \alpha a_2, \ldots, \alpha a_n),\\ (a_1, a_2, \ldots, a_n)(b_1, b_2, \ldots, b_n) &= (a_1b_1, a_2b_2, \ldots, a_nb_n).\end{aligned}$$

Clearly, in this way A becomes an algebra over K which is called the *direct product* of the algebras $A_1, A_2, \ldots, A_n$ and is denoted by $A_1 \times A_2 \times \ldots \times A_n$, or $\prod_{i=1}^{n} A_i$. The algebras $A_1, A_2, \ldots, A_n$ are said to be *direct factors* of the algebra A. Of course, the preceding example is a particular case of the present example, if $A_1 = A_2 = \ldots = A_n = K$.

An algebra is called *commutative* if the multiplication is commutative, i. e. if $ab = ba$ for all $a, b \in A$. The algebras of the Examples 2, 5 and 7 are commutative. The algebra of Example 6 is commutative if the group G is commutative. The algebra of Example 8 is commutative if all the direct factors $A_1, A_2, \ldots, A_n$ are commutative. The remaining algebras of the above examples are non-commutative.

A subset B of an algebra A is said to be a *subalgebra* if B itself is an algebra with respect to the operations in A, and has the same identity. In other words, B has to be a subspace of A such that $e \in B$ and if $a, b \in B$, then $ab \in B$.

Examples. 1. The set of triangular matrices, i. e. all matrices (a_{ij}) such that $a_{ij} = 0$ for $j < i$, form a subalgebra of the algebra $M_n(K)$ of all matrices. This algebra will be denoted by $T_n(K)$.

2. The diagonal matrices also form a subalgebra of $M_n(K)$; it will be denoted by $D_n(K)$.

3. The set of all matrices of the form

$$\begin{pmatrix} \alpha_1 & \alpha_2 & \alpha_3 & \ldots & \alpha_{n-1} & \alpha_n \\ 0 & \alpha_1 & \alpha_2 & \ldots & \alpha_{n-2} & \alpha_{n-1} \\ 0 & 0 & \alpha_1 & \ldots & \alpha_{n-3} & \alpha_{n-2} \\ \ldots & \ldots & \ldots & \ldots & \ldots & \ldots \\ 0 & 0 & 0 & \ldots & \alpha_1 & \alpha_2 \\ 0 & 0 & 0 & \ldots & 0 & \alpha_1 \end{pmatrix}$$

form a subalgebra of $M_n(K)$ of dimension n. This algebra will be called the *Jordan algebra* and denoted by $J_n(K)$.

4. If H is a subgroup of G, then KH is a subalgebra of KG.

5. The set of all elements c of an algebra A which commute with all elements of the algebra, i. e. such that $ca = ac$ for all $a \in A$, form, evidently, a subalgebra of A; it is called the *center* of the algebra A and is denoted by $C(A)$.

6. Consider, in an algebra A, the set of all scalar multiples of the identity, i. e. of all elements of the form αe with $\alpha \in K$. Since $(\alpha e)(\beta e) = \alpha\beta e$, this set forms a subalgebra denoted by Ke.

The fundamental goal of any theory is a classification of the objects under investigation. We are going to classify the algebras of small dimensions. If $[A : K] = 1$, then $A = Ke$ and the structure of A is fully determined; consequently, the first interesting case is when $[A : K] = 2$.

Choose a basis in a two-dimensional algebra A, taking e as the first basis element: $\{e, a\}$, $a \notin Ke$. Then the multiplication is uniquely determined by the product $aa = a^2$; clearly, associativity is automatically satisfied. Moreover, such an algebra is necessarily commutative. Let $a^2 = pa + qe$, where p and q are some (fixed) elements of the field K. Consider the polynomial $g(x) = x^2 - px - q$. The element a is a "root" of this polynomial. It turns out that the structure of A is essentially determined by the properties of the roots of $g(x)$ in K. There are 3 possible cases.

Case 1. $g(x)$ has in K two distinct roots $x_1 \neq x_2$. Then $p = x_1 + x_2$, $q = -x_1x_2$. Put

$$b = \frac{a - x_1e}{x_2 - x_1}.$$

Since $b \notin Ke$, $\{e, b\}$ is a basis of A; moreover,

$$b^2 = \frac{a^2 - 2x_1a + x_1^2e}{(x_2 - x_1)^2} = \frac{pa + qe - 2x_1a + x_1^2e}{(x_2 - x_1)^2} = \frac{(x_2 - x_1)a - (x_2 - x_1)x_1e}{(x_2 - x_1)^2} = \frac{a - x_1e}{x_2 - x_1} = b.$$

Case 2. $g(x)$ has in K a unique (double) root, i. e. $g(x) = (x - x_1)^2$, where $x_1 \in K$. Putting $b = a - x_1e$, we obtain the basis $\{e, b\}$ for which

$$b^2 = (a - x_1e)^2 = g(a) = 0.$$

Case 3. $g(x)$ has no roots in K, i. e. $g(x)$ is irreducible over the field K. We shall show that A is a field, i. e. that every non-zero element b has an inverse b^{-1} such that $bb^{-1} = e$. The easiest way to show this consists in "destroying irrationality in the denominator". Let $b = \alpha a + \beta e$. Then $g(x) = (\alpha x + \beta)f(x) + r$, where $r \in K$, the remainder of $g(x)$ when divided by $\alpha x + \beta$, is non-zero and

$f(x) = \alpha' x + \beta'$. But then it follows that $g(a) = 0 = (\alpha a + \beta e)(\alpha' a + \beta' e) + re$, i. e. the element $-\dfrac{(\alpha' a + \beta' e)}{r}$ is the inverse of b.

Consequently, we have obtained the following result.

Theorem 1.1.1. *A two-dimensional algebra A over a field K is either a field or it possesses a basis $\{e, b\}$ such that $b^2 = b$ or $b^2 = 0$.*

If the field K is algebraically closed (for example, if K is the field of complex numbers), then Case 3 above (that of the field) is not possible.

1.2 Isomorphisms and Homomorphisms. Division Algebras

In describing the two-dimensional algebras, we have seen that many of them (for instance, all algebras of Cases 1 and 2 above) have a "similar structure" in the sense that, for instance, it is possible to choose bases with identical tables of multiplication. Such algebras possess essentially the same properties and cannot be "internally differentiated", although they may be defined over distinct vector spaces. All such algebras will be identified and considered simply to be different copies of the same algebra. This leads to an important concept of the theory of algebras (and many other mathematical theories), namely to the concept of an isomorphism.

An *isomorphism* from an algebra A to an algebra B is a one-to-one linear map f of the space A onto the space B which preserves multiplication, i. e. such that $f(a_1 a_2) = f(a_1) f(a_2)$ for any elements a_1, a_2 from the algebra A. If there is an isomorphism from the algebra A to the algebra B, then the algebras A and B are called *isomorphic*. This will be denoted by $A \simeq B$ or, if the isomorphism f is to be indicated explicitly, by $f : A \xrightarrow{\sim} B$.

It is obvious that the existence of an isomorphism $f : A \xrightarrow{\sim} B$ is equivalent to the fact that one can choose bases in A and B with identical tables of multiplication. In particular, all two-dimensional algebras over the field K in Cases 1 or 2 of the preceding paragraph are mutually isomorphic.

In the theory of algebras, isomorphic algebras are, as a rule, identified. It is said that the algebras are studied "up to an isomorphism". For instance, up to an isomorphism, there are two two-dimensional algebras over an algebraically closed field. It is not difficult to verify that these algebras are K^2 and $J_2(K)$.

A classical example of an isomorphism, one which is very important for linear algebra, is the isomorphism between the algebra of the linear operators $E(V)$ of an n-dimensional space V and the matrix algebra $M_n(K)$; it is obtained by assigning to an operator its matrix with respect to a fixed basis.

Another example of isomorphic algebras is given by the algebras K^n and $D_n(K)$ (the algebra of diagonal matrices).

Finally, for an arbitrary algebra with identity e, the subalgebra Ke is isomorphic to the basis field. In what follows, we shall always identify the element α of the field K with the element $\alpha e \in A$ (its image in this isomorphism) and consider the field K as a subalgebra of the K-algebra A. In particular, the identity of the algebra A will be often denoted simply by 1.

The concept of a homomorphism plays also an important role in the theory of algebras.

A *homomorphism* from an algebra A to an algebra B is a linear map $f : A \to B$ which preserves multiplication and the identity, i. e. such that $f(a_1 a_2) = f(a_1)f(a_2)$ for any $a_1, a_2 \in A$ and $f(e_A) = e_B$, where e_A is the identity of the algebra A and e_B the identity of the algebra B.

If the homomorphism f is injective, i. e. if $a_1 \neq a_2$ implies $f(a_1) \neq f(a_2)$, then it is called a *monomorphism*. If f is surjective, i. e. for an arbitrary element $b \in B$, there is $a \in A$ such that $b = f(a)$, then it is called an *epimorphism*.

Obviously, if f is at the same time a monomorphism as well as an epimorphism, then it is an isomorphism. In this case (and only in this case) f possesses an inverse map f^{-1} which is an isomorphism of B to A.

Since the homomorphism f is a linear map, it is sufficient, in order that it is a monomorphism, that $f(a) = 0$ implies $a = 0$. Indeed, if this is the case, then it follows from $f(a_1) = f(a_2)$ that $f(a_1 - a_2) = 0$, and thus $a_1 - a_2 = 0$, i. e. $a_1 = a_2$.

As for maps, there is a product (or composition) defined for homomorphisms: if $f : A \to B$ and $g : B \to C$ are homomorphisms of algebras, then the map $gf : A \to C$ defined by $gf(a) = g(f(a))$ can easily be shown to be a homomorphism, as well. The multiplication of homomorphisms is associative: if one of the products $(gf)h$ and $g(fh)$ is defined, then the other is defined and they are equal.

Let us give the following example of a homomorphism. Let a be a fixed element of an algebra A. Consider the map $K[x] \to A$, assigning to each polynomial $f(x) = \alpha_0 x^n + \alpha_1 x^{n-1} + \ldots + \alpha_n$ the element $f(a) = \alpha_0 a^n + \alpha_1 a^{n-1} + \ldots + \alpha_n$. Clearly, it is a homomorphism and its range consists of all possible elements of the form $f(a)$. This image is usually denoted by $K[a]$ and is called a *monogenic subalgebra* generated by the element a. In the particular case that $K[a] = A$, the algebra A is called *monogenic*. The element $f(a)$ is said to be the value of $f(x)$ for $x = a$.

We shall now turn our attention to the study of internal properties of algebras and their elements.

An element a of an algebra A is called a *left (right) divisor of zero* if there is a non-zero element $b \in A$ such that $ab = 0$ ($ba = 0$, respectively).

Similarly, a is called a *left (right) divisor of identity* if there is an element $b \in A$ such that $ab = 1$ (or $ba = 1$, respectively).

Observe that if a is at the same time both left and right divisor of identity, i. e. if there are b, b' such that $ab = b'a = 1$, then $b' = b'(ab) = (b'a)b = b$, and if $ac = 1$, then $b = b(ac) = (ba)c = c$. Thus, b is a uniquely determined element satisfying $ab = 1$, and similarly a uniquely determined element satisfying

$ba = 1$. In this case the element is called *invertible*, and b is called the *inverse* of a and is denoted by $b = a^{-1}$.

Generally speaking, the relationship among the four properties introduced above is rather complex. However, in finite dimensional algebras the matter is very simple.

Theorem 1.2.1. *In a finite dimensional algebra*

1) *every left divisor of zero (identity) is a right divisor of zero (identity), and vice versa;*
2) *every element is either a divisor of zero or a divisor of identity;*
3) *a divisor of zero cannot be a divisor of identity.*

Proof. a) First, let us prove that in every algebra, a left (right) divisor of zero cannot be a right (left) divisor of identity. Indeed, let $ab = 0$ but $b \neq 0$, and at the same time $ca = 1$. Then $0 = c(ab) = (ca)b = b$, a contradiction to the assumption $b \neq 0$.

b) Now, let the algebra A be finite dimensional, and let $a \in A$ be an element which is not a left divisor of zero. Consider the map f of the vector space A into itself, given by the formula $f(x) = ax$. It turns out, in view of the algebra properties, that f is a linear map and that, since a is not a left divisor of zero, $f(x) = 0$ implies $x = 0$. But then, since A is finite dimensional, f is a non-singular map and its image coincides with the entire space A. In particular, $1 = f(b) = ab$ for some $b \in A$, and a is thus a left divisor of identity.

In a similar manner, if a is not a right divisor of zero, then it is a right divisor of identity.

c) We can now complete the proof of the theorem. If an element $a \in A$ is a left divisor of zero, then, in view of a), it cannot be a right divisor of identity. Thus, in view of b), it must be a right divisor of zero. The other assertions of 1) can be proved in a similar way. Furthermore, a) implies 3) and b) implies 2). □

An algebra in which every non-zero element is invertible is called a *division algebra*.

Corollary 1.2.2. *A finite dimensional algebra without divisors of zero is a division algebra.*

Corollary 1.2.3. *A subalgebra of a finite dimensional division algebra is a division algebra. In particular, the center of a finite dimensional division algebra is a field.*

Every element of a finite dimensional K-algebra A is a "root" of some non-zero polynomial $f(x) \in K[x]$ (i. e. $f(a) = 0$). Indeed, otherwise the subalgebra $K[a]$ would be isomorphic to $K[x]$; this is impossible, because the space $K[x]$ is infinite dimensional. The polynomial of the least degree with the leading coefficient 1 whose root is a is called the *minimal polynomial* of the element a and is denoted by $m_a(x)$.

Proposition 1.2.4. *Every polynomial having a as its root is divisible by $m_a(x)$. In particular, the minimal polynomial is uniquely determined.*

Proof. Let $f(a) = 0$. Divide $f(x)$ by $m_a(x)$:

$$f(x) = m_a(x)g(x) + r(x),$$

where $r(x) = 0$ or its degree is less than the degree of $m_a(x)$. But $f(a) = r(a) = 0$, and thus the latter case is impossible: $f(x)$ is divisible by $m_a(x)$. □

Proposition 1.2.5. *If A is a finite dimensional division algebra over a field K, then the minimal polynomial $m_a(x)$ of every element $a \in A$ is irreducible.*

Proof. If $m_a(x) = f(x)g(x)$, where f and g are polynomials of smaller degree, then $0 = m_a(a) = f(a)g(a)$; but, since $f(a) \neq 0$, $g(a) \neq 0$, this is impossible. □

Corollary 1.2.6. *If K is algebraically closed, then the only finite dimensional division algebra over K is K itself.*

Proof. If A is such a division algebra and a is an (arbitrary) element, then $m_a(x)$ is an irreducible polynomial and thus, since A is algebraically closed, it is linear: $m_a(x) = x - \alpha$. Therefore $a = \alpha \in K$. Consequently, $A = K$, completing the proof. □

1.3 Representations and Modules

The definition of an algebra given at the beginning of this chapter is useful and important in that it covers a rather large variety of objects. However, in investigations of the structure and properties of algebras, it is very often essential to have a concrete realization of a given algebra, for instance, as a suitable matrix algebra (or as an algebra of linear operators). Such realizations are studied in the theory of representations which in many ways will be our main tool of investigation in this book.

A *representation* of a K-algebra A is a homomorphism T of A into the algebra $E(V)$ of the linear operators on some K-space V. In other words, to define a representation T is to assign to every element $a \in A$ a linear operator $T(a)$ in such a way that

$$\begin{aligned} T(a+b) &= T(a)+T(b) \\ T(\alpha a) &= \alpha T(a) \\ T(ab) &= T(a)T(b) \\ T(1) &= E \quad \text{(the identity operator)} \end{aligned}$$

for arbitrary $a, b \in A$, $\alpha \in K$. If the space V is finite dimensional, then its dimension is called the *dimension* (or *degree*) of the representation T. Evidently, the image of the representation T, i. e. the set of all operators of the form $T(a)$, forms a subalgebra of $E(V)$. If T is a monomorphism, then this subalgebra is isomorphic to the algebra A. In this case, the representation is said to be *faithful*.

Theorem 1.3.1 (Cayley). *Every algebra admits a faithful representation. In other words, every algebra is isomorphic to a subalgebra of the algebra of linear operators.*

Proof. It follows from the axioms of an algebra that, for an arbitrary element $a \in A$, the map $T(a) : x \mapsto xa$, $x \in A$, is a linear operator on the space A and that $T(a+b) = T(a)+T(b)$, $T(\alpha a) = \alpha T(a)$, $T(ab) = T(a)T(b)$ and $T(1) = E$ (identity operator). Thus, T is a representation of the algebra A. If $a \neq b$, then $1a \neq 1b$. It follows that the operators $T(a)$ and $T(b)$ are distinct and that T is a faithful representation, as required. □

The representation constructed in the proof of Cayley's theorem is called *regular* and is of great importance in the theory of algebras (one may expect this because it provides a relatively simple and standard realization of the given algebra). The dimension of the regular representation equals the dimension of the algebra.

If the representation T is finite dimensional (and in what follows, we shall consider only such representations), then one may choose a basis in the space V and assign to each operator $T(a)$ its matrix $(T(a))$. Obviously, the correspondence $a \mapsto (T(a))$ is a homomorphism of the algebra A to the matrix algebra $M_n(K)$, where n is the dimension of the representation T. Such a homomorphism is called a *matrix representation* of the algebra A. If a new basis is chosen in the space V, then every matrix $(T(a))$ transforms into $C(T(a))C^{-1}$, where C is the matrix of the transformation. The matrix representations related this way are said to be *similar*.

The concept of similarity can be defined also for operator representations: two representations $T : A \to E(V)$ and $S : A \to E(W)$ are called *similar* if there is an isomorphism f of the space V onto the space W such that $T(a) = fS(a)f^{-1}$ for any element $a \in A$. From the above, it follows easily that one can choose bases of V and W in such a way that the matrices of the operators $T(a)$ and $S(a)$ coincide. Therefore, it is reasonable to study the representations up to a similarity, i. e. to identify similar representations.

In what follows, as a rule, we shall not distinguish between the representation and the corresponding matrix representation. Observe that Theorem 1.3.1

(Cayley's theorem) can be also formulated as follows: Every finite dimensional algebra is isomorphic to a subalgebra of a matrix algebra (of course, the fact that the algebra is finite dimensional is necessary in order that the regular representation be finite dimensional).

As a rule, it is convenient not to consider the space V and the homomorphism $T : A \to E(V)$ separately but to view the elements of the algebra as operators on V. This leads to the concept of a module.

A *right module* over a K-algebra A, or a right A-module, is a vector space M over the field K whose elements can be multiplied by the elements of the algebra, i. e. to every pair (m, a), $m \in M$, $a \in A$, there corresponds a uniquely determined element $ma \in M$ such that the following axioms are satisfied:

1) $(m_1 + m_2)a = m_1 a + m_2 a$;
2) $m(a_1 + a_2) = ma_1 + ma_2$;
3) $(\alpha m)a = m(\alpha a) = \alpha(ma)$ where $\alpha \in K$;
4) $m(ab) = (ma)b$;
5) $m1 = m$.

We shall show that, for any representation of the algebra A, we can construct a right module over that algebra, and vice versa: for any right module, we can construct a representation.

Let $T : A \to E(V)$ be a representation of the algebra A. Define the product of the elements of V by the elements of the algebra by putting $va = vT(a)$ for any $v \in V$, $a \in A$. It follows immediately from the definition of a representation that, in this way, V becomes a right A-module. We say that this module corresponds to the representation T.

On the other hand, if M is a right module over A, then it follows from the axioms of a module that, for a fixed $a \in A$, the map $T(a) : m \mapsto ma$ is a linear transformation in the space M. Assigning to every a the operator $T(a)$ (or its matrix with respect to a basis), we obtain a representation of the algebra A corresponding to the module M.

In particular, to a regular representation, there corresponds a *regular module*. Here $M = A$ and ma is the product of the elements m and a in the algebra A.

In what follows, unless stated otherwise, all modules under consideration will be assumed finite dimensional (as vector spaces over K).

We introduce also the concepts of homomorphism and isomorphism for modules.

A *homomorphism* of a right A-module M into a right A-module N is a linear map $f : M \to N$ for which $(ma)f = (mf)a$ for arbitrary elements $m \in M$ and $a \in A$.

If, in addition, f is bijective, then it is called an *isomorphism*, and the modules M and N are called *isomorphic*. In this case, we write $f : M \xrightarrow{\sim} N$, or simply $M \simeq N$. Evidently, if $f : M \xrightarrow{\sim} N$, then $f^{-1} : N \xrightarrow{\sim} M$. Isomorphic modules have the same properties and are identified.

Theorem 1.3.2. *The representations corresponding to isomorphic modules are similar and conversely, the modules corresponding to similar representations are isomorphic.*

Proof. Let T and S be the representations corresponding to the modules M and N, and $f : M \xrightarrow{\sim} N$. Then, for any $m \in M$ and $a \in A$, we have

$$mT(a)f = (ma)f = (mf)a = mfS(a),$$

i.e. $T(a)f = fS(a)$, or $T(a) = fS(a)f^{-1}$.

Conversely, assume that the representations T and S are similar, i.e. that $T(a) = fS(a)f^{-1}$. Then, if M and N are the corresponding modules, f is a one-to-one linear transformation of M onto N satisfying

$$(ma)f = mT(a)f = mfS(a) = (mf)a$$

for any $m \in M$ and $a \in A$, i.e. f is an isomorphism of the modules. □

In this way the concept of a module isomorphism corresponds precisely to the concept of a representation similarity.

In the sequel, it will be convenient to write the homomorphisms of right modules on the left, i.e. to write fa instead of af. Thus, unless stated otherwise, we shall always keep to this system of notation.

As homomorphisms of algebras, homomorphisms of modules can also be multiplied defining the product $gf : M \to L$ of the homomorphisms $f : M \to N$ and $g : N \to L$ by $gf(m) = g(f(m))$ for all elements $m \in M$ (it is easy to verify that gf is again a homomorphism).

However, for homomorphisms of modules we have also other operations: additon and scalar multiplication. If f and g are homomorphisms of a module M into a module N, then $f + g : M \to N$ is defined by $(f + g)(m) = f(m) + g(m)$ and $\alpha f : M \to N$, where $\alpha \in K$, by $(\alpha f)(m) = \alpha f(m)$ for all $m \in M$.

One sees immediately that both $f + g$ and αf defined above are homomorphisms and that the set of all homomorphisms from M to N forms with respect to these operations a vector space over the field K. We shall denote this space by $\mathrm{Hom}_A(M, N)$.

The multiplication of homomorphisms behaves in the usual way: it is associative, i.e. $(gf)h = g(fh)$ whenever these products are defined (obviously, they are defined simultaneously), and bilinear, i.e. $g(f + h) = gf + gh$; $(g + f)h = gh + fh$; $g(\alpha f) = (\alpha g)f = \alpha(gf)$ whenever these expressions have meanings. The proofs are easy and are left to the reader.

If a homomorphism $f : M \to N$ is injective, i.e. if $m_1 \neq m_2$ implies that $f(m_1) \neq f(m_2)$, then it is called a *monomorphism*. If f is surjective, i.e. if every element of N is of the form $f(m)$, then f is called an *epimorphism*. Clearly, if f is both a monomorphism and an epimorphism, then it is an isomorphism. As in the case of algebras, in order that f be a monomorphism, it is sufficient that $f(m) = 0$ implies $m = 0$.

Analogously to the concept of a right module, one can define a *left module* over the algebra A as a vector space L together with a multiplication $a\ell$ ($a \in A$, $\ell \in L$, $a\ell \in L$) satisfying the following axioms:

$$\begin{aligned}
&1)\ a(\ell_1 + \ell_2) = a\ell_1 + a\ell_2\,,\\
&2)\ (a_1 + a_2)\ell = a_1\ell + a_2\ell\,,\\
&3)\ a(\alpha\ell) = (\alpha a)\ell = \alpha(a\ell),\ \ \alpha \in K\,,\\
&4)\ (ab)\ell = a(b\ell)\,,\\
&5)\ 1\ell = \ell\,.
\end{aligned}$$

To left modules, there correspond the *antirepresentations* of the algebra A, i.e. the linear transformations $T : A \to E(V)$ such that $T(ab) = T(b)T(a)$; $T(1) = E$. The concepts of similarity, homomorphism and isomorphism can be defined for the antirepresentations and left modules in a similar way as for representations and right modules. Also the theorems corresponding to Theorems 1.3.1 and 1.3.2 hold. In particular, by considering the algebra A as a left module over itself, we obtain the concept of the *regular left module* and the *regular antirepresentation.*

In what follows, we shall consider, as a rule, just the right modules and shall simply speak about *modules* over an algebra A, or A-modules. The reader may verify easily that all results which will be proved, hold also for left modules. Therefore, whenever convenient, we shall use them without any particular notice.

1.4 Submodules and Factor Modules. Ideals and Quotient Algebras

It is well-known that, in linear algebra, the concept of an invariant subspace of an operator plays a very important role. If we have a representation $T : A \to E(V)$ of an algebra A, then it is natural to consider the subspaces of V which are invariant with respect to all operators of the representation. This leads to the concept of a submodule.

A *submodule* of an A-module M is a subspace $N \subset M$ such that $na \in N$ for all elements $n \in N$ and $a \in A$.

Choose a basis $\{e_1, \ldots, e_k\}$ in the subspace N and complete it to a basis of M: $\{e_1, \ldots, e_k, e_{k+1}, \ldots, e_m\}$. Then, with respect to this basis, the representation T corresponding to the module M has the form

$$T(a) = \begin{pmatrix} T_1(a) & 0 \\ X(a) & T_2(a) \end{pmatrix}. \tag{1.4.1}$$

Such a representation (and any one similar to it) is called *reducible.* Clearly, T_1 is the representation corresponding to the module N.

On the other hand, let a representation be reducible, i.e. have a form (1.4.1), where T_1 is a representation of dimension k. Then the subspace N

spanned by the first k elements of the basis, is invariant with respect to all operators $T(a)$, i.e. it is a submodule of M.

It follows from the properties of operations with matrices partitioned into blocks, that the map $a \mapsto T_2(a)$ is also a representation of the algebra A. The corresponding module can be interpreted as follows.

Let $m \in M$. Consider the set $m + N$ consisting of all elements of the form $m+n$, where n runs through all N. Such sets are called the *congruence classes* of M by N (clearly, the congruence class $m + N$ is a linear variety defined by the subspace N through the vector m). If an element x belongs to the class $m + N$, then we say that x *is congruent to* m *modulo* N and write $x \equiv m \,(\mathrm{mod}\, N)$. We are going to show that two congruent classes either coincide or are disjoint.

Indeed, if $(m_1+N)\cap(m_2+N) \neq \emptyset$, then there are two elements n_1 and n_2 in N such that $m_1+n_1 = m_2+n_2$. From here, $m_1 - m_2 = n_2 - n_1 = n_0 \in N$, and for every element $n \in N$,

$$m_1 + n = m_2 + n_0 + n \in m_2 + N$$

and

$$m_2 + n = m_1 + n - n_0 \in m_1 + N\,,$$

i.e. $m_1 + N = m_2 + N$.

One can see easily that if $x \in m + N$ and $y \in m' + N$, then $x + y \in (m + m') + N$ and also $\alpha x \in \alpha m + N$ and $xa \in ma + N$ for all elements $\alpha \in K$, $a \in A$. Consequently, one can define on the set of the congruence classes an A-module structure, defining

$$\begin{aligned} (m+N)+(m'+N) &= (m+m')+N\,,\\ \alpha(m+N) &= \alpha m + N\,,\\ (m+N)a &= ma+N\,. \end{aligned} \tag{1.4.2}$$

The fact that all axioms are satisfied is clear because the operations with the classes are determined by means of their "representatives", i.e. by the operations in the module M.

The set of congruence classes of M by N together with the module structure defined by (1.4.2) is called the *factor module* of the module M by the submodule N and is denoted by M/N.

Observe that the factor module defines a canonical map $\pi : M \to M/N$ assigning to each element $m \in M$ the class $m + N$. Moreover, the formulae (1.4.2) imply that π is a homomorphism (and obviously an epimorphism). We shall call this epimorphism the *projection of* M *onto the factor module* M/N.

It is trivial to verify that if $\{e_1, \ldots, e_k\}$ is a basis of N and $\{e_{k+1}, \ldots, e_m\}$ its completion to a basis of M, then the classes $\pi(e_{k+1}), \ldots, \pi(e_m)$ form a basis of M/N and the corresponding representation coincides with T_2.

The submodules of the regular module are called the *right ideals* of A. Thus, a right ideal is a space $I \subset A$ such that, if $x \in I$ and $a \in A$, then $xa \in I$. The submodules of the left regular module are called the *left ideals*.

Let us point out that in the term "right ideal" we shall never omit the adjective "right" because the term "ideal" alone is used with quite a different meaning.

Important examples of submodules and factor modules occur in the study of homomorphisms.

Let $f : M_1 \to M_2$ be a homomorphism of A-modules. The set of all elements $m \in M_1$ for which $f(m) = 0$ is its *kernel* $\operatorname{Ker} f$. The *image* $\operatorname{Im} f$ of the homomorphism f is the set of all elements of M_2 of the form $f(m)$.

Theorem 1.4.1 (Homomorphism Theorem). *For any homomorphism $f : M_1 \to M_2$ the kernel and the image are submodules of M_1 and M_2, respectively, and $\operatorname{Im} f \simeq M_1/\operatorname{Ker} f$.*

Proof. If $f(m) = f(m') = 0$, then $f(m+m') = f(m) + f(m') = 0$, $f(\alpha m) = \alpha f(m) = 0$ and $f(ma) = f(m)a = 0$, i. e. $\operatorname{Ker} f = N_1$ is a submodule of M_1. Similarly, since $f(m) + f(m') = f(m+m')$, $\alpha f(m) = f(\alpha m)$ and $f(m)a = f(ma)$, $\operatorname{Im} f = N_2$ is a submodule of M_2.

Let $m + N_1$ be an element of M_1/N_1 and $x \in m + N_1$. Then $x = m + n$, where $f(n) = 0$ which yields $f(x) = f(m)$. Thus, putting $g(m + N_1) = f(m)$, we define a map $g : M_1/N_1 \to N_2$; moreover, from the fact that f is a homomorphism and from the definition of the operations (1.4.2) in a factor module it follows that g is a homomorphism.

Assume that $g(m + N_1) = 0$. Then $f(m) = 0$, i. e. $m \in N_1$, and therefore $m + N_1 = 0 + N_1$ is the zero class of the factor module M_1/N_1, and thus g is a monomorphism. Since every element from N_2 has a form $f(m) = g(m + N_1)$, g is an epimorphism, and hence an isomorphism of $M_1/\operatorname{Ker} f$ onto $\operatorname{Im} f$. □

Although it is very simple, the homomorphism theorem plays an important role in the study of modules. We shall illustrate this with an example.

A module M is said to be *cyclic*, if it contains an element m_0 such that every element of M is of the form $m_0 a$, where $a \in A$. The element m_0 is called a *generator* of the module M.

Corollary 1.4.2. *Every cyclic module is isomorphic to a factor module of the regular module by a suitable right ideal.*

Proof. Let M be a cyclic module with a generator m_0. It follows from the module axioms that the map $f : A \to M$ defined by $f(a) = m_0 a$ is a module homomorphism and that, since m_0 is a generator, $\operatorname{Im} f = M$. But then $M \simeq A/\operatorname{Ker} f$, where $\operatorname{Ker} f$ is a right ideal. □

We shall also often use the following result which refines the homomorphism theorem.

Theorem 1.4.3 (Noether). *Let N be a submodule of M and π the projection of M onto $\bar{M} = M/N$. For any submodule $L \subset M$, set $\pi(L) = \{\pi(x) \mid x \in L\}$ and for any submodule $\bar{L} \subset \bar{M}$, set $\pi^{-1}(\bar{L}) = \{x \in M \mid \pi(x) \in \bar{L}\}$. Then*

1) *$\pi^{-1}(\bar{L})$ is a submodule in M containing N;*

2) *$\pi(\pi^{-1}(\bar{L})) = \bar{L}$ and if $L \supset N$, then $\pi^{-1}(\pi(L)) = L$;*
3) *if $L = \pi^{-1}(\bar{L})$, then $L/N \simeq \bar{L}$ and $M/L \simeq \bar{M}/\bar{L}$.*

In this way, we obtain a bijective correspondence between the submodules of $\bar{M}$ and the submodules of M containing N; moreover, this correspondence preserves the operation of forming factor modules.

Proof. The assertion 1) is trivial. Furthermore, every element $\bar{x} \in \bar{L}$ is of the form $\pi(x)$, where $x \in M$, and also $x \in \pi^{-1}(\bar{L})$ because $\pi(x) = \bar{x} \in \bar{L}$, from where we get the formula $\bar{L} = \pi(\pi^{-1}(\bar{L}))$.

Now, let L be a submodule of M containing N. If we restrict π to L, we obtain a homomorphism $\bar{\pi} : L \to \bar{M}$ whose kernel is N and whose image is $\pi(L) = \bar{L}$. Obviously, $\pi^{-1}(\bar{L}) \supset L$. We show the converse inclusion. If $m \in \pi^{-1}(\bar{L})$, then $\pi(m) \in \bar{L}$ and therefore it has the form $\pi(x)$ with $x \in L$. From $\pi(m) = \pi(x)$ it follows that $\pi(m - x) = 0$, i.e. $m - x = n \in N$. However, $N \subset L$ and thus also $m = x + n \in L$. As a result, $\pi^{-1}(\bar{L}) = L$ and $\bar{L} = \operatorname{Im}\bar{\pi} \simeq L/\operatorname{Ker}\bar{\pi} = L/N$.

Denote by τ the projection of $\bar{M}$ onto $\bar{M}/\bar{L}$ and consider the homomorphism $\tau\pi : M \to \bar{M}/\bar{L}$. Since τ and π are epimorphisms, $\tau\pi$ is an epimorphism, too.

Let us determine $\operatorname{Ker}\tau\pi$. The fact that $\tau\pi(m) = 0$ implies that $\pi(m) \in \bar{L}$, i.e. $m \in \pi^{-1}(\bar{L}) = L$. Hence $\operatorname{Ker}\tau\pi = L$ and, by the homomorphism theorem, $\bar{M}/\bar{L} \simeq M/L$. □

Let us investigate what will happen if L is a submodule of M which does not contain N. As before, we can consider the restriction $\bar{\pi} : L \to M/N$ of the projection $M \to M/N$. Here, $\pi(x) = 0$ means that $x \in N$, and thus $\operatorname{Ker}\bar{\pi} = L \cap N$ and $\bar{L} = \operatorname{Im}\bar{\pi} \simeq L/L \cap N$. But we have seen that $\bar{L} \simeq \pi^{-1}(\bar{L})/N$. At the same time, $m \in \pi^{-1}(\bar{L})$ if and only if $\pi(m) = \pi(x)$ for a certain $x \in L$, i.e. $m = x + n$, where $n \in N$. Therefore, denoting by $L + N$, as usually, the subspace of M consisting of all possible sums $x + n$, we see that $L + N = \pi^{-1}(\bar{L})$ is a submodule of M (this can easily be seen directly) and $(L + N)/N \simeq \bar{L} \simeq L/L \cap N$.

The following theorem has been proved.

Theorem 1.4.4 (Noether). *For any submodules L and N of a module M, $(L + N)/N \simeq L/(L \cap N)$.*

If we wish to illustrate the position of the submodules L, N, $L + N$ and $L \cap N$ in the module M, we obtain a "parallelogram".

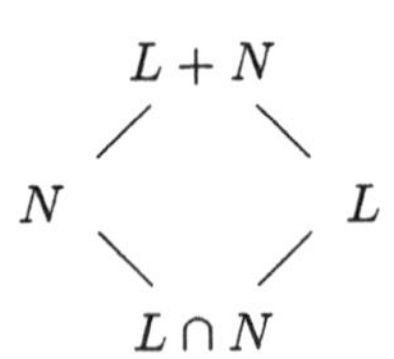

The factor modules $(L+N)/N$ and $L/(L\cap N)$ are the "opposite sides of the parallelogram". Therefore we shall occasionally refer to the second Noether theorem as the "parallelogram rule".

A natural process of translating the above results to the homomorphisms of algebras leads to the concept of an ideal (or, as one often says, of a two-sided ideal).

Let A and B be two algebras over a field K and $\Phi : A \to B$ a K-algebra homomorphism. Its image $\operatorname{Im}\Phi = \{\Phi(a) \mid a \in A\}$ is, of course, a subalgebra of B. But the kernel $\operatorname{Ker}\Phi = \{a \in A \mid \Phi(a) = 0\}$ is not a subalgebra because it does not contain the identity. Since Φ is a linear map, $\operatorname{Ker}\Phi$ is a subspace of A. In addition, if $x \in \operatorname{Ker}\Phi$, then for any $a \in A$, $\Phi(ax) = \Phi(a)\Phi(x) = \Phi(a)0 = 0$, and similarly $\Phi(xa) = 0$, i. e. ax and xa both belong to $\operatorname{Ker}\Phi$. In other words, $\operatorname{Ker}\Phi$ is simultaneously a right and a left ideal.

A subspace which is at the same time a right and a left ideal of an algebra is called an *ideal.*

Given an ideal $I \subset A$, one can construct a new algebra as follows.

Again, consider the set of all congruence classes of A by I. If $a + I$ and $b + I$ are two such classes, then, for any $x \in a + I$ and $y \in b + I$, the element xy lies in the class $ab + I$. Therefore the set of all congruence classes forms an algebra over the field K if we put

$$\begin{aligned}(a+I)+(b+I) &= (a+b)+I\,,\\ \alpha(a+I) &= \alpha a + I\,, \quad \alpha \in K\,,\\ (a+I)(b+I) &= ab+I\,.\end{aligned}$$

This algebra is called the *quotient algebra* of the algebra A by the ideal I and is denoted by A/I. The zero of this algebra is the class $0 + I = I$, and the identity is the class $1 + I$.

The map $\pi : A \to A/I$ for which $\pi(a) = a + I$, is an epimorphism of the algebra A onto the quotient algebra A/I. It is called the *projection* of A onto A/I.

The following results are completely analogous to the corresponding theorems proved for modules. Their proofs, also similar to those given above, are left to the reader as a simple exercise.

Theorem 1.4.5 (Homomorphism Theorem). *For an algebra homomorphism $\Phi : A \to B$, we have $\operatorname{Im}\Phi \simeq A/\operatorname{Ker}\Phi$.*

Corollary 1.4.6. *If $A = K[a]$ is a monogenic algebra, then $A \simeq K[x]/I$, where I is the ideal consisting of all multiples of the polynomial $m_a(x)$.*

Theorem 1.4.7 (Noether). *Let π be a projection of an algebra A onto its quotient algebra $\bar{A} = A/I$. For any subspace $B \subset A$, put $\pi(B) = \{\pi(b) \mid b \in B\}$, and for any subspace $\bar{B} \subset \bar{A}$, put $\pi^{-1}(\bar{B}) = \{b \in A \mid \pi(b) \in \bar{B}\}$. Then*

1) *if B is an ideal (subalgebra) of A, then $\pi(B)$ is an ideal (subalgebra) of $\bar{A}$; if $\bar{B}$ is an ideal (subalgebra) of $\bar{A}$, then $\pi^{-1}(B)$ is an ideal (subalgebra) of A;*
2) *for any ideal (subalgebra) $\bar{B} \subset \bar{A}$, $\pi(\pi^{-1}(\bar{B})) = \bar{B}$; for any ideal (subalgebra) $B \subset A$ containing I, $\pi^{-1}(\pi(B)) = B$;*
3) *if $\bar{B}$ is an ideal (subalgebra) of $\bar{A}$, $B = \pi^{-1}(\bar{B})$, then $A/B \simeq \bar{A}/\bar{B}$ ($B/I \simeq \bar{B}$, respectively).*

Theorem 1.4.8 (Noether). *If I is an ideal and B a subalgebra of an algebra A, then $(B+I)/I \simeq B/(B\cap I)$.*

If N is a submodule of a module M and I is a right ideal of an algebra A, define NI as the set of all sums of the form $\sum n_i a_i$, where $n_i \in N$, $a_i \in I$. It is easy to see that NI is also a submodule of M. It may happen that $NI = 0$, i. e. that $na = 0$ for any $n \in N$ and $a \in I$. We say in this case that I *annihilates* the submodule N. For any module M, one can determine the greatest right ideal of A which annihilates M. Put $\operatorname{Ann} M = \{a \in A \mid ma = 0 \text{ for all } m \in M\}$. Obviously, $\operatorname{Ann} M$ is a right ideal, and also an ideal, of the algebra A. It is called the *annihilator* of the module M.

If an ideal I annihilates a module M, it is possible to view M as a module over the quotient algebra A/I, setting $m(a+I) = ma$ (verify that this definition does not depend on the choice of a representative in the class $a+I$). Clearly, in this case, since I annihilates every submodule and every factor module of the module M, the "structure" of the module M does not depend on whether we consider it as an A- or as an A/I-module.

Conversely, every A/I-module M can be considered as an A-module if we set $ma = m(a+I)$ (here, I automatically annihilates M).

In what follows, we shall always identify the modules over A/I and the modules over A which are annihilated by the ideal I. In particular, we shall often consider the regular A/I-module as an A-module. Obviously, its annihilator is the ideal I.

Moreover, let us remark that, for any element $m \in M$ and any right ideal $I \subset A$, the subspace $mI = \{ma \mid a \in I\}$ is a submodule of M. If $mI = 0$, we say that I *annihilates* m. Among the right ideals which annihilate a given element, there is also a greatest one which is called the annihilator of m: $\operatorname{Ann} m = \{a \in A \mid ma = 0\}$. In difference to the annihilator of a module, $\operatorname{Ann} m$ may not be an ideal (cf. Exercise 9 at the end of this chapter).

1.5 The Jordan-Hölder Theorem

In every non-zero module M, there are evidently always at least two submodules: M itself and the zero subspace (these submodules are said to be *trivial*). If there are no other submodules of M, the module M is called *simple*. The corresponding representation is *irreducible*, i. e. it is not of the form (1.4.1) in any basis.

Let us assume that the module M is not simple. Then it contains a non-trivial submodule N: $N \neq 0$ and $N \neq M$ (i.e. $M/N \neq 0$). We say that the module M is an *extension* of the module $L = M/N$ by the kernel N. In view of the homomorphism theorem, this is equivalent to the existence of an epimorphism $M \to L$ with kernel N.

If the modules L and N are not simple, then we can choose a non-trivial submodule L_1 of L and a non-trivial submodule N_1 of N. By Theorem 1.4.3, $L_1 \simeq N_2/N$, where N_2 is a submodule of M containing N. As a result, we obtain the following chain of submodules of M: $M \supset N_2 \supset N \supset N_1 \supset 0$. If in this chain the module N_1 or any of the factor modules M/N_2, N_2/N, N/N_1 is still non-simple, then it is possible to insert in it, in the same way as above, yet another submodule. Since the space M is finite dimensional, this process cannot be repeated indefinitely. This means that in the end we obtain a chain $M = M_0 \supset M_1 \supset M_2 \supset \ldots \supset M_{s-1} \supset M_s = 0$ such that all factor modules M_i/M_{i+1} are simple. Such a chain is called a *composition series* of the module M. The factor modules M_i/M_{i+1} are called the *factors* of this series and their number s is the *length* of the series.

We could say, that the factors of a composition series are "bricks" from which the module M is constructed by subsequent extensions. Of course, these factors do not determine, in general, the module M but they carry rather significant information on its structure. It is natural to ask the question to what extent are they determined by the module M. The answer is given in the following theorem.

Theorem 1.5.1 (Jordan-Hölder). *If $M = M_0 \supset M_1 \supset \ldots \supset M_s = 0$ and $M = N_0 \supset N_1 \supset \ldots \supset N_t = 0$ are two composition series, then their lengths are equal and there is a bijection between the factors of these series such that the corresponding factors are isomorphic.*

Proof. We shall give a proof by induction on s. If $s = 1$, then the module $M = M_0/M_1$ is simple. Therefore, $t = 1$ and $N_0/N_1 = M = M_0/M_1$. Assume that $s > 1$ and that for any series of length $s - 1$ the theorem holds.

If $M_1 = N_1$, then $M_0/M_1 = N_0/N_1$ and the theorem follows immediately from the induction hypothesis. If $M_1 \neq N_1$, then $M_1 + N_1 \neq M_1$, and since there are no intermediate submodules between M and M_1, $M_1 + N_1 = M$, and by the parallelogram rule

$$N_1/M_1 \cap N_1 \simeq M/M_1\,; \quad M_1/M_1 \cap N_1 \simeq M/N_1\,.$$

Now we construct a composition series in the module $M_1 \cap N_1$:

$$M_1 \cap N_1 = L_2 \supset L_3 \supset \ldots \supset L_k = 0\,.$$

Then $M_1 \supset L_2 \supset L_3 \supset \ldots \supset L_k = 0$ is a composition series of M_1. In comparison to the series $M_1 \supset M_2 \supset \ldots \supset M_s = 0$, it turns out, by the induction hypothesis, that $s = k$ and that there is a bijection such that the factors of the series are pairwise isomorphic.

Furthermore, compare the series $M \supset M_1 \supset L_2 \supset L_3 \supset \ldots \supset L_s = 0$ and $M \supset N_1 \supset L_2 \supset L_3 \supset \ldots \supset L_s = 0$. Their factors coincide from the third position, and the isomorphisms $M/M_1 \simeq N_1/L_2$ and $M/N_1 \simeq M_1/L_2$ were established earlier. It follows that all factors of these series are pairwise isomorphic.

Finally, comparing the series $N_1 \supset L_2 \supset L_3 \supset \ldots \supset L_s = 0$ and $N_1 \supset N_2 \supset \ldots \supset N_t = 0$, we get, by the induction hypothesis, that $s = t$ and that the factors of these series are pairwise isomorphic (in a certain bijection). The proof of the theorem is completed. □

The length of a composition series is called the *length of the module M* and is denoted by $\ell(M)$, and the factors of a composition series are called the *simple factors of the module M*. In view of the Jordan-Hölder theorem, the definition of the length and the simple factors does not depend on the choice of the series.

Let us remark that the order of the simple factors in a composition series is, in general, not determined in a unique way. For instance, for the semisimple modules which will be studied in the next chapter, it is quite arbitrary.

Corollary 1.5.2. *If a module M is an extension of a module L by a kernel N, then $\ell(M) = \ell(L) + \ell(N)$.*

Proof. Consider a composition series of the module $L \simeq M/N$: $L = L_0 \supset L_1 \supset \ldots \supset L_k = 0$ and take the preimages M_i of the modules L_i in M. Then, by Theorem 1.4.3, $M_i/M_{i+1} \simeq L_i/L_{i+1}$ are simple modules. Now, construct a composition series of the module N: $N = N_0 \supset N_1 \supset \ldots \supset N_t = 0$. Then

$$M = M_0 \supset M_1 \supset \ldots \supset M_k = N = N_0 \supset N_1 \supset \ldots \supset N_t = 0$$

is a composition series of the module M of length $k + t$, as required. □

Corollary 1.5.3 (Grassmann's Rule). *If L and N are submodules of M, then*

$$\ell(L + N) + \ell(L \cap N) = \ell(L) + \ell(N).$$

Proof. It is an immediate consequence of Corollary 1.5.2 and the parallelogram rule. □

A submodule N of a module M is said to be *maximal* if $N \neq M$ and there is no submodule L, different from M and N such that $M \supset L \supset N$. This is, obviously, equivalent to M/N being simple. In a composition series, every subsequent submodule is maximal in the immediately preceding one.

1.6 Direct Sums

The knowledge of a submodule and factor module provides rather significant information on the structure of the entire module. This is quite evident if one takes into account the matrix expression (1.4.1) of the respective representation. However, the "gluing block" $X(a)$ which appears in the left lower corner of the matrix is not, in general, determined by the submodule and the factor module and its structure can be rather complex.

The most favourable case is, of course, the one when the additional information carried by the gluing is absent, i. e. when $X(a) = 0$ and the representation has the form

$$T(a) = \begin{pmatrix} T_1(a) & 0 \\ 0 & T_2(a) \end{pmatrix} . \tag{1.6.1}$$

Such representations (and all similar ones) are called *decomposable.*

In the language of modules, the concept of a decomposable representation leads to the definition of the direct sum of modules.

Let $M_1, M_2, \dots, M_n$ be modules over an algebra A. Consider the set M of the n-tuples $(m_1, m_2, \dots, m_n)$, where $m_i \in M_i$, and define the operations coordinatewise:

$$(m_1, m_2, \dots, m_n) + (m'_1, m'_2, \dots, m'_n) = (m_1 + m'_1, m_2 + m'_2, \dots, m_n + m'_n) ,$$
$$\alpha(m_1, m_2, \dots, m_n) = (\alpha m_1, \alpha m_2, \dots, \alpha m_n) , \quad \alpha \in K ,$$
$$(m_1, m_2, \dots, m_n)a = (m_1 a, m_2 a, \dots, m_n a) , \quad a \in A .$$

Obviously, M becomes an A-module which is called the *direct sum* of the modules $M_1, M_2, \dots, M_n$ and is denoted by $M_1 \oplus M_2 \oplus \dots \oplus M_n$, or $\bigoplus_{i=1}^{n} M_i$. As a vector space, M is the direct sum of the spaces $M_1, M_2, \dots, M_n$.

If $n = 2$, and $\{e_1, e_2, \dots, e_k\}$, $\{f_1, f_2, \dots, f_\ell\}$, are bases of M_1 and M_2, respectively, then $\{(e_1, 0), (e_2, 0), \dots, (e_k, 0), (0, f_1), (0, f_2), \dots, (0, f_\ell)\}$ is a basis of $M_1 \oplus M_2$ and the corresponding representation has the form (1.6.1), where $T_1(a)$ and $T_2(a)$ are the representations corresponding to M_1 and M_2.

A module M which is isomorphic to $M_1 \oplus M_2$, where M_1 and M_2 are non-zero modules, is said to be *decomposable.* We shall give an internal characterization of decomposable modules.

Let N and L be two submodules of a module M. Define the map $f : N \oplus L \to M$ by $f(x, y) = x + y$, where $x \in N$, $y \in L$. It is trivial to verify that f is a homomorphism and $\operatorname{Im} f = L + N$. We shall calculate $\operatorname{Ker} f$.

If $(x, y) \in \operatorname{Ker} f$, then $x + y = 0$, i. e. $x = -y$. Therefore, $x \in N \cap L$. Conversely, if $x \in N \cap L$, then the element $(x, -x)$ of the module $N \oplus L$ belongs to $\operatorname{Ker} f$. Thus $\operatorname{Ker} f \simeq N \cap L$, and we get the following proposition.

Proposition 1.6.1. *The homomorphism $f : N \oplus L \to M$ (N, L are submodules of M) defined by the formula $f(x, y) = x + y$ is an isomorphism if and only if $N + L = M$ and $N \cap L = 0$.*

If the above conditions are satisfied, we say that M is *decomposable into a direct sum* of its submodules N and L, and we write $M = N \oplus L$. The submodule L is called in this case the *complement* of the submodule N (and vice versa). Furthermore, we say that the submodule N is a *direct summand* of the module M.

The same submodule N of M can possess different complements (even in the simple case when $A = K$, i.e. when the modules are just vector spaces). However, all complements are mutually isomorphic: it is easy to see that each of them is isomorphic to M/N.

Proposition 1.6.2. *The following conditions are equivalent:*

1) *the submodule N of the module M is a direct summand;*
2) *there is a homomorphism $p : M \to N$ such that $p(x) = x$ for every $x \in N$;*
3) *there is a homomorphism $i : M/N \to M$ such that $i(\bar{y}) \in \bar{y}$ for every class $\bar{y} \in M/N$.*

Proof. 1) $\Rightarrow$ 2). If $M = N \oplus L$, then every element $m \in M$ can be uniquely expressed in the form $m = x + y$, where $x \in N$, $y \in L$. Put $p(m) = x$. One gets immediately that $p(m + m') = p(m) + p(m')$ for every $m' \in M$, and $p(\alpha m) = \alpha p(m)$, $p(ma) = p(m)a$ for every $\alpha \in K$, $a \in A$, i.e. p is a homomorphism. If $x \in N$, then $x = x + 0$ and thus $p(x) = x$.

2) $\Rightarrow$ 3). Define the value of i at the class $\bar{m} = m + N$ by the rule $i(\bar{m}) = m - p(m)$. If m' is another element of the same class, then $m' = m + x$, where $x \in N$, and thus $m' - p(m') = m + x - p(m) - p(x) = m - p(m)$, i.e. our definition does not depend on the choice of the representative in the class $\bar{m}$. It is easy to verify that i is a homomorphism and that, since $p(m) \in N$, $i(\bar{m}) \in m + N = \bar{m}$.

3) $\Rightarrow$ 1). Denote $L = \operatorname{Im} i$. Since $i(\bar{m}) \in \bar{m}$, where $\bar{m} = m + N$, $m - i(\bar{m}) \in N$ and the expression $m = (m - i(\bar{m})) + i(\bar{m})$ shows that $N + L = M$. If $x \in N \cap L$, then $x = i(y)$, where $y = x + N = N$, i.e. $y = 0$ in M/N, and therefore $x = 0$. Consequently, $N \cap L = 0$ and $M = N \oplus L$. □

The homomorphism p is often called a *projector* onto the submodule N. Like complements, projectors are not determined uniquely.

The direct sum of several modules also allows an internal formulation.

Theorem 1.6.3. *Let $M_1, M_2, \ldots, M_k$ be submodules of a module M and let $f : M_1 \oplus M_2 \oplus \ldots \oplus M_k \to M$ be the homomorphism defined by the formula $f(m_1, m_2, \ldots, m_k) = m_1 + m_2 + \ldots + m_k$. Then the following conditions are equivalent:*

1) *f is an isomorphism;*
2) *$M_1 + M_2 + \ldots + M_k = M$ and $M_i \cap \big(\sum_{j \neq i} M_j\big) = 0$ for any i;*
3) *$M_1 + M_2 + \ldots + M_k = M$ and $M_i \cap \big(\sum_{j < i} M_j\big) = 0$ for any $i > 1$.*

Proof. 1) $\Rightarrow$ 2). The fact that f is an epimorphism yields immediately that $M = M_1 + M_2 + \ldots + M_k$. Furthermore, if $x \in M_i \cap (\sum_{j\neq i} M_j)$, then $x = \sum_{j\neq i} m_j$, where $m_j \in M_j$. If we put $m_i = -x$, then $f(m_1, m_2, \ldots, m_k) = 0$, and since f is a monomorphism, we get that $m_1 = m_2 = \ldots = m_k = 0$ and $x = 0$.

2) $\Rightarrow$ 3). Trivial.

3) $\Rightarrow$ 1). From the condition that $M_1 + M_2 + \ldots + M_k = M$, we get that f is an epimorphism. Moreover, if $f(m_1, m_2, \ldots, m_k) = 0$ and i is the last position for which $m_i \neq 0$, then $m_i = -\sum_{j<i} m_j \in M_i \cap (\sum_{j<i} M_j)$, a contradiction. Therefore, $m_1 = m_2 = \ldots = m_k = 0$ and f is a monomorphism. □

If any of the equivalent conditions of Theorem 1.6.3 holds, then we say that M *decomposes into a direct sum* of the submodules $M_1, M_2, \ldots, M_k$ and we write $M = M_1 \oplus M_2 \oplus \ldots \oplus M_k$.

The external and internal definitions of the direct sum are equivalent: If $M = M_1 \oplus M_2 \oplus \ldots \oplus M_k$ is an external direct sum, then the set of the elements $(0, \ldots, 0, m_i, 0, \ldots, 0)$ (all the coordinates but the ith one are 0) forms a submodule M_i' in M and one can see easily that $M_i' \simeq M_i$ and $M = M_1' \oplus M_2' \oplus \ldots \oplus M_k'$ (as the internal direct sum).

Obviously, every (finite dimensional) module can be decomposed into a direct sum of indecomposable modules. We shall see in Chapter 3 that such a decomposition is unique (up to isomorphism and a permutation of the summands). Therefore, if we know all indecomposable modules over an algebra A, then we can describe all A-modules. However, in many cases the description of the indecomposable modules is a very difficult problem which is unaccessible by presently known methods.

1.7 Endomorphisms. The Peirce Decomposition

In this section we shall prove some fundamental theorems, establishing a connection between the theory of representations and the structure theory of algebras. These results will play the main role in the following chapters of the book.

Let us recall that in Sect. 1.2 we have defined operations over the homomorphisms of modules and showed that, for two given A-modules M and N, the set $\mathrm{Hom}_A(M, N)$ can be considered as a vector space over the field K.

A particularly important case is when $M = N$. The homomorphisms of $\mathrm{Hom}_A(M, M)$ can always be multiplied. Hence the space $\mathrm{Hom}_A(M, M)$ is also a K-algebra. This algebra is called the *algebra of endomorphisms* of the module M and is denoted by $E_A(M)$. Its elements (homomorphisms into itself) are called the *endomorphisms* of M. The invertible elements of this algebra, i. e. the isomorphisms of M onto M, are called the *automorphisms* of M.

Let us see the meaning of the above introduced concepts in the case of the regular module. Let $f : A \to M$ be a homomorphism of the regular module into a module M. Then, for every element $a \in A$, $f(a) = f(1)a = m_0 a$, where $m_0 = f(1)$ is a fixed element of M. Conversely, if we fix an arbitrary element $m_0 \in M$ and put $f(a) = m_0 a$, then, as one can easily see, we obtain a homomorphism $f : A \to M$. In this way we establish a bijection between the elements of M and the homomorphisms from $\mathrm{Hom}_A(A, M)$. If f and g are two such homomorphisms whereby $f(1) = m_0$, $g(1) = m_1$, then $(f+g)(1) = m_0 + m_1$ and $(\alpha f)(1) = \alpha m_0$ for arbitrary $\alpha \in K$. Consequently, our bijection is an isomorphism of the vector spaces.

Now, let $M = A$ and f and g be endomorphisms of A with $f(1) = a$, $g(1) = b$. Then $(fg)(1) = f(g(1)) = f(b) = f(1)b = ab$. Therefore the bijection between A and $E_A(A)$ is an algebra isomorphism. Hence, we have proved the following theorem.

Theorem 1.7.1. *The map $f \mapsto f(1)$ is an isomorphism of the vector spaces $\mathrm{Hom}_A(A, M)$ and M. If $M = A$, this map is an isomorphism of the algebras $E_A(A)$ and A.*

The endomorphism algebra of M is a subalgebra of the algebra of all linear operators of the space M. It consists of those transformations which commute with all transformations $T(a)$, $a \in A$, where T is the representation defined by the module M. In this way, the presentation of the algebra $E_A(M)$ provides its faithful representation (more precisely, anti-representation because we have agreed to write the endomorphisms on the left and the linear transformations on the right of the elements). The corresponding left module can be canonically identified with the vector space M endowed by the endomorphisms acting on it by the rule $fm = f(m)$ (the value of f at the element $m \in M$).

Thus, every A-module M can be considered as a left module over the algebra $E_A(M)$. It is easy to verify that if $M = A$ is the regular module, then the corresponding left module over the algebra $E_A(A) \simeq A$ is simply the left regular module.

The analogous assertions hold also for the left modules. Here, it is convenient to write the homomorphisms of the left modules on the right of the elements. The image of m in the homomorphism f will be denoted by mf; correspondingly, the product fg is defined by $m(fg) = (mf)g$. In this notation, a left A-module becomes a right module over its endomorphism algebra. Also, the analogue of Theorem 1.7.1 holds.

What is the relation between the structure of the endomorphism algebra and the structure of the module, in particular the decompositions into direct sums?

Let M decompose into a direct sum of its submodules: $M = M_1 \oplus M_2 \oplus \ldots \oplus M_s$. This means that every element $m \in M$ is uniquely represented in the form of the sum $m = m_1 + m_2 + \ldots + m_s$, where $m_i \in M_i$. Write $m_i = e_i m$. It follows from the fact that the sum is unique that $e_i(m+n) = e_i m + e_i n$, $e_i(\alpha m) = \alpha e_i(m)$ and $e_i(ma) = (e_i m)a$ for every $m, n \in M$, $\alpha \in K$, $a \in A$, i.e. e_i is an endomorphism of M. Besides, if $m \in M_i$, then $e_i m = m$ and $e_j m = 0$ for $j \neq i$ which yields that $e_j e_i = \delta_{ij} e_i$, where δ_{ij} is the Kronecker symbol ($\delta_{ij} = 1$ for $i = j$ and $\delta_{ij} = 0$ for $i \neq j$). Finally, from the definition of e_i, it follows that $m = e_1m + e_2m + \ldots + e_sm$, i.e. $e_1 + e_2 + \ldots + e_s = 1$.

An element e of an algebra A is said to be an *idempotent* if $e^2 = e$. Two idempotents e and f such that $ef = fe = 0$ are called *orthogonal*. The equality $1 = e_1 + e_2 + \ldots + e_s$, where $e_1, e_2, \ldots, e_s$ are pairwise orthogonal idempotents, will be called a *decomposition of the identity* of the algebra A.

Theorem 1.7.2. *There is a bijective correspondence between the decompositions of an A-module M into a direct sum of submodules and the decompositions of the identity of the algebra $E = E_A(M)$.*

Proof. We have already attached to every decomposition of the module M a decomposition of the identity of the algebra E. Now, let $1 = e_1 + e_2 + \ldots + e_s$ be a decomposition of the identity of the algebra E. Put $M_i = \operatorname{Im} e_i$. Then, for every element $m \in M$, $m = (e_1 + e_2 + \ldots + e_s)m = e_1m + e_2m + \ldots + e_sm$, where $e_im \in M_i$. If $m = m_1 + m_2 + \ldots + m_s$ is a decomposition of the element m in the form of the sum of the elements $m_i \in M_i$, then $m_i = e_ix_i$ for some $x_i \in M$. Therefore,

$$e_im = \sum_{j=1}^{s} e_im_j = \sum_{j=1}^{s} e_ie_jx_j = e_ix_i = m_i$$

(since e_i and e_j are for $i \neq j$ orthogonal). Consequently, such a form is unique, i.e. $M = M_1 \oplus M_2 \oplus \ldots \oplus M_s$. □

Corollary 1.7.3. *A module M is indecomposable if and only if there are no non-trivial (i. e. different from 0 and 1) idempotents in the algebra $E_A(M)$.*

Proof. If e is a non-trivial idempotent, then $f = 1 - e$ is also a non-trivial idempotent which is orthogonal to e, and thus $1 = e + f$ is a decomposition of the identity. □

Combining Theorems 1.7.1 and 1.7.2, we obtain the following corollary.

Corollary 1.7.4. *There is a bijective correspondence between the decompositions of a module M and the decompositions of the regular module over the algebra $E_A(M)$.*

Observe that if $1 = e_1 + e_2 + \ldots + e_s$ is a decomposition of the identity of an algebra A, then the corresponding decomposition of the regular A-module has the form $A = e_1A \oplus e_2A \oplus \ldots \oplus e_sA$. This decomposition is called the *right Peirce decomposition* of the algebra A. Similarly one can define the *left Peirce decomposition*: $A = Ae_1 \oplus Ae_2 \oplus \ldots \oplus Ae_s$ (this is a decomposition of the left regular module).

Moreover, if M is an arbitrary module, then a given decomposition of the identity of the algebra A induces a decomposition of M as a vector space

$M = Me_1 \oplus Me_2 \oplus \ldots \oplus Me_s$. Since $f(me_i) = (fm)e_i$ for every endomorphism $f \in E_A(M)$, this decomposition is also a decomposition of M as a left $E_A(M)$-module (but, in general, it is not a decomposition of M as an A-module). We shall call this decomposition the *Peirce decomposition of the module* M.

The summands e_iA of the right Peirce decomposition of an algebra A are the right ideals, i. e. the A-modules. If we apply the Peirce decomposition of modules, we obtain the following decomposition of the vector space A:

$$A = \bigoplus_{i,j=1}^{s} e_i A e_j \,. \tag{1.7.1}$$

This decomposition is called the *two-sided Peirce decomposition*, or simply the *Peirce decomposition of the algebra* A. The components of the Peirce decomposition $A_{ij} = e_iAe_j$ are, in general, no longer right nor left ideals. Nevertheless, this decomposition allows a convenient interpretation of the elements of the algebra A in the form of some matrices.

Let a and b be two elements of the algebra A. We shall decompose them in accordance with the Peirce decomposition (1.7.1): $a = \sum_{i,j} a_{ij}$, $b = \sum_{i,j} b_{ij}$, where $a_{ij} = e_iae_j$, $b_{ij} = e_ibe_j$. Then $a + b = \sum_{i,j}(a_{ij} + b_{ij})$ and

$$ab = \sum_{i,k} \sum_{\ell,j} a_{ik} b_{\ell j} = \sum_{i,j} \Big(\sum_{k} a_{ik} b_{kj} \Big) \,,$$

because, for $k \neq \ell$, $a_{ik}b_{\ell j} = e_iae_ke_\ell be_j = 0$. Thus, $e_i(ab)e_j = \sum_k a_{ik}b_{kj}$. This allows the element a to be written in the form of a matrix of its Peirce components

$$a = \begin{pmatrix} a_{11} & a_{12} & \ldots & a_{1s} \\ a_{21} & a_{22} & \ldots & a_{2s} \\ \vdots & \vdots & \ddots & \vdots \\ a_{s1} & a_{s2} & \ldots & a_{ss} \end{pmatrix} , \quad a_{ij} = e_iae_j \in A_{ij} \,. \tag{1.7.2}$$

We have just established that the addition and multiplication of these elements translates in this interpretation into the addition and multiplication of the matrices defined in the usual way. In what follows, we shall often use this description. In particular, the Peirce decomposition (1.7.1) will be written in the form

$$A = \begin{pmatrix} A_{11} & A_{12} & \ldots & A_{1s} \\ A_{21} & A_{22} & \ldots & A_{2s} \\ \vdots & \vdots & \ddots & \vdots \\ A_{s1} & A_{s2} & \ldots & A_{ss} \end{pmatrix} .$$

Let us apply the Peirce decomposition to the endomorphism algebra of a module M decomposed into the direct sum $M = M_1 \oplus M_2 \oplus \ldots \oplus M_s$. Let $1 = e_1 + e_2 + \ldots + e_s$ be the corresponding decomposition of the identity of the algebra $E = E_A(M)$; $E_{ij} = e_iEe_j$. The Peirce decomposition of the element $f \in E$ is of the form

$$f = \begin{pmatrix} f_{11} & f_{12} & \dots & f_{1s} \\ f_{21} & f_{22} & \dots & f_{2s} \\ \vdots & \vdots & \ddots & \vdots \\ f_{s1} & f_{s2} & \dots & f_{ss} \end{pmatrix}, \quad f_{ij} = e_i f e_j .$$

Let $m = m_1 + m_2 + \ldots + m_s$ be an element of M ($m_i = e_i m$). Then

$$fm = \sum_{i,j} e_i f e_j m = \sum_{i,j} f_{ij} m_j ,$$

i. e. if one writes m in the form of a column of the elements $m_1, m_2, \ldots, m_s$, then

$$fm = \begin{pmatrix} f_{11} & f_{12} & \dots & f_{1s} \\ f_{21} & f_{22} & \dots & f_{2s} \\ \vdots & \vdots & \ddots & \vdots \\ f_{s1} & f_{s2} & \dots & f_{ss} \end{pmatrix} \begin{pmatrix} m_1 \\ m_2 \\ \vdots \\ m_s \end{pmatrix},$$

where the multiplication is to be again the matrix multiplication.

Note that $f_{ij}m$ always belongs to M_i. Besides, the value of $f_{ij}m$ is determined uniquely by the component $m_j = e_j m$, because $f_{ij}m = f_{ij}m_j$. Therefore, f_{ij} can be interpreted as a homomorphism $M_j \to M_i$. Conversely, if $g : M_j \to M_i$ is a homomorphism, then one can define the homomorphism $\bar{g} : M \to M$ by $\bar{g}m = gm_j$ (where $m_j = e_j m$) and, obviously, $\bar{g}$ will belong to E_{ij}. Consequently $E_{ij} \simeq \mathrm{Hom}_A(M_j, M_i)$ and we shall always identify E_{ij} and $\mathrm{Hom}_A(M_j, M_i)$ by means of this isomorphism.

In particular, if we turn our attention to the regular A-module, then we see that, for any decomposition of the identity $1 = e_1 + e_2 + \ldots + e_s$ of the algebra A the components of the Peirce decomposition $A_{ij} = e_i A e_j$ can be canonically identified with $\mathrm{Hom}_A(e_j A, e_i A)$ and this identification agrees with the matrix form (1.7.2) of the elements of the algebra.

Finally, we obtain an interesting result if all summands $M_1, M_2, \ldots, M_s$ are mutually isomorphic: $M_1 \simeq M_2 \simeq \ldots \simeq M_s \simeq L$. In this case, we shall write $M \simeq sL$. Obviously, $E_{ij} \simeq E_A(L)$ and the matrix form of the endomorphisms yields the following conclusion.

Theorem 1.7.5. *If $M \simeq sL$, then the algebra $E_A(M)$ is isomorphic to the algebra of the matrices of degree s with coefficients from $E_A(L)$.*

In what follows, the algebra of all matrices of degree n with coefficients from an algebra A will be denoted by $M_n(A)$.

Corollary 1.7.6. $M_n(A) \simeq E_A(nA)$.

In conclusion, let us now consider the relation between the idempotents and the direct products of algebras.

Let $A = A_1 \times A_2 \times \ldots \times A_k$. Put $e_i = (0, \ldots, 1, \ldots, 0)$, the identity of the algebra A_i at the ith position, zeroes elsewhere. Obviously, $e_1, e_2, \ldots, e_k$

are pairwise orthogonal idempotents and $1 = e_1 + e_2 + \ldots + e_k$. But the idempotents e_i have an additional property: they belong to the center of the algebra, i. e. $e_i a = a e_i$ for any $a \in A$.

Idempotents which lie in the center are said to be *central.* If, in a decomposition of the identity, all idempotents are central, then the decomposition is called itself *central.*

For every central decomposition of the identity $1 = e_1 + e_2 + \ldots + e_k$, we have $e_i A = A e_i = e_i A e_i$ and $e_i A e_j = e_i e_j A = 0$ for $i \neq j$. Thus, the right, left and two-sided Peirce decompositions coincide in this case and its components are ideals of the algebra A.

Theorem 1.7.7. *There is a bijective correspondence between*

1) *the decompositions of the algebra A into a direct product of algebras;*
2) *the central decompositions of the identity of the algebra A;*
3) *the decompositions of A into a direct sum of ideals.*

Proof. We have already constructed, for a given decomposition into a direct product, the central decomposition of the identity and then, the decomposition into a direct sum of ideals.

Conversely, let $A = I_1 \oplus I_2 \oplus \ldots \oplus I_k$, where I_i are ideals, and $1 = e_1 + e_2 + \ldots + e_k$ the corresponding decomposition of the identity. Then $I_i = e_i A$, and thus $e_i a e_j \in I_i \cap I_j$ for any $a \in A$; from here, $e_i a e_j = 0$, i. e. $e_i A e_j = 0$ for $i \neq j$, and the Peirce decomposition has the form

$$A = \begin{pmatrix} A_1 & & & 0 \\ & A_2 & & \\ & & \ddots & \\ 0 & & & A_k \end{pmatrix},$$

where $A_i = e_i A e_i = e_i A$. But then $A \simeq A_1 \times A_2 \times \ldots \times A_k$, as required. □

Corollary 1.7.8. *There is a bijective correspondence between the direct product decompositions of the algebra A and those of its center.*

It is clear from the proof of Theorem 1.7.7 that the decomposition of the identity $1 = e_1 + e_2 + \ldots + e_s$ is central if and only if $e_i A e_j = 0$ for $i \neq j$. Taking into account the interpretation of the Peirce components of the endomorphism algebra, we obtain the following result.

Corollary 1.7.9. *If $M = M_1 \oplus M_2 \oplus \ldots \oplus M_s$ and if $\mathrm{Hom}_A(M_j, M_i) = 0$ for $i \neq j$, then $E_A(M) \simeq E_A(M_1) \times E_A(M_2) \times \ldots \times E_A(M_s)$.*

Exercises to Chapter 1

1. Compute the regular matrix representation of the field $\mathbb{C}$ of complex numbers and of the algebra $\mathbb{H}$ of quaternions (over the field of reals) with respect to the natural bases ($\{1, i\}$ for $\mathbb{C}$ and $\{1, i, j, k\}$ for $\mathbb{H}$).
2. Using the regular representation, prove that the algebra of quaternions is a division algebra.
3. Let A be the algebra over the field of complex numbers with basis $\{e, i, j, k\}$ whose multiplication table is the same as that of quaternions. Find the divisors of zero. Establish an isomorphism $A \simeq M_2(\mathbb{C})$.
4. Find the center and the ideals of the algebra $M_n(K)$.
5. Compute a regular matrix representation of the Jordan algebra $J_n(K)$; find all ideals of $J_n(K)$.
6. Prove that every monogenic algebra over an algebraically closed field is isomorphic to a direct product of Jordan algebras. (Hint: Use Cayley's theorem and the Jordan normal form of a matrix.)
7. Let M be a vector space of n-tuples considered as a module over the algebra $A = T_n(K)$ of triangular matrices (the image of a matrix X in the respective representation is X).
   Find the submodules of the module M and the respective representations. Construct a composition series in M. Show that the module M is indecomposable. Compute $E_A(M)$.
8. Let $M_1, M_2, \ldots, M_k$ be submodules of M and f the natural homomorphism $M_1 \oplus M_2 \oplus \ldots \oplus M_k \to M$ defined in Theorem 1.6.3. Prove that $\operatorname{Ker} f$ has a filtration with the factors $N_i \simeq M_i \cap (\sum_{j<i} M_j)$, $2 \le i \le k$.
9. Let M be a cyclic module over an algebra A, generated by m. Prove that $M \simeq A/\operatorname{Ann} m$. Verify that, if $\operatorname{Ann} m$ is an ideal of A (for instance in case that A is commutative), then $\operatorname{Ann} m = \operatorname{Ann} M$ and for an arbitrary generator m' of M, $\operatorname{Ann} m' = \operatorname{Ann} m$.
   Show that, if A is not commutative, then $\operatorname{Ann} m$ may depend on the choice of m. (Hint: Take $A = M_n(K)$ and M the space of all n-tuples.)
10. (Peirce decomposition of an ideal) Let I be an ideal of an algebra A and $1 = e_1 + e_2 + \ldots + e_s$ a decomposition of the identity of this algebra. Prove that the element $a = \sum_{i,j} a_{ij}$, where $a_{ij} \in e_i A e_j$, belongs to I if and only if $a_{ij} \in e_i I e_j$.
11. Using the statement of Exercise 10, describe the ideals of the algebra of triangular matrices.
12. Let A be an algebra not necessarily with identity. Let $\tilde{A}$ be the algebra obtained from A by adjoining an identity (see Sect. 1.1):
$$\tilde{A} = \{(a, \alpha) \mid a \in A, \alpha \in K\}.$$
   a) Prove that the elements of the form $(a, 0)$ generate an ideal of $\tilde{A}$ which is isomorphic to A (as an algebra without an identity).
   b) If A has an identity, show that $\tilde{A} \simeq A \times K$.
   c) Show that every homomorphism $f : A \to B$ between two algebras without an identity extends uniquely to a homomorphism $\tilde{f} : \tilde{A} \to \tilde{B}$ between the algebras with identity, and that every homomorphism $g : \tilde{A} \to \tilde{B}$ maps A into B. In particular, $A \simeq B$ if and only if $\tilde{A} \simeq \tilde{B}$.

13. A representation of an algebra A without an identity is a homomorphism $f : A \to \mathrm{End}(V)$.
   a) Define a module over an algebra without an identity and establish a relation between the modules and representations.
   b) Let M be a module over an algebra A without an identity. Put $m(a, \alpha) = ma + \alpha m$ ($a \in A$, $\alpha \in K$). Prove that M is an $\tilde{A}$-module.
   c) Verify that, for any A-modules M and N, $\mathrm{Hom}_A(M, N) \simeq \mathrm{Hom}_{\tilde{A}}(M, N)$, considering M and N as $\tilde{A}$-modules (see b)). In particular, $M \simeq N$ (as A-modules) if and only if $M \simeq N$ considered as $\tilde{A}$-modules.

2. Semisimple Algebras

The classical theory of semisimple algebras is one of the most striking examples how "module theoretical" methods produce deep structural results. Moreover, semisimple algebras and their representations play a very important role in many parts of mathematics. In this chapter, we establish the most fundamental properties of semisimple algebras and their modules, and prove the Wedderburn-Artin theorem which gives complete classification of such algebras. The results of Chapter 1 (in particular, of Sect. 1.7) and a description of the homomorphisms of simple modules, the so-called Schur's lemma, will play a fundamental role in this process.

2.1 Schur's Lemma

Let us recall that a non-zero module M is said to be simple if it has no non-trivial (i. e. different from 0 and M) submodules. We have explained their significance in the module theory: every module is obtained by subsequent extensions of simple modules. The simple modules play also a very important role in the structure theory; this aspect is to large extent a consequence of the following results.

Theorem 2.1.1 (Schur). *If U and V are simple A-modules, then every non-zero homomorphism $f : U \to V$ is an isomorphism.*

Proof. This follows from the fact that $\operatorname{Ker} f$ and $\operatorname{Im} f$ are submodules of U and V, respectively, and that $f \neq 0$ implies that $\operatorname{Ker} f \neq U$ and $\operatorname{Im} f \neq 0$. Consequently, $\operatorname{Ker} f = 0$ and $\operatorname{Im} f = V$, i. e. f is both a monomorphism and an epimorphism, thus an isomorphism. □

Corollary 2.1.2 (Schur). *The endomorphism algebra of a simple module is a division algebra.*

Indeed, every non-zero element of such an algebra is an isomorphism and therefore invertible.

Corollary 2.1.3. *A regular A-module is simple if and only if the algebra A is a division algebra.*

Proof. This follows from the isomorphism $A \simeq E_A(A)$ (Theorem 1.7.1) and the fact that there are no non-trivial right ideals in a division algebra. □

Let us remark that the converse of Schur's lemma does not hold: There are non-simple modules M such that $E_A(M)$ is a division algebra (see Exercise 7 of Chap. 1).

2.2 Semisimple Modules and Algebras

A module M is called *semisimple* if it is isomorphic to a direct sum of simple modules.

To a semisimple module, there corresponds a *completely reducible* representation, i. e. a representation T of the form

$$T(a) = \begin{pmatrix} T_1(a) & & & 0 \\ & T_2(a) & & \\ & & \ddots & \\ 0 & & & T_n(a) \end{pmatrix},$$

where T_i are irreducible representations.

Proposition 2.2.1. *The following conditions are equivalent:*

1) *the module M is semisimple;*
2) $M = \sum_{i=1}^{m} M_i$, *where M_i are simple submodules of M;*
3) *every submodule $N \subset M$ has a complement;*
4) *every simple submodule $N \subset M$ has a complement.*

Proof. 1) ⇒ 2) by definition.

2) ⇒ 3). If $M = \sum_{i=1}^{m} M_i$, then trivially $M = N + \sum_{i=1}^{m} M_i$. Observe that, from the fact that M_j is simple, it follows that either $M_j \cap (N + \sum_{i=1}^{j-1} M_i) = 0$ or $M_j \subset N + \sum_{i=1}^{j-1} M_i$. Omitting those M_j for which $M_j \subset N + \sum_{i=1}^{j-1} M_i$, we obtain a family of submodules N_k such that $N + \sum_k N_k = M$ and $N_\ell \cap (N + \sum_{k<\ell} N_k) = 0$. Consequently, M is the direct sum of the submodules N and N_k (Theorem 1.6.3) and $N' = \sum_k N_k$ is a complement of N.

3) ⇒ 4) is trivial.

4) ⇒ 1) can be proved easily by induction on the length $\ell(M)$ of the module M. If $\ell(M) = 1$, then the module M is simple. Let $\ell(M) > 1$ and U be a simple submodule of M. Then $M = U \oplus U'$, where U' is a complement of U in

M, and $\ell(M) = \ell(U)+\ell(U')$, i. e. $\ell(U') = \ell(M)-1$. If N is a simple submodule of U' and N' its complement in M, then every element $x \in U'$ can be written in the form $x = n+n'$, where $n \in N$, $n' \in N'$, and thus $n' = x - n \in N' \cap U'$. As a result, $U' = N \oplus (N' \cap U')$, i.e. every simple submodule of U' has a complement in U'. By the induction hypothesis, $U' \simeq \bigoplus_{i=2}^{n} U_i$ with simple U_i's. Therefore, $M \simeq \bigoplus_{i=1}^{n} U_i$, where $U_1 = U$, as required. □

Corollary 2.2.2. *Every submodule and every factor module of a semisimple module is semisimple.*

Proof. Let N be a submodule of a semisimple module M, and $\pi : M \to M/N$ be the projection of M onto the factor module M/N. If $M = \sum_{i=1}^{m} M_i$, where M_i are simple modules, then $M/N = \sum_{i=1}^{m} \pi(M_i)$. But $\pi(M_i)$ is a factor module of M_i and therefore either $\pi(M_i) = 0$, or $\pi(M_i) \simeq M_i$. Hence, M/N is a sum of simple submodules and therefore semisimple. Now, the fact that N is semisimple follows immediately: if N' is a complement of N in M, then $N \simeq M/N'$. □

If M is a semisimple module, and $M = \bigoplus_{i=1}^{n} U_i$ a decomposition into a direct sum of simple submodules, then $M_j = \bigoplus_{i>j} U_i$ $(j = 0, 1, \dots, n)$ are submodules of M and $M_{j-1} \supset M_j$ with $M_{j-1}/M_j \simeq U_j$. Consequently, $M = M_0 \supset M_1 \supset M_2 \supset \dots \supset M_n = 0$ is a composition series of M and $U_1, U_2, \dots, U_n$ are its simple factors. Then, by the Jordan-Hölder theorem (1.5.1), we obtain the following proposition.

Proposition 2.2.3. *If M is a semisimple module and $M \simeq U_1 \oplus U_2 \oplus \dots \oplus U_n \simeq V_1 \oplus V_2 \oplus \dots \oplus V_m$ two decompositions of M into a direct sum of simple modules, then $n = m$ and by a suitable permutation of the summands, $U_i \simeq V_i$ for all i.*

An algebra is said to be *semisimple* if its regular module is semisimple. The simple submodules of the regular right (left) A-module are called *minimal right (left) ideals* of the algebra A.

The following lemma, together with Proposition 2.2.1, provides an "internal" characterization of the semisimple algebras.

Lemma 2.2.4 (Brauer). *If I is a minimal right ideal of an algebra A, then either $I^2 = 0$, or $I = eA$, where e is an idempotent.*

Proof. Assume that $I^2 \neq 0$, i.e. there are elements x and y in I such that $xy \neq 0$. Then the map $f : I \to I$ given by $f(a) = xa$ is a non-zero homomorphism, and since I is a simple module, it is an isomorphism (Theorem 2.1.1). Therefore, there is an element $e \in I$ such that $x = xe$. But then $xe = xe^2$, i.e.

$f(e) = f(e^2)$ and therefore $e = e^2$. It follows that e is an idempotent. Finally, eA is a non-zero submodule of I and therefore $I = eA$. □

A right (left, two-sided) ideal of I of an algebra A is called *nilpotent* if $I^m = 0$ for some m.

Corollary 2.2.5. *The following conditions for an algebra A are equivalent:*

1) *A is semisimple;*
2) *every right ideal of A is of the form eA, where e is an idempotent;*
3) *every non-zero ideal of A contains a non-zero idempotent;*
4) *A has no non-zero nilpotent ideals;*
5) *A has no non-zero nilpotent right ideals.*

Proof. 1) ⇒ 2). If I is a right ideal, then by Proposition 2.2.1, $A = I \oplus I'$, where I' is a complement of I. Hence $I = eA$, where $1 = e + e'$ is the corresponding decomposition of the identity (Theorem 1.7.2).

2) ⇒ 3) is trivial.

3) ⇒ 4) follows from the fact that if e is a non-zero idempotent, then $e^k = e \neq 0$ for every k.

4) ⇒ 5). If $I \neq 0$ is a nilpotent right ideal, then AI is a two-sided ideal of A and $(AI)^m = AI^m$ implies that AI is nilpotent, too.

5) ⇒ 1). If I is a simple submodule of the regular module, i. e. a minimal right ideal of the algebra A, then $I^2 \neq 0$ and, by Lemma 2.2.4, $I = eA$. Therefore, there is a complement $I' = (1-e)A$ of I and, by Proposition 2.2.1, the algebra A is semisimple. □

Note that the conditions 3) and 4) of Corollary 2.2.5 are symmetric with respect to the notion of "right" and "left". Therefore "left semisimplicity", i. e. semisimplicity of the left regular module, is equivalent to semisimplicity, and one may add to the conditions of Corollary 2.2.5 all those obtained by replacing "right" with "left".

The above criterion can be easily reformulated in terms of elements. To do so, one assigns to an element $a \in A$ the right ideal $aA = \{ax \mid x \in A\}$ (the right ideals of this form are called *principal*) and expresses the meaning of the nilpotency of this ideal. Every element of $(aA)^m$ is a sum of elements of the form $ax_1ax_2 \dots ax_m$, where $x_1, x_2, \dots, x_m$ are some elements of the algebra. Therefore $(aA)^m = 0$ if and only if every product of such a form is equal to zero. In other words, whenever the element a appears m times in a product $a_1a_2 \dots a_t$, the product is zero. Such elements are called *strongly nilpotent.*

Corollary 2.2.6. *An algebra is semisimple if and only if it contains no non-zero strongly nilpotent elements.*

Corollary 2.2.7. *A commutative algebra is semisimple if and only if it contains no nilpotent elements (i. e. $a \neq 0$ such that $a^m = 0$).*

Indeed, in a commutative algebra every nilpotent element is strongly nilpotent.

Note that in a non-commutative algebra there may be nilpotent elements which are not strongly nilpotent. For example, in the matrix algebra $M_2(K)$, the matrix e_{12} is nilpotent: $e_{12}^2 = 0$; however $e_{21}e_{12} = e_{22}$ is an idempotent, and therefore e_{12} is not strongly nilpotent.

It is not difficult to see that there are no non-zero strongly nilpotent elements in $M_2(K)$ (the reader should verify this statement).

Corollary 2.2.8. *The center of a semisimple algebra is semisimple.*

Proof. The statement follows from the fact that an element of the center is nilpotent if and only if it is strongly nilpotent. □

Finally, our considerations yield the following important criterion of semisimplicity in terms of representation theory.

Theorem 2.2.9. *An algebra A is semisimple if and only if there is a faithful semisimple A-module.*[1]

Proof. The necessity of the condition follows immediately from the fact that the regular module is faithful. We are going to prove the sufficiency.

Let $M = \bigoplus_{i=1}^{n} U_i$ be a faithful module, where U_i are simple A-modules. If $I \neq 0$ is an ideal of the algebra A, then $U_iI \neq 0$ for some i because M is faithful. But then $U_iI = U_i$, and thus $U_iI^m = U_i$. Therefore $I^m \neq 0$ for every m. In view of Corollary 2.2.5, this means that the algebra A is semisimple. □

2.3 Vector Spaces and Matrices

Before we embark on the general theory of semisimple algebras and their representations, we shall consider vector spaces and matrix algebras over division algebras.

Let D be a (finite dimensional) division algebra over K and V a (finite dimensional) left D-module. Then V is called a (finite dimensional) *vector space* over the division algebra D.

Proposition 2.3.1. *A vector space is a semisimple module. Every simple left D-module is isomorphic to the regular one.*

Proof. Let V be a vector space over D and v_1 a non-zero element (vector) of V. The homomorphism $f : D \to V$ mapping $x \in D$ into $xv_1 \in V$ is non-zero,

[1] Recall that a module M is said to be faithful if the respective representation is faithful, i. e. if $\operatorname{Ann} M = 0$.

and thus $\operatorname{Ker} f = 0$ (since the regular left D-module is simple). Therefore $Dv_1 = \operatorname{Im} f \simeq D$.

If V is a simple module, then $V = Dv_1 \simeq D$. Otherwise, there is an element $v_2 \notin Dv_1$. But then $Dv_2 \simeq D$ and $Dv_1 + Dv_2$ contains properly Dv_1. Continuing in this way, we get the equality $V = \sum_{i=1}^{n} Dv_i$, where $Dv_i \simeq D$ are simple modules. By Proposition 2.2.1, V is a semisimple module, as required. □

Corollary 2.3.2. *Every vector space over a division algebra D is isomorphic to nD (direct sum of n copies of the regular module). The number n is determined uniquely.*

The fact that n is an invariant is a consequence of n being the length of the module V. It is usually called the *dimension* of the vector space V and denoted by $[V : D]$. Evidently, $[V : K] = [V : D][D : K]$.

In accordance with Corollary 1.7.6, the endomorphism algebra of an n-dimensional vector space V over a division algebra D is isomorphic to the matrix algebra $M_n(D)$ with entries from the division algebra D. Consequently, it is natural to consider the space V as a right module over $M_n(D)$. In what follows, the elements of V will be identified with n-tuples $(x_1, x_2, \dots, x_n)$, $x_i \in D$. Then the action of the matrices is the usual multiplication of vectors by matrices.

Proposition 2.3.3. *The module V over the algebra $A = M_n(D)$ is simple. The algebra $M_n(D)$ is semisimple.*

Proof. Let U be a non-zero A-submodule of V, and $u = (u_1, u_2, \dots, u_n)$ a non-zero element of U. Assume, without loss of generality, that $u_1 \neq 0$. Then every vector $x = (x_1, x_2, \dots, x_n)$ can be represented in the form $x = uX$, where

$$X = \begin{pmatrix} u_1^{-1}x_1 & u_1^{-1}x_2 & \dots & u_1^{-1}x_n \\ 0 & 0 & \dots & 0 \\ \dots & \dots & \dots & \dots \\ 0 & 0 & \dots & 0 \end{pmatrix}.$$

Therefore $uA = V$, the A-module V is simple and the algebra $M_n(D)$ is semisimple by Theorem 2.2.9. □

Let us describe an explicit decomposition of the regular A-module. Denote by I_i the right ideal of the algebra $M_n(D)$ consisting of all matrices of the form

$$\begin{pmatrix} 0 & 0 & \dots & 0 \\ \dots & \dots & \dots & \dots \\ x_1 & x_2 & \dots & x_n \\ \dots & \dots & \dots & \dots \\ 0 & 0 & \dots & 0 \end{pmatrix}$$

(the non-zero row is the ith one). Making $(x_1, x_2, \dots, x_n) \in V$ correspond to the matrix, we obtain, obviously, an isomorphism of I_i onto V.

In the sequel, we shall denote by e_{ij} the *elementary matrices*, i. e. the matrices such that the (ij)th entry is 1 and all the other entries are 0. In particular, $1 = e_{11} + e_{22} + \dots + e_{nn}$ is a decomposition of the identity, whereby $e_{ii}A = I_i$. It follows that $E_A(V) \simeq e_{ii}Ae_{ii} \simeq D$.

Proposition 2.3.4. *Every module over the algebra $A = M_n(D)$ is semisimple. Every simple A-module is isomorphic to V, and the regular A-module is isomorphic to nV.*

Proof. Clearly, $A = I_1 \oplus I_2 \oplus \dots \oplus I_n$ is a decomposition of the regular A-module and, as we have seen, $I_i \simeq V$. If M is an arbitrary A-module and m a non-zero element of M, then mI_i is either the zero module or is isomorphic to I_i, since I_i is simple. Besides, $m = \sum_{i=1}^{n} me_{ii} \in \sum_{i=1}^{n} mI_i$ and thus some of the modules mI_i is non-zero. If M is simple, then $M \simeq mI_i \simeq V$. Otherwise, we proceed as in the proof of Proposition 2.3.1 to represent M as a sum of simple modules, and thus to complete the proof of our proposition. □

Corollary 2.3.5. *Two modules M and N over the algebra $M_n(D)$ are isomorphic if and only if $[M : K] = [N : K]$.*

Proof. If $M = mV$ and $N = kV$, then $[M : K] = m[V : K]$ and $[N : K] = k[V : K]$, i. e. $m = k$ if and only if $[M : K] = [N : K]$. □

Proposition 2.3.6. *There are no proper ideals (different from 0 and A) in the algebra $A = M_n(D)$.*

Proof. Let I be a non-zero ideal of A, $X = (x_{ij})$ a non-zero matrix from I and $x_{k\ell}$ a non-zero entry of this matrix. Then $e_{ik}X \neq 0$ and belongs to $I_i \cap I$. Therefore $I_i \cap I \neq 0$, and since I_i is simple, $I \supset I_i$ (for all i!). Consequently, $I \supset \sum_{i=1}^{n} I_i = A$, as required. □

An algebra which has no ideals different from 0 and the algebra itself is called *simple*. Proposition 2.3.6 shows that the algebra $M_n(D)$ is simple. In the following section, we shall show that there are no other (finite dimensional) simple algebras.

2.4 The Wedderburn-Artin Theorem

The results of Sect. 1.7 and those of the preceding section of Chapter 2 allow us to obtain the fundamental structure theorems of the semisimple algebras.

First of all, Schur's lemma yields immediately a description of the commutative semisimple algebras.

Theorem 2.4.1 (Weierstrass-Dedekind). *A commutative semisimple algebra is isomorphic to a direct product of fields. Conversely, a direct product of fields is a semisimple algebra.*

Proof. Let A be a commutative semisimple algebra, $A = U_1 \oplus U_2 \oplus \ldots \oplus U_n$ a decomposition of the regular A-module into a direct sum of simple modules and $1 = e_1 + e_2 + \ldots + e_n$ the corresponding decomposition of the identity. Because of commutativity, all idempotents are central. Then $A \simeq A_1 \times A_2 \times \ldots \times A_n$, where $A_i = e_i A \simeq E_A(U_i)$ (see Theorem 1.7.7). By Schur's lemma, A_i are division algebras and since they are commutative, they are fields.

Conversely, if $A \simeq A_1 \times A_2 \times \ldots \times A_n$, where A_i are fields, then the regular A-module is of the form $A = U_1 \oplus U_2 \oplus \ldots \oplus U_n$, where U_i are regular A_i-modules which are simple. □

Corollary 2.4.2. *If K is algebraically closed, then every commutative semisimple K-algebra is isomorphic to K^n.*

The general structure theorem is obtained by a combination of Schur's lemma and the matrix form of endomorphisms.

Theorem 2.4.3 (Wedderburn-Artin). *Every semisimple algebra is isomorphic to a direct product of matrix algebras over division algebras. Conversely, a direct product of matrix algebras over division algebras is a semisimple algebra.*

Proof. Let A be a semisimple algebra and $A \simeq n_1 U_1 \oplus n_2 U_2 \oplus \ldots \oplus n_s U_s$ be a decomposition of the regular A-module into a direct sum of simple modules with $U_i \not\simeq U_j$ for $i \neq j$. Denoting $n_i U_i$ by M_i, we get a decomposition $A \simeq M_1 \oplus M_2 \oplus \ldots \oplus M_s$, where, for $i \neq j$, $\mathrm{Hom}_A(M_i, M_j) = 0$ because $\mathrm{Hom}_A(U_i, U_j) = 0$ by Theorem 2.1.1. But then $A \simeq A_1 \times A_2 \times \ldots \times A_s$, where $A_i = E_A(M_i)$ (see Corollary 1.7.9). Since $M_i = n_i U_i$, Theorem 1.7.5 yields $A_i \simeq M_{n_i}(D_i)$ with a division algebra $D_i = E_A(U_i)$ (by Schur's lemma).

Conversely, if $A = A_1 \times A_2 \times \ldots \times A_s$, where $A_i = M_{n_i}(D_i)$, then the regular A-module can be decomposed into a direct sum $A = I_1 \oplus I_2 \oplus \ldots \oplus I_s$ with the regular A_i-modules I_i (see Theorem 1.7.7). In view of Proposition 2.3.4, I_i are semisimple modules. Thus, A is a semisimple algebra, too. □

Corollary 2.4.4 (Molien). *If K is algebraically closed, then every semisimple K-algebra is isomorphic to the algebra of the form $M_{n_1}(K) \times M_{n_2}(K) \times \ldots \times M_{n_s}(K)$.*

Corollary 2.4.5. *Every simple K-algebra is isomorphic to an algebra of the form $M_n(D)$, where D is a division algebra.*

Proof. We have already seen that $M_n(D)$ is a simple algebra. Conversely, if A is simple, then the unique non-zero two-sided ideal (A itself) contains a non-zero idempotent 1, i. e. the algebra is semisimple (by Corollary 2.2.5). In addition, the algebra A is indecomposable (into a direct product). This means that $A \simeq M_n(D)$. □

Corollary 2.4.6. *Every simple algebra over an algebraically closed field K is isomorphic to $M_n(K)$ for some n.*

2.5 Uniqueness of the Decomposition

The Wedderburn-Artin theorem assigns to every semisimple algebra A a system $(D_1, D_2, \ldots, D_s; n_1, n_2, \ldots, n_s)$, where D_i are division algebras and n_i the degrees of the matrices, so that

$$A \simeq M_{n_1}(D_1) \times M_{n_2}(D_2) \times \ldots \times M_{n_s}(D_s). \tag{2.5.1}$$

There is a natural question: Are the division algebras D_i and the degrees n_i determined uniquely? We are going to answer this question in the affirmative. Moreover, we will show that a decomposition (2.5.1) is, in fact, unique.

First, we are going to establish a general result concerning the uniqueness of the decomposition of algebras into direct products.

Theorem 2.5.1. *Let $A \simeq A_1 \times A_2 \times \ldots \times A_s \simeq B_1 \times B_2 \times \ldots \times B_t$ be two decompositions of the algebra A into a direct product of indecomposable algebras, and $1 = e_1 + e_2 + \ldots + e_s = f_1 + f_2 + \ldots + f_t$ the corresponding central decompositions of the identity. Then $s = t$ and, by a suitable permutation of the idempotents, $e_i = f_i$ and $A_i \simeq B_i$ for all i.*

Proof. In view of Theorem 1.7.7, the indecomposability of $A_i \simeq e_i A$ and $B_j \simeq f_j B$ means that, if $e_i = e_i' + e_i''$, $f_j = f_j' + f_j''$, where e_i' and e_i'', as well as f_j' and f_j'' are central idempotents, then either $e_i' = 0$ or $e_i'' = 0$ and similarly, $f_j' = 0$ or $f_j'' = 0$. But in view of the fact that e_i and f_j are central, $e_i f_j$ and $e_i - e_i f_j$ are central orthogonal idempotents. Therefore, either $e_i f_j = 0$ or $e_i f_j = e_i$. Similarly, either $e_i f_j = 0$ or $e_i f_j = f_j$. Since $e_i = e_i(f_1 + f_2 + \ldots + f_t)$, there is an index j such that $e_i f_j \neq 0$, i. e. $e_i = e_i f_j = f_j$. But there can be only one such index; for, $k \neq j$ implies $e_i f_k = f_j f_k = 0$. It follows immediately that $s = t$ and that, by a suitable relabelling, $e_i = f_i$ for all i. Then $A_i \simeq e_iA = f_iA \simeq B_j$. □

Now we can establish the uniqueness in the formulation of the Wedderburn-Artin theorem.

Theorem 2.5.2. *If $A \simeq M_{n_1}(D_1) \times M_{n_2}(D_2) \times \ldots \times M_{n_s}(D_s) \simeq M_{k_1}(F_1) \times M_{k_2}(F_2) \times \ldots \times M_{k_t}(F_t)$, where $D_1, D_2, \ldots, D_s, F_1, F_2, \ldots, F_t$ are division algebras then $s = t$ and, by a suitable permutation, $n_i = k_i$ and $D_i \simeq F_i$ for all i.*

Proof. Since there are no ideals in the algebra $M_n(D)$, it is indecomposable. Therefore, by Theorem 2.5.1, it follows immediately that $s = t$ and that, by a suitable renumbering, $M_{n_i}(D_i) \simeq M_{k_i}(F_i)$. It remains to be shown that if $A \simeq M_n(D) \simeq M_k(F)$, where D and F are division algebras, then $n = k$ and $D \simeq F$.

According to Proposition 2.3.4, A possesses a simple module V such that $A \simeq nV \simeq kV$; from here $k = n$. Besides, $D \simeq E_A(V) \simeq F$, which completes the proof of the theorem. □

The simple algebras $A_i = M_{n_i}(D_i)$ are called *simple components* of the semisimple algebra A. In view of Theorem 2.5.2, they are determined uniquely. Thus, the classification of semisimple algebras is completely reduced to the classification of finite dimensional division algebras.

2.6 Representations of Semisimple Algebras

The structure theorem of Wedderburn-Artin, together with the results of Sect. 2.3, enables us to give a complete description of the modules over semisimple algebras.

Proposition 2.6.1. *Let M be a module over an algebra $A = A_1 \times A_2 \times \ldots \times A_s$ and $1 = e_1 + e_2 + \ldots + e_s$ be the corresponding central decomposition of the identity of A. Then $M = \bigoplus_{i=1}^{s} Me_i$, where Me_i are modules over A_i.*

Proof. Clearly, Me_i is a submodule of M because e_i belongs to the center of A. Furthermore, $m = me_1 + me_2 + \ldots + me_s$ and, moreover, if $m = m_1 + m_2 + \ldots + m_s$, where $m_i \in Me_i$, then $me_i = m_i$, i.e. this representation is unique. Consequently, $M = \bigoplus_{i=1}^{s} Me_i$. But $Me_ie_j = 0$ for $i \neq j$ and thus $e_jA \subset \mathrm{Ann}\,(Me_i)$; hence, Me_i is a module over $A/\bigoplus_{j\neq i} e_jA \simeq A_i$, which was to be proved. □

Theorem 2.6.2. *Let A be a semisimple algebra, and $A_i \simeq M_{n_i}(D_i)$ its simple components ($i = 1, 2, \ldots, s$). Then every A-module is semisimple and can be* *uniquely written in the form* $\bigoplus_{i=1}^{s} k_i V_i$, *where* V_i *is the simple* A_i*-module. In particular, the simple* A*-modules are in a bijective correspondence with the simple components of the algebra.*

Proof. In view of Proposition 2.6.1, every A-module decomposes into a direct sum $M_1 \oplus M_2 \oplus \ldots \oplus M_s$, where M_i are A_i-modules. By Proposition 2.3.4, $M_i \simeq k_i V_i$, from where the theorem follows (the uniqueness follows from Proposition 2.2.3). □

In representation theory of groups, the following consequence of the preceding theorem plays an important role.

Corollary 2.6.3. *Let* T *and* S *be two representations of a semisimple algebra* A *over a field of characteristic 0. These representations are similar if and only if* $\operatorname{tr} T(a) = \operatorname{tr} S(a)$ *for all* $a \in A$.[2]

Proof. If the representations T and S are similar, then the matrices $T(a)$ and $S(a)$ are similar and therefore have equal traces.

Conversely, assume that $\operatorname{tr} T(a) = \operatorname{tr} S(a)$ for all $a \in A$. Let $A = A_1 \times A_2 \times \ldots \times A_s$, where A_i are simple components of A, furthermore let $1 = e_1 + e_2 + \ldots + e_s$ be the central decomposition of the identity and V_i the simple A_i-modules. Decompose the modules M and N corresponding to the representations T and S: $M \simeq \bigoplus_{i=1}^{s} m_i V_i$, $N \simeq \bigoplus_{i=1}^{s} k_i V_i$. If $v \in V_i$, then $ve_i = v$ and $ve_j = 0$ for $j \neq i$; from here, $\operatorname{tr} T(e_i) = m_i d_i$ and $\operatorname{tr} S(e_i) = k_i d_i$, where $d_i = [V_i : K]$. Therefore, since the traces are equal, $m_i = k_i$ for all i, i. e. $M \simeq N$ and the representations T and S are similar.[3] □

Theorem 2.6.2 allows us to describe the endomorphism algebra of a module over a semisimple algebra A. In order to do that, we recall that, for the simple module V_i over the algebra $A_i = M_{n_i}(D_i)$ (the simple component of the algebra A), $E_A(V_i) \simeq D_i$. In addition, by Schur's lemma, $\operatorname{Hom}_A(V_i, V_j) = 0$ for $i \neq j$. Therefore, the matrix form of the endomorphism yields the following result.

Theorem 2.6.4. *If* M *is a module over a semisimple algebra* A *then the algebra* $E_A(M)$ *is also semisimple. More precisely, if* $M = \bigoplus_{i=1}^{s} k_i V_i$, *where* V_i *is the simple module over the simple component* $A_i = M_{n_i}(D_i)$ *of the algebra* A, *then* $E_A(M) \simeq E_1 \times E_2 \times \ldots \times E_s$, *where* $E_i \simeq M_{k_i}(D_i)$.

---

[2] The symbol $\operatorname{tr} X$ denotes the trace of the matrix X: If $X = (x_{ij})$, then $\operatorname{tr} X = x_{11} + x_{22} + \ldots + x_{nn}$.

[3] The proof shows that it is sufficient to assume that the traces coincide only on the elements of the center.

Two semisimple algebras A and B are said to be *isotypic* if they have the same number of simple components and the corresponding division algebras are isomorphic.

Corollary 2.6.5. *The algebras A and B are isotypic if and only if $B \simeq E_A(M)$, where M is a faithful A-module.*

Proof. This follows from Theorem 2.6.4 and the fact that, for a faithful module $M = \bigoplus_{i=1}^{s} k_i V_i$, $k_i > 0$ for all i. □

Corollary 2.6.6. *If M is a semisimple module over an algebra A, then $E_A(M)$ is a semisimple algebra.*

Indeed, $E_A(M) = E_{\bar{A}}(M)$, where $\bar{A} = A/\operatorname{Ann} M$. But M is a faithful $\bar{A}$-module and thus $\bar{A}$, and consequently $E_{\bar{A}}(M)$, are semisimple algebras (Theorem 2.2.9).

In conclusion, from the above results, we derive the so-called *density theorem* for semisimple algebras.

Theorem 2.6.7 (Burnside). *Let M be a semisimple module over an algebra A, $B = E_A(M)$, $\bar{A} = E_B(M)$ (here, M is considered as a left B-module). Let us attach to every element $a \in A$ the endomorphism $\bar{a} \in \bar{A}$, defined by $x\bar{a} = xa$. Then the map $a \mapsto \bar{a}$ is an epimorphism of the algebras.*

Proof. Consider $A/\operatorname{Ann} M$ instead of A; this has no effect on the algebras B or $\bar{A}$. Since $a \in \operatorname{Ann} M$ implies $\bar{a} = 0$, we thus assume that M is a faithful module. Then the algebra A is semisimple (by Theorem 2.2.9). In this case, $a \neq 0$ implies $\bar{a} \neq 0$ and we need only to verify that the above map is an isomorphism.

Let $A = A_1 \times A_2 \times \ldots \times A_s$, where $A_i \simeq M_{n_i}(D_i)$ and $M = \bigoplus_{i=1}^{s} k_i V_i$, where V_i is the simple A_i-module. Then $B = B_1 \times B_2 \times \ldots \times B_s$, where $B_i \simeq M_{k_i}(D_i)$ and the corresponding decomposition of the identity $1 = e_1 + e_2 + \ldots + e_s$ satisfies $e_i M = k_i V_i$. Denote the simple B_i-module by U_i. Then, as a B-module, $e_i M \simeq m_i U_i$. But $[V_i : K] = n_i d_i$, where $d_i = [D_i : K]$, and $[U_i : K] = k_i d_i$. Therefore, $[e_i M : K] = k_i n_i d_i = m_i k_i d_i$, from where $m_i = n_i$, $\bar{A} = E_B(M) \simeq A$ and the monomorphism $A \to \bar{A}$ has to be an isomorphism. □

Corollary 2.6.8. *If U is a simple A-module, $D = E_A(U)$ its division algebra of endomorphisms and $[U : D] = n$, then A has a quotient algebra isomorphic to $M_n(D)$.*

Proof. This is an immediate consequence of Theorem 2.6.7 and the homomorphism theorem.

Exercises to Chapter 2

1. Prove that an irreducible representation of a commutative algebra over an algebraically closed field is one-dimensional.

2. Let A be a commutative semisimple subalgebra of the algebra $M_n(K)$, where K is an algebraically closed field. Prove that then there is a matrix S such that, for every matrix $X \in A$, SXS^{-1} is a diagonal matrix. (Two subalgebras A and A' of an algebra B are said to be *conjugate* if there is an element $b \in B$ such that $bab^{-1} \in A'$ for every $a \in A$ and, furthermore, every element $a' \in A'$ has such a form. Exercise 2 states that the subalgebra $A \subset M_n(K)$ is conjugate to a subalgebra of the algebra of the diagonal matrices.)

3. Under what conditions is a monogenic algebra semisimple?

4. Prove that a matrix $X \in M_n(K)$, where K is an algebraically closed field, is conjugate to a diagonal matrix if and only if the monogenic subalgebra generated by the matrix X is semisimple.

5. Let G be a cyclic group of order n. Prove that the group algebra KG is semisimple if and only if the characteristic of the field K does not divide n.

6. Prove that a commutative algebra A is semisimple if and only if every monogenic subalgebra of A is semisimple.

7. Describe those algebras whose monogenic subalgebras are semisimple.

8. Describe the algebras without nilpotent elements.

9. Prove that every non-zero idempotent of the algebra $M_n(D)$ is conjugate to $\sum_{i=1}^{k} e_{ii}$ for some k $(1 \le k \le n)$.

10. Let X and Y be two matrices of the algebra $A = M_n(K)$, $XA = \{XS \mid S \in A\}$, $YA = \{YS \mid S \in A\}$. Clearly, XA and YA are right ideals. When are they isomorphic as A-modules?

11. Prove that isomorphic simple subalgebras of the algebra $M_n(K)$ are conjugate.

12. Show that, if the algebra A is semisimple, then the length of the left regular module equals the length of the right regular module (for algebras which are not semisimple, the statement is, in general, false).

13. Let A be a semisimple algebra over a field K of characteristic 0; let T and S be two representations of A of the same dimension with the property that for every $a \in A$, there is a matrix C_a such that $C_aT(a)C_a^{-1} = S(a)$. Prove that the representations T and S are similar.[4]

[4] This exercise has been proposed by A.V. Rojter.

3. The Radical

Theorems 2.4.3 (Wedderburn-Artin) and 2.6.2 give a complete description of semisimple algebras and their representations. In comparison, we know very little on the structure of non-semisimple algebras and modules over them, even in the case when K is algebraically closed. The fundamental concept here is the notion of a *radical*: the least ideal such that the respective quotient algebra is semisimple. An essential property of the radical is its nilpotency. It allows to "lift the idempotents modulo the radical". In this way, the class of *projective modules*, related to semisimple modules, appears in a natural way. Their decomposition into the indecomposable ones can be shown to be unique, and by means of the endomorphism algebras, this result can be extended to arbitrary modules. Finally, in the last section of this chapter, we introduce the concept of a *diagram of an algebra* and of a *universal algebra over a diagram*; making use of them we obtain a description (of course, by no means complete) of algebras, at least in the algebraically closed case.[5] In particular, we obtain the classification of so-called *hereditary algebras* (over an algebraically closed field).

Recall that, unless stated otherwise, all algebras are finite dimensional (infinite dimensional algebras will appear in Sect. 3.6 as universal algebras over diagrams).

3.1 The Radical of a Module and of an Algebra

Let M be a semisimple module: $M = \bigoplus_{i=1}^{s} U_i$, where U_i are simple modules and let π_i be the projection of M on U_i. Then, for any non-zero element $m \in M$, $\pi_i(m) \neq 0$ for at least one index i. One can say that the homomorphisms of M into all possible simple modules "distinguish" the elements of the module M. Conversely, it is not difficult to verify that if the homomorphisms of M into simple modules distinguish the elements, then M is semisimple. Indeed, if N is a minimal submodule of M and $n \in N$ a non-zero element, then $f(n) \neq 0$ for a suitable homomorphism $f : M \to U$, where U is a simple module. But then $N \cap \operatorname{Ker} f = 0$ because N is minimal. Besides, $\operatorname{Im} f = U$, i.e. $M/\operatorname{Ker} f \simeq U$

[5] In Chapter 8 these concepts will be generalized to arbitrary algebras whose quotients by the radical are separable.

and $\operatorname{Ker} f$ is a maximal submodule of M. Therefore, $N + \operatorname{Ker} f = M$ and $\operatorname{Ker} f$ is a complement of N in M. Thus M is a semisimple module (see Proposition 2.2.1).

For a given module M, we shall introduce "a measure of how far it is from being semisimple": the set of all elements $m \in M$ such that $f(m) = 0$ for any homomorphism f of M to a simple module. Evidently, these elements form a submodule of M, which will be called the *radical of the module* M and be denoted by $\operatorname{rad} M$.

Since, for any non-zero homomorphism $f : M \to U$, where U is a simple module, $\operatorname{Ker} f$ is a maximal submodule of M, and conversely, since every maximal submodule $M' \subset M$ is a kernel of the projection $\pi : M \to M/M'$, and the module M/M' is simple, the radical is the intersection of all maximal submodules of the module M.

Theorem 3.1.1. *A module M is semisimple if and only if* $\operatorname{rad} M = 0$. *The factor module $M/\operatorname{rad} M$ is always semisimple.*

Proof. The first assertion has already been proved above. Therefore, it is sufficient to prove that $\operatorname{rad}(M/\operatorname{rad} M) = 0$. But, by Theorem 1.4.3, the maximal submodules of $M/\operatorname{rad} M$ are of the form $M'/\operatorname{rad} M$, where M' is maximal submodule of M (since always $M' \supset \operatorname{rad} M$). Clearly, $\cap(M'/\operatorname{rad} M) = (\cap M')/\operatorname{rad} M = 0$, i. e. $\operatorname{rad}(M/\operatorname{rad} M) = 0$ and the theorem is proved. □

Proposition 3.1.2. $\operatorname{rad}(\bigoplus_{i=1}^{s} M_i) = \bigoplus_{i=1}^{s} \operatorname{rad} M_i$.

Proof. Every homomorphism $f : M \to U$, where $M = \bigoplus_{i=1}^{s} M_i$, is determined uniquely by a family of homomorphisms $f_i : M_i \to U$ according to the formula $f(m_1, m_2, \ldots, m_s) = \sum_{i=1}^{s} f_i(m_i)$ (see Sect. 1.7). Therefore, if $m_i \in \operatorname{rad} M_i$ for all i, then $f(m_1, m_2, \ldots, m_s) = 0$ and thus $(m_1, m_2, \ldots, m_s) \in \operatorname{rad} M$.

Conversely, if $(m_1, m_2, \ldots, m_s) \in \operatorname{rad} M$, consider the homomorphisms $f : M \to U$ for which $f_j = 0$ for $j \neq i$. We get that $f_i(m_i) = 0$ for any homomorphism $f_i : M_i \to U$. Consequently, $m_i \in \operatorname{rad} M_i$, as required. □

Proposition 3.1.3. $f(\operatorname{rad} M) \subset \operatorname{rad} N$ *for any module homomorphism* $f : M \to N$.

Proof. If $m \in \operatorname{rad} M$, then, for any homomorphism $g : N \to U$, where U is a simple module, $gf(m) = 0$, i. e. $f(m) \in \operatorname{rad} N$. □

This result implies that, for any homomorphism $f : M \to N$, one can construct the induced homomorphism $\tilde{f} : M/\operatorname{rad} M \to N/\operatorname{rad} N$ by setting $\tilde{f}(m + \operatorname{rad} M) = f(m) + \operatorname{rad} N$.

Lemma 3.1.4 (Nakayama). *A homomorphism $f : M \to N$ is an epimorphism if and only if the induced homomorphism $\tilde{f} : M/\operatorname{rad} M \to N/\operatorname{rad} N$ is an epimorphism.*

Proof. If f is an epimorphism, then clearly $\tilde{f}$ is an epimorphism. Conversely, the fact that $\tilde{f}$ is an epimorphism means that $\operatorname{Im} f + \operatorname{rad} N = N$. But if $\operatorname{Im} f \neq N$, then $\operatorname{Im} f$ is contained in a maximal submodule N' of N. Since $\operatorname{rad} N \subset N'$, $\operatorname{Im} f + \operatorname{rad} N \subset N'$ and thus cannot be N. Consequently, $\operatorname{Im} f = N$, i.e. f is an epimorphism. □

It follows that the fact that a homomorphism is an epimorphism is sufficient to be verified "modulo the radical". Occasionally, the following form of this result can be found useful.

Corollary 3.1.5. *If N and L are submodules of M such that $N + L = M$ and $N \subset \operatorname{rad} M$, then $L = M$.*

Proof. An exercise. □

The radical of a regular module of an algebra A is called the *radical of the algebra*. By definition, $\operatorname{rad} A$ is a right ideal. It follows from Theorem 3.1.1 that the semisimplicity of the algebra A is equivalent to the equality $\operatorname{rad} A = 0$.

Theorem 3.1.6. *For any A-module M, $\operatorname{rad} M = MR$, where $R = \operatorname{rad} A$. In particular, the radical of an algebra is a two-sided ideal and the corresponding quotient algebra is semisimple.*

Proof. Consider the homomorphism $A \to M$ which maps every element $a \in A$ into ma, where m is a fixed element of M (see Theorem 1.7.1). By Proposition 3.1.3, it maps the radical into the radical, i.e. $mR \subset \operatorname{rad} M$. Therefore $MR \subset \operatorname{rad} M$.

In particular, $AR \subset R$, i.e. R is a two-sided ideal and the quotient algebra $A/R = \bar{A}$ is semisimple according to Theorem 3.1.1.

Consider the factor module $\bar{M} = M/MR$. It is annihilated by the radical R and can be therefore considered as an $\bar{A}$-module which is semisimple in view of Theorem 2.6.2. Consequently, for every non-zero class $\bar{m} = m + MR$, there is a homomorphism $f : \bar{M} \to U$, with a simple module U such that $f(\bar{m}) \neq 0$. In combination with the projection $\pi : M \to \bar{M}$, we obtain the homomorphism $f\pi : M \to U$ satisfying $f\pi(m) \neq 0$, i.e. $m \notin \operatorname{rad} M$. It follows that $\operatorname{rad} M \subset MR$ and the proof of our theorem is completed. □

Let us remark that we have, in fact, proved that $MR = \operatorname{rad} M$ is the least submodule of M such that the respective factor module is semisimple.

Corollary 3.1.7. *Every semisimple A-module is a module over the semisimple quotient algebra $\bar{A} = A/radA$. In particular, the number of simple A-modules is equal to the number of the simple components of the algebra $\bar{A}$.*

Corollary 3.1.8. *The radical of an algebra is the intersection of all maximal ideals.*

Proof. If I is a maximal ideal of an algebra A, then A/I is a simple algebra, and hence it is a semisimple A-module. Therefore, $(A/I)R = 0$, i.e. $R \subset I$ and I/R is a maximal ideal of the quotient algebra $\bar{A} = A/R$ (Theorem 1.4.7). Denote by J the intersection of all maximal ideals of the algebra A. Then $J \supset R$ and J/R is the intersection of all maximal ideals of the algebra $\bar{A}$. But $\bar{A}$ is a semisimple algebra, and thus $J/R = 0$, i.e. $J = R$. □

Corollary 3.1.8 provides a symmetric characterization of the radical (free of the concepts "right" and "left"). It follows that the radical of the left regular module coincides with the radical of the algebra.

Proposition 3.1.9. *The radical R of an algebra A is a nilpotent ideal containing all nilpotent right and left ideals.*

Proof. Theorem 3.1.6 together with Nakayama's lemma (Corollary 3.1.5) implies that $R^2 = \operatorname{rad} R \neq R$ and, in general, $R^{m+1} = \operatorname{rad} R^m \neq R^m$ whenever $R^m \neq 0$. However, the chain of the subspaces $R \supset R^2 \supset \ldots \supset R^m \supset R^{m+1} \supset \ldots$ must become stationary, and thus $R^m = 0$ for some m.

Conversely, if I is a nilpotent right ideal, then $(I+R)/R$ is a nilpotent right ideal of the semisimple algebra A/R. Hence, by Corollary 2.2.5, $(I+R)/R = 0$, i.e. $I + R = R$. □

Corollary 3.1.10. *The radical of an algebra is the set of all strongly nilpotent elements.*

An important property of the radical relates to the following concept. A right (left) ideal I of an algebra A is called *quasiregular* if every element $1 - x$, where $x \in I$, is invertible.

Proposition 3.1.11. *The radical is a quasiregular ideal containing all right and left quasiregular ideals.*

Proof. If $x \in R$, then $x^k = 0$ for some k and therefore $(1 - x)(1 + x + x^2 + \ldots + x^{k-1}) = 1 - x^k = 1$, i.e. $1 - x$ is invertible. Conversely, let $I \not\subset R$, where I is a right ideal. Then $I \not\subset M$ for some maximal right ideal M. Thus $I + M = A$. In particular, $1 = x + m$, where $x \in I$ and $m \in M$. Consequently, $m = 1 - x$ is not invertible because $mA \neq A$. Hence, I is not quasiregular. □

Thus, in finite dimensional algebras, the concepts of being nilpotent and quasiregular coincide.

Corollary 3.1.12. *Every right (left) nil ideal (i. e. such whose elements are nilpotent) is contained in the radical and is therefore nilpotent.*

Proof. The statement follows from the fact that, for a nil right ideal I, all elements $1-x$ $(x \in I)$ are invertible; for, if $x^m = 0$, then $(1-x)(1+x+x^2+\dots+x^{m-1}) = 1$. □

In what follows, we shall often make use of the following characteristic of the radical.

Proposition 3.1.13. *The radical of an algebra is the unique nil ideal such that the respective quotient algebra is semisimple.*

Proof. The radical has all these properties, as we have seen above. Conversely, since A/I is semisimple, it follows that $I \supset \operatorname{rad} A$ and, from the fact that I is a nil ideal, we deduce that $I \subset \operatorname{rad} A$. □

Corollary 3.1.14. $\operatorname{rad}(A/I) = (R+I)/I$, *where* $R = \operatorname{rad} A$.

Proof. Since R is nilpotent, it follows that $(R+I)/I$ is nilpotent. At the same time, $(A/I)/((R+I)/I) \simeq A/(R+I) \simeq (A/R)/((R+I)/R)$. Since A/R is semisimple, its quotient algebra is also semisimple; thus $(R+I)/I = \operatorname{rad}(A/I)$. □

3.2 Lifting of Idempotents. Principal Modules

The fact that the radical is nilpotent provides a powerful method of investigation of non-semisimple algebras, viz. lifting of idempotents modulo the radical.

Lemma 3.2.1. *Let I be a nil ideal of an algebra A and u an element of the algebra such that $u^2 \equiv u \pmod{I}$. Then there is an idempotent e in A such that $e \equiv u \pmod{I}$.*

Proof. Put $v = u + r - 2ur$, where $r = u^2 - u$. Then $ur = ru$, $r^2 \in I^2$ and thus $v^2 \equiv u^2 + 2ur - 4u^2r \equiv u + r + 2ur - 4ur \equiv v \pmod{I^2}$. Besides, evidently, $v \equiv u \pmod{I}$. Proceeding in this way with the element v in place of u, we can construct an element v_1 such that $v_1^2 \equiv v_1 \pmod{I^4}$ and $v_1 \equiv v \pmod{I^2}$. Thus $v_1 \equiv u \pmod{I}$. Finally, continuing this process, and taking into account that the ideal I is nilpotent, we can construct a required idempotent e. □

Lemma 3.2.1 above yields a characterization of the algebras whose regular module is indecomposable.

Theorem 3.2.2. *The following conditions are equivalent:*

1) *the regular A-module is indecomposable;*
2) *A/R, where $R = \operatorname{rad} A$, is a division algebra;*

3) *there is a unique maximal right ideal in the algebra A;*
4) *the non-invertible elements of the algebra A form a right ideal.*[6]

Proof. 1) $\Rightarrow$ 2). The condition 1) means that there are no non-trivial idempotents in the algebra $E_A(A) = A$ (Corollary 1.7.3). But then, in view of Lemma 3.2.1, there are no non-trivial idempotents in the semisimple algebra A/R, and thus A/R is a division algebra.

2) $\Rightarrow$ 4). If A/R is a division algebra and an element $a \in A$ does not belong to R, then the class $a + R$ is invertible in A/R, i. e. there is an element $b \in A$ such that $ab \equiv 1 \pmod R$. It follows from Proposition 3.1.11 that the element ab, and therefore also a, is invertible. Since all elements of the radical are not invertible, the radical R is just the set of all non-invertible elements, as required.

4) $\Rightarrow$ 3). Let I be the right ideal consisting of all non-invertible elements of the algebra A. Then every right ideal $J \neq A$ must be contained in I because the ideal J cannot contain invertible elements, and thus I is the unique maximal right ideal.

3) $\Rightarrow$ 1) follows from the fact that if M is decomposable, $M = N \oplus L$, where $N \neq 0$, $L \neq 0$, then M contains at least two maximal submodules $N' \oplus L$ and $N \oplus L'$, where N' and L' are maximal submodules of N and L, respectively. □

The algebras satisfying the conditions of Theorem 3.2.2 are called *local.* Corollary 1.7.4 implies the following assertion.

Corollary 3.2.3 (Fitting). *A module is indecomposable if and only if its endomorphism algebra is local.*

We shall apply the above results to direct summands of the regular module. The modules which are isomorphic to indecomposable direct summands of the regular module are called *principal indecomposable* modules, or simply *principal* modules. In other words, a principal A-module is of the form eA, where e is a *minimal idempotent*: it cannot be represented in the form $e = e' + e''$, where e' and e'' are non-zero orthogonal idempotents. By corollary 3.2.3, this is equivalent to the fact that the algebra $E_A(eA) \simeq eAe$ is local.

Proposition 3.2.4. *For an idempotent e of A,* $\operatorname{rad}(eAe) = eRe$. *An idempotent $e \in A$ is minimal if and only if the idempotent $\bar{e} = e + R$ of the algebra $\bar{A} = A/R$ is minimal.*

Proof. Evidently, eRe is an ideal of eAe and is nilpotent (because R is nilpotent). On the other hand, $eAe/eRe \simeq \bar{e}\bar{A}\bar{e}$ is a semisimple algebra. By Proposition 3.1.13, $eRe = \operatorname{rad}(eAe)$. Now, the second assertion follows from Theorem 3.2.2, since $\bar{e}$ is minimal if and only if $\bar{e}\bar{A}\bar{e}$ is a division algebra. □

[6] Since condition 2) is symmetric, one can replace right modules and ideals in 1), 3) and 4) by left ideals.

Corollary 3.2.5. *A principal module contains a unique maximal submodule.*

Proof. If eA is a principal module, then $\bar{e} = e + R$ is a minimal idempotent; thus $\bar{e}\bar{A} \simeq eA/eR$ is a simple module and eR is a maximal submodule of eA. But, by Theorem 3.1.6, $eR = \operatorname{rad}(eA)$ is contained in all maximal submodules and is therefore the unique maximal submodule of eA. □

Proposition 3.2.6. *If $f : eA \to M$ is a homomorphism of A-modules, then $f(e) \in Me$ and, making the element $f(e)$ correspond to f, we establish an isomorphism of the vector spaces $\operatorname{Hom}_A(eA, M) \simeq Me$.*

Proof. $f(e) = f(e^2) = f(e)e \in Me$, and if $f(e) = g(e)$, then $f(ea) = f(e)a = g(e)a = g(ea)$ for any $a \in A$, and thus the map $\operatorname{Hom}_A(eA, M) \to Me$ is a monomorphism. On the other hand, restricting the homomorphism $A \to M$ which maps a into ma, to eA, we obtain the homomorphism $f : eA \to M$ which maps e into me and therefore the map $\operatorname{Hom}_A(eA, M) \to Me$ is an epimorphism. □

Corollary 3.2.7. *For any homomorphism $f : eA \to M$ and any epimorphism $g : N \to M$, there is a homomorphism $\varphi : eA \to N$ such that $f = g\varphi$.*

Proof. Let $f(e) = me$ and n be a preimage of m in N. Then $\varphi : eA \to N$ can be defined by mapping e into ne. □

Corollary 3.2.8. *If a module M has a unique maximal submodule, then M is isomorphic to a factor module of a principal module.*

Proof. Let M' be the unique maximal submodule of M. Then $M' = MR$ and M/MR is a simple module (here $R = \operatorname{rad} A$). Thus $M/MR \simeq \bar{e}\bar{A}$ for some minimal idempotent $\bar{e}$ of the algebra $\bar{A} = A/R$. By Lemma 3.2.1, $\bar{e} = e + R$ with an idempotent e of A which is by Proposition 3.2.4 minimal and for which $\bar{e}\bar{A} \simeq eA/eR$. Denote by π the projection of M onto $\bar{e}\bar{A}$ and by f the projection of eA onto $\bar{e}\bar{A}$. By Corollary 3.2.7, there is a homomorphism $\varphi : eA \to M$ such that $\pi\varphi = f$ and, moreover, the induced homomorphism $\bar{\varphi} : eA/eR \to M/MR$ is an isomorphism. By Nakayama's lemma (Lemma 3.1.4), φ is an epimorphism and $M \simeq eA/\operatorname{Ker}\varphi$. □

Corollary 3.2.9. *The principal modules eA and fA are isomorphic if and only if the simple modules $\bar{e}\bar{A}$ and $\bar{f}\bar{A}$ are isomorphic ($\bar{A} = A/R$ and $\bar{a} = a + R$).*

Proof. If $eA \simeq fA$, then $\bar{e}\bar{A} \simeq eA/eR \simeq fA/fR \simeq \bar{f}\bar{A}$. Conversely, let $\bar{e}\bar{A} \simeq \bar{f}\bar{A}$. Combining this isomorphism with the projection $eA \to \bar{e}\bar{A}$, we obtain an epimorphism $g : eA \to \bar{f}\bar{A}$. By Corollary 3.2.7, $g = \pi\varphi$, where $\varphi : eA \to fA$ and π is a projection $fA \to \bar{f}\bar{A}$. Since φ induces an isomorphism $\bar{e}\bar{A} \simeq \bar{f}\bar{A}$, it is an epimorphism (by Lemma 3.1.4). In a similar way, one can define an epimorphism $\psi : fA \to eA$. But then both $\varphi\psi$ and $\psi\varphi$ are epimorphisms and therefore isomorphisms (since eA and fA are finite dimensional spaces). Therefore φ and ψ are isomorphisms and the proof of the corollary is completed. □

In this way, a natural bijective correspondence has been established between the principal and the simple modules. Let us remark that the same results hold for the left modules. Now one can see easily that in a semisimple algebra $\bar{A}$, $\bar{e}\bar{A} \simeq \bar{f}\bar{A}$ if and only if $\bar{A}\bar{e} \simeq \bar{A}\bar{f}$ (any of these isomorphisms means that e and f belong to the same simple component). Hence, one obtains the following result.

Corollary 3.2.10. *The left principal modules Ae and Af are isomorphic if and only if the modules eA and fA are isomorphic.*

3.3 Projective Modules and Projective Covers

Corollary 3.2.7 expresses the most important property of principal modules, viz. their projectivity.

A module P over an algebra A is called *projective* if for every epimorphism $g : M \to N$ and every homomorphism $f : P \to N$ there is a homomorphism $\varphi : P \to M$ such that $f = g\varphi$. We say that f *can be lifted* to φ or that φ is a *lifting of* f to M.

Proposition 3.3.1. *Two projective modules P and Q are isomorphic if and only if the semisimple modules $\bar{P} = P/PR$ and $\bar{Q} = Q/QR$ are isomorphic (here $R = \operatorname{rad} A$).*

Proof. Every isomorphism $P \xrightarrow{\sim} Q$ induces an isomorphism $\bar{P} \simeq \bar{Q}$. Conversely, if $\bar{P} \simeq \bar{Q}$, then there is an epimorphism $f : P \to \bar{Q}$ which can be lifted to a homomorphism $\varphi : P \to Q$ (since P is a projective module); by Nakayama's lemma, φ is an epimorphism. In a similar way, one obtains an epimorphism $\psi : Q \to P$. Comparing the dimensions, it turns out that φ and ψ are isomorphisms. □

Proposition 3.3.2. *A direct sum of modules is projective if and only if every direct summand is projective.*

Proof. Every homomorphism $f : P \oplus Q \to N$ is uniquely determined by a pair of homomorphisms $f_1 : P \to N$ and $f_2 : Q \to N$; indeed, $f(p,q) = f_1(p) + f_2(q)$. Now, the solution of the equation $f = g\varphi$, where g is an epimorphism $M \to N$ and φ the required homomorphism $P \oplus Q \to M$, is a pair $\varphi_1 : P \to M$ and $\varphi_2 : Q \to M$ such that $f_1 = g\varphi_1$ and $f_2 = g\varphi_2$. Consequently, f can be lifted if and only if each of the homomorphisms f_1 and f_2 can be lifted. □

Among the fundamental examples of projective modules are the free modules. A *free module* is a module which is isomorphic to a direct sum of regular modules, i. e. is of the form nA for some n. The elements of a free module can be viewed as the "vectors" $(a_1, a_2, \dots, a_n)$ with the components from A and componentwise operations. The number n is called the *rank* of the free module.

Proposition 3.3.3. *Every homomorphism $f : nA \to M$ is given uniquely by a choice of the elements $\{m_1, m_2, \dots, m_n\}$ of the module M, according to the formula*

$$f(a_1, a_2, \dots, a_n) = \sum_{i=1}^{n} m_i a_i \,. \tag{3.3.1}$$

Hence, $\mathrm{Hom}_A(nA, M) \simeq nM$.

Proof. It is trivial to verify that the map $f : nA \to M$ given by the formula (3.3.1) for arbitrary $m_1, m_2, \dots, m_n$ is a homomorphism. Conversely, if f is a homomorphism $nA \to M$, we put $m_i = f(u_i)$, where $u_i = (0, \dots, 1, \dots, 0)$ (with 1 at the ith position). Since $(a_1, a_2, \dots, a_n) = \sum_{i=1}^{n} u_i a_i$, $f(a_1, a_2, \dots, a_n) = \sum_{i=1}^{n} f(u_i) a_i = \sum_{i=1}^{n} m_i a_i$ and the proposition is proved. □

A set of elements $\{m_1, m_2, \dots, m_n\}$ of a module M is called a *generating set* (a set of generators) if every element $m \in M$ can be expressed in the form $m = \sum_{i=1}^{n} m_i a_i$ for some $a_i \in A$. This means that the homomorphism $f : nA \to M$ defined by the formula (3.3.1) is an epimorphism. The homomorphism theorem then yields the following corollary.

Corollary 3.3.4. *If a module M has a generating set consisting of n elements, then M is isomorphic to a factor module of the free module of rank n.*

Observe that a finite dimensional module M always has a finite generating set (for example, a basis). The least number of elements in any of the generating sets of M is called the *number of generators* of the module M and is denoted by $\mu_A(M)$.

Free modules as well as principal ones are projective (by Proposition 3.3.2). The following theorem describes the relation between free, principal and projective modules.

Theorem 3.3.5. *The following conditions for a module P are equivalent:*

1) *P is projective;*
2) *P is isomorphic to a direct sum of principal modules;*
3) *P is isomorphic to a direct summand of a free module;*
4) *the kernel of every epimorphism $f : M \to P$ has a complement in M (and then, evidently, $M \simeq P \oplus \mathrm{Ker}\, f$).*

Proof. 2) $\Rightarrow$ 1) and 3) $\Rightarrow$ 1) by Proposition 3.3.2.

4) $\Rightarrow$ 3) by Corollary 3.3.4.

1) $\Rightarrow$ 4). There exists a homomorphism $\varphi : P \to M$ such that $f\varphi = 1$, and by Proposition 1.6.2, this is equivalent to the fact that $\operatorname{Ker} f$ has a complement in M.

1) $\Rightarrow$ 2). Decompose the semisimple module $\bar{P} = P/PR$ into a direct sum $\bar{P} = U_1 \oplus U_2 \oplus \ldots \oplus U_t$ of simple modules U_i. Let P_i be a principle module such that $P_i/P_iR \simeq U_i$. Write $Q = \bigoplus_{i=1}^{t} P_i$. Then Q is a projective module and $Q/QR \simeq P/PR$. Consequently, $P \simeq Q$ by Proposition 3.3.1. □

It turns out that the projective modules are in a bijective correspondence with the semisimple ones, in a similar manner as the principal modules are in a bijective correspondence with the simple ones.

Theorem 3.3.6. *The map assigning to every projective module P the semisimple module $\bar{P} = P/PR$, is a bijective correspondence between the projective and semisimple modules. If $P = P_1 \oplus P_2 \oplus \ldots \oplus P_n = Q_1 \oplus Q_2 \oplus \ldots \oplus Q_m$ are two decompositions of a projective module P into a direct sum of principal modules, then $n = m$ and, under a suitable relabelling, $P_i \simeq Q_i$ for all i.*

Proof. In view of Proposition 3.3.1, we need to verify only that every semisimple module M is isomorphic to P/PR, where P is a projective module. To this end, it is sufficient to decompose M into a direct sum of simple modules $M = U_1 \oplus U_2 \oplus \ldots \oplus U_n$ and to put $P = P_1 \oplus P_2 \oplus \ldots \oplus P_n$, where P_i is a principal modules satisfying $U_i \simeq P_i/P_iR$.

If $P_1 \oplus P_2 \oplus \ldots \oplus P_n \simeq Q_1 \oplus Q_2 \oplus \ldots \oplus Q_m$, where P_i and Q_i are principal modules, then for the factors modulo the radical we obtain an isomorphism $\bar{P}_1 \oplus \bar{P}_2 \oplus \ldots \oplus \bar{P}_n \simeq \bar{Q}_1 \oplus \bar{Q}_2 \oplus \ldots \oplus \bar{Q}_m$ with the simple modules $\bar{P}_i = P_i/P_iR$ and $\bar{Q}_j = Q_j/Q_jR$. By Proposition 2.2.3, $n = m$ and $\bar{P}_i \simeq \bar{Q}_i$ for all i (after an appropriate reindexing), and then, by Proposition 3.2.9, also $P_i \simeq Q_i$. □

It follows from Proposition 3.3.4 that every module is isomorphic to a factor module of a projective module. Of course, there are many different ways of representing a given module M in the form P/N with a projective module P. However, we are going to show that there is, in a certain sense, a unique minimal representation of this form.

A projective module P is called a *projective cover* of a module M and is denoted by $P(M)$ if there is an epimorphism $f : P \to M$ which induces an isomorphism $P/\operatorname{rad} P \xrightarrow{\sim} M/\operatorname{rad} M$. Evidently, this is equivalent to the fact that $M \simeq P/N$ with $N \subset \operatorname{rad} P$.

Theorem 3.3.7. 1) *For every module M, there is a projective cover and it is unique up to an isomorphism.*

2) *$P(M) \simeq P(\bar{M})$, where $\bar{M} = M/\operatorname{rad} M$.*

3) *If g is an epimorphism of a projective module Q onto a module M, then $Q \simeq P_1 \oplus P_2$, where $P_1 \simeq P(M)$, the restriction of g to P_1 is an epi-*

morphism, and $P_2 \subset \operatorname{Ker} g$. *Moreover, the isomorphism* $P(M) \xrightarrow{\sim} P_1$ *can be chosen in such a way that the composition with the epimorphism* g *is equal to a fixed epimorphism* $f : P(M) \to M$.

Proof. By Theorem 3.3.6, $\bar{M} \simeq P/\operatorname{rad} P$ for some projective module P. Consequently, the epimorphism $P \to \bar{M}$ can be extended to a homomorphism $f : P \to M$, whereby, by Nakayama's lemma, f is an epimorphism. Thus, P is a projective cover of M and the assertions 1) and 2) are proved (the fact that the cover is unique follows from Proposition 3.3.1).

Now, let Q be a projective module and $g : Q \to M$ an epimorphism. Let us lift g to a homomorphism $\varphi : Q \to P$ such that $f\varphi = g$. Then the induced map $\bar{\varphi} : Q/\operatorname{rad} Q \to P/\operatorname{rad} P$ is an epimorphism and, by Nakayama's lemma, φ is an epimorphism. By Theorem 3.3.5, $Q = P_1 \oplus \operatorname{Ker} \varphi$, where $P_1 \simeq P$. Since $P_2 = \operatorname{Ker} \varphi \subset \operatorname{Ker} g$, we have obtained the assertion 3), as well. □

Corollary 3.3.8. $P(M \oplus N) \simeq P(M) \oplus P(N)$.

The proof is immediate.

Corollary 3.3.9. *Let* $R = \operatorname{rad} A$, $\bar{A} = A/R$ *and* $1 = \bar{e}_1 + \bar{e}_2 + \ldots + \bar{e}_n$ *be a decomposition of the identity of the algebra* $\bar{A}$. *Then there is a decomposition of the identity* $1 = e_1 + e_2 + \ldots + e_n$ *of the algebra* A *such that* $\bar{e}_i = e_i + R$.

Proof. Evidently, $A = P(\bar{A})$. On the other hand, if $U_i = \bar{e}_i\bar{A}$ and $P_i = P(U_i)$, then $P(\bar{A}) \simeq P_1 \oplus P_2 \oplus \ldots \oplus P_n$. Consequently, $A \simeq P_1 \oplus P_2 \oplus \ldots \oplus P_n$ and the isomorphism can be chosen in such a way that the composition with the natural epimorphism $P_1 \oplus P_2 \oplus \ldots \oplus P_n \to U_1 \oplus U_2 \oplus \ldots \oplus U_n = \bar{A}$ would give the projection $\pi : A \to \bar{A}$. Let $1 = e_1 + e_2 + \ldots + e_n$ be a decomposition of the identity of the algebra A corresponding to the decomposition $A \simeq P_1 \oplus P_2 \oplus \ldots \oplus P_n$ of the regular module. Then $e_iA/e_iR \simeq \bar{e}_i\bar{A}$, i.e. the idempotents $e_i + R$ determine a decomposition of the identity corresponding to the decomposition $\bar{A} = \bar{e}_1\bar{A} \oplus \bar{e}_2\bar{A} \oplus \ldots \oplus \bar{e}_n\bar{A}$. In view of the fact that there is a bijective correspondence between the decompositions of the identity and the decompositions of the module (Theorem 1.7.2), $e_i + R = \bar{e}_i$, as required. □

In conclusion of this section, we are going to apply the above results to the study of a particular class of algebras.

An algebra A is called *primary* if $A/\operatorname{rad} A$ is a simple algebra.

Theorem 3.3.10. *The following statements for an algebra* A *are equivalent:*

1) *A is primary;*
2) *there is a unique maximal ideal in A;*
3) *every proper ideal of the algebra A is nilpotent;*
4) *A has a single simple module;*

5) *A has a single principal module;*
6) $A \simeq M_n(B)$, *where B is a local algebra.*

Proof. We are going to give a proof following the implications 3) ⇒ 2) ⇒ 1) ⇒ 4) ⇒ 5) ⇒ 6) ⇒ 1) ⇒ 3).

3) ⇒ 2). If all ideals are nilpotent, then they are all contained in $R = \operatorname{rad} A$ (Proposition 3.1.9) and thus R is the unique maximal ideal.

2) ⇒ 1). This follows from Corollary 3.1.8. Also, 1) ⇒ 4) follows from Corollary 3.1.7, and 4) ⇒ 5) follows from Corollary 3.2.9.

5) ⇒ 6). The regular module A is isomorphic to nP, where P is the unique principal module; from here, $A \simeq E_A(A) \simeq M_n(B)$, where $B = E_A(P)$ is a local algebra (by Corollary 3.2.3).

6) ⇒ 1). Let $A = M_n(B)$. We shall prove that $\operatorname{rad} A = M_n(J)$, where $J = \operatorname{rad} B$. Indeed, $M_n(J)$ is a nilpotent ideal in A and $A/M_n(J) \simeq M_n(B/J)$ is a simple algebra, because B/J is a division algebra. Therefore $M_n(J)$ is $\operatorname{rad} A$ (by Proposition 3.1.13) and A is a primary algebra.

1) ⇒ 3). If I is an ideal of a primary algebra A, then $(I+R)/R$ is an ideal of the simple algebra A/R. From here, either $(I+R)/R = 0$, i. e. $I \subset R$, or $(I+R)/R = A/R$, i. e. $I + R = A$, and by Nakayama's lemma, $I = A$. This means that if $I \neq A$, then the ideal $I \subset R$ and therefore it is nilpotent. □

Observe that, by proving the implication 6) ⇒ 1), we have also proved the following proposition.

Proposition 3.3.11. $\operatorname{rad} M_n(B) = M_n(\operatorname{rad} B)$.

3.4 The Krull-Schmidt Theorem

Theorem 3.3.6 implies, in particular, that a decomposition of the regular module into a direct sum of indecomposable modules is unique. In this section, using the results of Sect. 1.7, we shall extend this fact to arbitrary modules. First, we shall express it in terms of the idempotents.

Theorem 3.4.1. *Let* $1 = e_1 + e_2 + \ldots + e_n = f_1 + f_2 + \ldots + f_m$ *be two decompositions of the identity of an algebra A with minimal idempotents* e_i *and* f_j*. Then* $n = m$ *and there is an invertible element a in the algebra A such that, up to a suitable reindexing,* $f_i = a e_i a^{-1}$ *for all i.*

Proof. Let $A = e_1A \oplus e_2A \oplus \ldots \oplus e_nA = f_1A \oplus f_2A \oplus \ldots \oplus f_mA$ be two decompositions of the regular module into a direct sum of principal modules. By Theorem 3.3.6, $n = m$ and $e_iA \simeq f_iA$ for all i (up to a suitable relabelling). But the isomorphism $e_iA \xrightarrow{\sim} f_iA$ is given by a suitable element $a_i \in f_iAe_i$ such that $f_ia_i = a_ie_i = a_i$. Put $a = a_1 + a_2 + \ldots + a_n$. Then $ae_i = a_ie_i = a_i$ and $f_ia = f_ia_i = a_i$ for all i. We show that a is invertible. To that end, we choose elements $b_i \in e_iAf_i$ defining the isomorphism $f_iA \simeq e_iA$ reciprocal to a_i and put $b = b_1 + b_2 + \ldots + b_n$. Since $a_ib_i = f_i$ and $e_ib = b_i = bf_i$, $ab = \sum_{i=1}^{n} a_ib_i = \sum_{i=1}^{n} f_i = 1$ and $b = a^{-1}$. Consequently, the equality $ae_i = f_ia$ yields $f_i = ae_ia^{-1}$. The theorem is proved. □

Now, in order to prove the uniqueness of module decompositions, we apply Theorem 3.4.1 to endomorphism algebras.

Theorem 3.4.2 (Krull-Schmidt). *If* $M = M_1 \oplus M_2 \oplus \ldots \oplus M_n = N_1 \oplus N_2 \oplus \ldots \oplus N_m$ *are two decompositions of the module* M *into a direct sum of indecomposable modules, then* $n = m$ *and, after a suitable reindexing,* $M_i \simeq N_i$ *for all* i.

Proof. By Theorem 1.7.2, we have two decompositions of the identity of the algebra $E = E_A(M)$ corresponding to the two decompositions of M into a direct sum of indecomposable modules: $1 = e_1+e_2+\ldots+e_n = f_1+f_2+\ldots+f_m$, where e_i and f_j are minimal idempotents, $M_i = e_iM$ and $N_j = f_jM$. By Theorem 3.4.1, $n = m$ and, after a suitable renumbering, $f_i = ae_ia^{-1}$, where a is an invertible element of the algebra E, i. e. an automorphism of the module M. Let a_i be the restriction of a to M_i. Since $ae_i = f_ia$, $ae_i(m) \in f_iM = N_i$, and thus a_i maps M_i into N_i. Since a is a monomorphism, a_i is also a monomorphism. On the other hand, a is an epimorphism, and therefore any element y of N_i is of the form $y = a(x)$. But then $f_i(y) = y = f_ia(x) = ae_i(x)$ with $e_i(x) \in M_i$; thus a_i is an epimorphism of M_i onto N_i, and therefore an isomorphism. The theorem is proved. □

3.5 The Radical of an Endomorphism Algebra

We shall apply the above results to clarify the behaviour of the radical with respect to a Peirce decomposition.

Lemma 3.5.1. *Let* $f : M \to N$ *be a homomorphism between two indecomposable* A*-modules. Then either* f *is an isomorphism or, for any homomorphism* $g : N \to M$, $fg \in \operatorname{rad} E_A(N)$ *and* $gf \in \operatorname{rad} E_A(M)$.

Proof. Let $g \in \operatorname{Hom}_A(N, M)$ and $fg \notin \operatorname{rad} E_A(N)$. Then, since the algebra $E_A(N)$ is local (Corollary 3.2.3), fg is an invertible element and thus an automorphism of N. Consequently, f is an epimorphism. If $\varphi = (fg)^{-1}$, then $f(g\varphi) = 1$, and by Proposition 1.6.2, $M \simeq N \oplus \operatorname{Ker} f$. Since M is indecomposable, we get $\operatorname{Ker} f = 0$ and thus f is a monomorphism; therefore, f is an isomorphism. In a similar manner, one can show that $gf \notin \operatorname{rad} E_A(M)$ also implies that f is an isomorphism. □

Theorem 3.5.2. *Let $M = M_1 \oplus M_2 \oplus \ldots \oplus M_s$, where $M_i = n_i N_i$, the modules N_i are indecomposable and such that $N_i \not\simeq N_j$ for $i \neq j$. Write* $\operatorname{Hom}_A(M_i, M_j) = E_{ij}$ *and consider the two-sided Peirce decomposition of the algebra $E = E_A(M)$*

$$E = \begin{pmatrix} E_{11} & E_{12} & \ldots & E_{1s} \\ E_{21} & E_{22} & \ldots & E_{2s} \\ \ldots & \ldots & \ldots & \ldots \\ E_{s1} & E_{s2} & \ldots & E_{ss} \end{pmatrix}.$$

Then the radical of the algebra E has the form

$$R = \begin{pmatrix} R_{11} & E_{12} & \ldots & E_{1s} \\ E_{21} & R_{22} & \ldots & E_{2s} \\ \ldots & \ldots & \ldots & \ldots \\ E_{s1} & E_{s2} & \ldots & R_{ss} \end{pmatrix} \tag{3.5.1}$$

where $R_{ii} = \operatorname{rad} E_{ii}$. In other words, if $1 = e_1 + e_2 + \ldots + e_s$ is the corresponding decomposition of the identity of the algebra E, then $e_i R e_j = e_i E e_j$ for $i \neq j$ and $e_i R e_i = \operatorname{rad} e_i E e_i$.

Proof. According to Sect. 1.7, the elements of E_{ij} can be interpreted as matrices of dimension $n_i \times n_j$ with coefficients from $H_{ij} = \operatorname{Hom}_A(N_i, N_j)$. Therefore, $R_{ii} = M_{n_i}(R_i)$, where $R_i = \operatorname{rad} E_A(N_i)$ (see Proposition 3.3.11). Moreover, it follows from Lemma 3.5.1 that $E_{ij}E_{ji} \subset R_{ii}$ for $i \neq j$. Consequently, the set R defined by the formula (3.5.1) is an ideal in E and $E/R \simeq E_{11}/R_{11} \times E_{22}/R_{22} \times \ldots \times E_{ss}/R_{ss}$ is a semisimple algebra; therefore $R \supset \operatorname{rad} E$.

On the other hand, consider the right ideal

$$I_i = \begin{pmatrix} 0 & 0 & \ldots & 0 & \ldots & 0 \\ \ldots & \ldots & \ldots & \ldots & \ldots & \ldots \\ E_{i1} & E_{i2} & \ldots & R_{ii} & \ldots & E_{is} \\ \ldots & \ldots & \ldots & \ldots & \ldots & \ldots \\ 0 & 0 & \ldots & 0 & \ldots & 0 \end{pmatrix}$$

(the ith row of R). We shall show that it is nilpotent. Indeed,

$$I_i^2 = \begin{pmatrix} 0 & \ldots & 0 & \ldots & 0 \\ \ldots & \ldots & \ldots & \ldots & \ldots \\ R_{ii}E_{i1} & \ldots & R_{ii}^2 & \ldots & R_{ii}E_{is} \\ \ldots & \ldots & \ldots & \ldots & \ldots \\ 0 & \ldots & 0 & \ldots & 0 \end{pmatrix},$$

and in general,

$$I_i^{k+1} = \begin{pmatrix} 0 & \ldots & 0 & \ldots & 0 \\ \ldots & \ldots & \ldots & \ldots & \ldots \\ R_{ii}^k E_{i1} & \ldots & R_{ii}^{k+1} & \ldots & R_{ii}^k E_{is} \\ \ldots & \ldots & \ldots & \ldots & \ldots \\ 0 & \ldots & 0 & \ldots & 0 \end{pmatrix}.$$

Since $R_{ii}^k = 0$ for some k, $I_i^{k+1} = 0$. Therefore $I_i \subset \operatorname{rad} E$ and we obtain $R \subset I_1 + I_2 + \ldots + I_s \subset \operatorname{rad} E$, as required. □

If we apply Theorem 3.5.2 to the algebra $A \simeq E_A(A)$, we get the following result.

Theorem 3.5.3. *Let $1 = e_1 + e_2 + \ldots + e_s$ be a decomposition of the identity of an algebra A such that the idempotents $\bar{e}_i = e_i + R$ are central in the quotient algebra $\bar{A} = A/R$, where $R = \operatorname{rad} A$. Then $e_i R e_j = e_i A e_j$ for $i \neq j$ and $e_i R e_i = \operatorname{rad}(e_i A e_i)$.*

Proof. Since $\bar{e}_i$ are central idempotents, the A-modules $\bar{e}_i \bar{A}$ and $\bar{e}_j \bar{A}$ have no isomorphic simple direct summands for $i \neq j$. Thus, by Corollary 3.2.9, $e_i A$ and $e_j A$ have no isomorphic principal direct summands and, in view of Theorem 3.5.2, $e_i A e_j = \operatorname{Hom}_A(e_j A, e_i A) \subset \operatorname{rad} A$. Since, by Proposition 3.2.4, $\operatorname{rad}(e_i A e_i) = e_i R e_i$, the theorem follows. □

Now we are going to introduce a class of algebras which plays a fundamental role in the theory of finite dimensional algebras.

Theorem 3.5.4. *The following conditions for an algebra A are equivalent:*

1) *the quotient algebra $\bar{A} = A/R$, where $R = \operatorname{rad} A$, is isomorphic to a product of division algebras;*
2) *if $A = P_1 \oplus P_2 \oplus \ldots \oplus P_s$ is a decomposition of the regular module into a direct sum of principal modules, then $P_i \not\simeq P_j$ for $i \neq j$;*
3) *there is an algebra B and a B-module M such that $A \simeq E_B(M)$ and $M = M_1 \oplus M_2 \oplus \ldots \oplus M_s$, where M_i are indecomposable modules and $M_i \not\simeq M_j$ for $i \neq j$.*

Proof. 1) ⇒ 2). If $\bar{A} \simeq D_1 \times D_2 \times \ldots \times D_s$, where D_i are division algebras, then, by Corollary 3.3.9, $A \simeq P_1 \oplus P_2 \oplus \ldots \oplus P_s$, where $P_i/P_i R \simeq U_i$ are simple D_i-modules, P_i are principal modules and $P_i \not\simeq P_j$ for $i \neq j$ because $U_i \not\simeq U_j$ for $i \neq j$.

2) ⇒ 3) follows from Theorem 1.7.1. It is sufficient to put $B = A$ and take the regular module for M.

3) ⇒ 1) follows immediately from Theorem 3.5.2. □

An algebra A satisfying the conditions of Theorem 3.5.4 is called *basic*.

Let A be an arbitrary algebra and $A \simeq n_1 P_1 \oplus n_2 P_2 \oplus \ldots \oplus n_s P_s$ a decomposition of the regular A-module into a direct sum of the principal modules, whereby $P_i \not\simeq P_j$ for $i \neq j$. Write $P = P_1 \oplus P_2 \oplus \ldots \oplus P_s$ and $B = E_A(P)$. Then B is basic and is called the *basic algebra* of the algebra A. If $1 = e_1 + e_2 + \ldots + e_s$ is a decomposition of the identity of the algebra B corresponding to the given decomposition of the module P, we say that the principal B-module $Q_i = e_i B$ *corresponds* to the principal A-module P_i and that the projective B-module $Q = k_1 Q_1 \oplus k_2 Q_2 \oplus \ldots \oplus k_s Q_s$ *corresponds* to the projective A-module $k_1 P_1 \oplus k_2 P_2 \oplus \ldots \oplus k_s P_s$.

Lemma 3.5.5. *If B is the basic algebra of an algebra A and Q_1, Q_2 are projective B-modules corresponding to the projective A-modules P_1, P_2, then* $\mathrm{Hom}_B(Q_1, Q_2) \simeq \mathrm{Hom}_A(P_1, P_2)$.

Proof. In view of the matrix form of the endomorphisms introduced in Sect. 1.7, the proof can be reduced to verifying that $\mathrm{Hom}_B(Q_j, Q_i) \simeq \mathrm{Hom}_A(P_j, P_i)$ for principal B-modules Q_i corresponding to the principal A-modules P_i. Since both $\mathrm{Hom}_B(Q_j, Q_i)$ and $\mathrm{Hom}_A(P_j, P_i)$ can be identified with e_iBe_j, the proof of the lemma follows. □

Theorem 3.5.6. *The following conditions are equivalent:*

1) *the basic algebras of the algebras A_1 and A_2 are isomorphic;*
2) *$A_2 \simeq E_{A_1}(P)$, where P is a projective A_1-module having among its direct summands all principal A_1-modules;*
3) *$A_2 \simeq E_{A_1}(P_1)$ and $A_1 \simeq E_{A_2}(P_2)$, where P_i are projective A_i-modules $(i = 1, 2)$.*

Proof. 1) ⇒ 2). Let B be a common basic algebra of the algebras A_1 and A_2, Q be a projective B-module corresponding to the regular A_2-module and P be a projective A_1-module corresponding to Q. Then Q contains all principal B-modules as direct summands. This means that P contains all principal A_1-modules and by Lemma 3.5.5, $A_2 \simeq E_{A_2}(A_2) \simeq E_B(Q) \simeq E_{A_1}(P)$.

Observe that, in view of the fact that condition 1) is symmetric, we have at the same time proved the implication 1) ⇒ 3).

2) ⇒1). Let $P = k_1P_1 \oplus k_2P_2 \oplus \ldots \oplus k_sP_s$, where $P_1, P_2, \ldots, P_s$ are all pairwise non-isomorphic A_1-modules and $A_2 = E_{A_1}(P)$. It follows from Theorem 3.5.2 that $A_2/\mathrm{rad}\, A_2 \simeq M_{k_1}(D_1) \times M_{k_2}(D_2) \times \ldots \times M_{k_s}(D_s)$, where $D_i = B_i/\mathrm{rad}\, B_i$ and $B_i = E_{A_1}(P_i)$. Therefore, there are s simple A_2-modules and s principal A_2-modules $P'_1, P'_2, \ldots, P'_s$ (Corollary 3.2.9), and moreover, as one can easily see using Lemma 3.5.5, $\mathrm{Hom}_{A_2}(P'_j, P'_i) \simeq \mathrm{Hom}_{A_1}(P_j, P_i)$. Thus $E_{A_2}(P'_1 \oplus P'_2 \oplus \ldots \oplus P'_s) \simeq E_{A_1}(P_1 \oplus P_2 \oplus \ldots \oplus P_s)$, as required.

3) ⇒ 2). If $A_2 \simeq E_{A_1}(P_1)$, where $P_1 = k_1Q_1 \oplus k_2Q_2 \oplus \ldots \oplus k_sQ_s$ with pairwise non-isomorphic principal A_1-modules Q_i, then, by Theorem 3.5.2, A_2 has s simple, and thus s principal modules. Similarly, if $A_1 \simeq E_{A_2}(P_2)$, where $P_2 = m_1Q'_1 \oplus m_2Q'_2 \oplus \ldots \oplus m_tQ'_t$ with pairwise non-isomorphic principal A_2-modules Q'_j, the A_1 has t principal modules. From here we conclude that $s \leq t$ and $t \leq s$, i.e. $t = s$ and that $Q_1, Q_2, \ldots, Q_s$ are all principal A_1-modules. □

The algebras satisfying the conditions of Theorem 3.5.6 are called *isotypic*. Evidently, for semisimple algebras this concept coincides with the one introduced in Sect. 2.6.

Corollary 3.5.7. *Every algebra A is isomorphic to the endomorphism algebra of a projective module P over a basic algebra B. The algebra B is determined* *(up to an isomorphism) uniquely and is isomorphic to the basic algebra of the algebra A.*

Let us remark that the module P is, in general, not uniquely determined (see Exercise 16 of this chapter).

3.6 Diagram of an Algebra

The preceding results allow us to outline a certain method of investigation of non-semisimple algebras. Taking into account Corollary 3.5.7, we can restrict ourselves to basic algebras.

Let $P_1, P_2, \ldots, P_s$ be pairwise non-isomorphic principal modules over an algebra A (by Corollary 3.2.9, their number equals the number of simple components of the algebra $\bar{A} = A/R$ where $R = \operatorname{rad} A$). Write $R_i = P_iR$ and $V_i = R_i/R_iR$. Here V_i is a semisimple module, and thus $V_i \simeq \bigoplus_{j=1}^{s} t_{ij}U_j$, where $U_j = P_j/R_j$ are simple modules (in view of Theorem 3.3.7, this is equivalent to the isomorphism $P(R_i) \simeq \bigoplus_{j=1}^{s} t_{ij}P_j$). Now, to each module P_i assign a point of the plane which will be denoted by i, and join the point i with the point j by t_{ij} arrows. The set of points and arrows which will be obtained in this way will be called a *diagram of the algebra A* and will be denoted by $\mathcal{D}(A)$.[7]

Observe that isotypic algebras have the same diagrams. Besides, since $V_i = P_iR/P_iR^2$, the diagrams of the algebras A and A/R^2 coincide.

Examples. 1. If the algebra A is semisimple, then $R_i = 0$ and $\mathcal{D}(A)$ is a set of points without any arrows.

2. Let $A = T_n(K)$ be the algebra of triangular matrices of degree n. The matrix units e_{ii} are minimal idempotents and $1 = e_{11} + e_{22} + \ldots + e_{nn}$ is a decomposition of the identity. Since $[e_{ii}A : K] = n - i + 1$, the principal A-modules $P_i = e_{ii}A$ are pairwise non-isomorphic. By Theorem 3.5.3, we get easily that $R_i = e_{i\,(i+1)}K + e_{i\,(i+2)}K + \ldots + e_{in}K \simeq P_{i+1}$. Therefore, the diagram $\mathcal{D}(A)$ looks as follows:

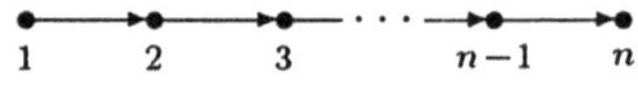

3. The Jordan-algebra $J_n(K)$ is local and its radical is cyclic. Therefore the only principal module is the regular one and it is the projective cover of its radical. Consequently, the diagram of the algebra $J_n(K)$ has the form

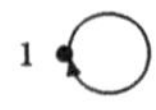

[7] Compare this concept to that of a K-species $\mathcal{S}(A)$ of an algebra A and its valued graph as presently used in literature and the Appendix. For split algebras, one usually speaks about a "quiver" of an algebra in the sense of P. Gabriel. See also Sect. 8.5. (*Translator's note*)

In general, by a *diagram* $\mathcal{D}$ we shall understand an arbitrary finite set of points together with arrows between them. Usually, the points will be denoted by the numbers $1, 2, \dots, s$. Then, the diagram is given by its *incidence matrix*

$$[\mathcal{D}] = \begin{pmatrix} t_{11} & t_{12} & \dots & t_{1s} \\ t_{21} & t_{22} & \dots & t_{2s} \\ \dots & \dots & \dots & \dots \\ t_{s1} & t_{s2} & \dots & t_{ss} \end{pmatrix},$$

where t_{ij} is the number of arrows from the point i to the point j. If an arrow σ of the diagram $\mathcal{D}$ joins the point i with the point j, then i is called the *tail* (origin) and j the *head* (top) of the arrow σ. This fact will be recorded as follows: $\sigma : i \to j$.

Two diagrams $\mathcal{D}_1$ and $\mathcal{D}_2$ are called *isomorphic* if there is a bijective correspondence between their points and arrows such that the tails and the heads of the corresponding arrows map one to the other. It is not difficult to see that $\mathcal{D}_1 \simeq \mathcal{D}_2$ if and only if the incidence matrix $[\mathcal{D}_1]$ can be transformed into the incidence matrix $[\mathcal{D}_2]$ by simultaneous permutations of the rows and columns. In particular, the diagram of an algebra is determined uniquely up to an isomorphism.

A *path of a diagram* $\mathcal{D}$ is an ordered sequence of arrows $\{\sigma_1, \sigma_2, \dots, \sigma_k\}$ such that the head of the arrow σ_ℓ coincides with the tail of the arrow $\sigma_{\ell+1}$ $(\ell = 1, 2, \dots, k-1)$. The number of the arrows k is called the *length of the path.* The tail of the arrow σ_1 is called the *tail of the path* and the head of the arrow σ_k the *head of the path.* We shall say that the path connects the point i with the point j and write $\sigma_1\sigma_2 \dots \sigma_k : i \to j$.

We shall assume that the algebra A is basic. Then $\bar{A} \simeq D_1 \times D_2 \times \dots \times D_s$, where $D_i = E_A(U_i)$ and U_i can be considered as the regular D_i-module. Let $1 = \bar{e}_1 + \bar{e}_2 + \dots + \bar{e}_s$ be the decomposition of the identity of the algebra $\bar{A}$ such that $\bar{e}_i\bar{A} \simeq D_i$ and $1 = e_1 + e_2 + \dots + e_s$ the corresponding decomposition of the identity of the algebra A (see Corollary 3.3.9). In this case, $P_i = e_iA$, $R_i = e_iR$ and $V_i = e_iV$, where $V = R/R^2$. Write $V_{ij} = e_iVe_j$. Now, V_{ij} is a right D_j-module and as such $V_{ij} \simeq t_{ij}U_j$. Thus, $\mu_{D_j}(V_{ij}) = t_{ij}$. Let us choose in V_{ij} a generating set of t_{ij} elements and index them by the arrows of the diagram $\mathcal{D}(A)$ which point from i to j (their number is also t_{ij}). Let v_σ be the generator corresponding to the arrow $\sigma : i \to j$ and r_σ its preimage in $R_{ij} = e_iRe_j$. The set of all elements $\{v_\sigma\}$ (over all arrows of the diagram $\mathcal{D}$) is a generating set of the module V. By Nakayama's lemma (Corollary 3.1.5), $\{r_\sigma\}$ is a generating set of R (as a right module). Note that if $\sigma : i \to j$, then $e_ir_\sigma = r_\sigma e_j = r_\sigma$ and $r_\sigma r_\tau = 0$ if the head of the arrow σ does not coincide with the tail of τ.

Lemma 3.6.1. *If there is no path in the diagram $\mathcal{D}(A)$ which connects the point i with the point j $(i \neq j)$, then* $\mathrm{Hom}_A(P_j, P_i) = 0$. *If the algebra A is basic, then every element $r \in R_{ij}$ can be represented in the form $r = \sum r_{\sigma_1} r_{\sigma_2} \dots r_{\sigma_k} a_{\sigma_1\sigma_2\dots\sigma_k}$, where the summation runs over all paths $\sigma_1\sigma_2 \dots \sigma_k : i \to j$ and $a_{\sigma_1\sigma_2\dots\sigma_k} \in A_{jj}$ with $A_{jj} = e_jAe_j$.*

Proof. By Lemma 3.5.5, the algebra A can be assumed to be basic and therefore, by Theorem 3.5.3, $\mathrm{Hom}_A(P_j, P_i) \simeq e_iAe_j = R_{ij}$ (for $i \neq j$). Therefore, it suffices to prove only the second assertion. The considerations introduced above show that if $r \in R_{ij}$, then $r \equiv \sum r_\sigma a_\sigma \pmod{R^2}$, where $a_\sigma \in A_{jj}$ and the summation runs over all arrows $\sigma : i \to j$. Then the element $r' : r - \sum r_\sigma a_\sigma$ belongs to $e_iR^2e_j$. However, $R = \sum\limits_{i,j} R_{ij}$ and therefore $e_iR^2e_j = \sum\limits_k R_{ik}R_{kj}$, i.e. $r' = \sum\limits_k x_k y_k$, where $x_k \in R_{ik}$, $y_k \in R_{kj}$. Again, $x_k \equiv \sum\limits_\tau r_\tau a_\tau \pmod{R^2}$, where $\tau : i \to k$, $a_\tau \in A_{kk}$ and $a_\tau y_k \equiv \sum\limits_p r_p a_{\tau p} \pmod{R^2}$, where $p : k \to j$, $a_{\tau p} \in A_{jj}$. Therefore, $r' \equiv \sum r_\tau r_p a_{\tau p} \pmod{R^3}$, where $\tau p : i \to j$. Continuing this process and taking into account that the radical is nilpotent, we obtain the required expression for r.

Let us remark here that $\mathrm{Hom}_A(P_j, P_i) = 0$ is possible even if there is a path from i to j (see Exercise 12). □

A diagram $\mathcal{D}$ is called *connected* if it cannot be divided into two non-empty disjoint subsets which are not connected by any arrows.

Theorem 3.6.2. *An algebra A is a non-trivial direct product if an only if the diagram $\mathcal{D}(A)$ is disconnected.*

Proof. Let the diagram $\mathcal{D} = \mathcal{D}(A)$ be disconnected: $\mathcal{D} = \mathcal{D}_1 \cup \mathcal{D}_2$, $\mathcal{D}_1 \cap \mathcal{D}_2 = \emptyset$, $\mathcal{D}_1 \neq \emptyset$, $\mathcal{D}_2 \neq \emptyset$ and there are no arrows between the points of $\mathcal{D}_1$ and $\mathcal{D}_2$. Thus, by Lemma 3.6.1, if $i \in \mathcal{D}_1$, $j \in \mathcal{D}_2$, then $\mathrm{Hom}_A(P_i, P_j) = 0$ and $\mathrm{Hom}_A(P_j, P_i) = 0$. By Corollary 1.7.9, the algebra A is decomposable. Conversely, if A decomposes, $A = A_1 \times A_2$, then, for any principal A_1-module P_i and any principal A_2-module P_j, $\mathrm{Hom}_A(P_i, P_j) = 0$ and $\mathrm{Hom}_A(P_j, P_i) = 0$; it follows that the points i and j are not connected and the diagram $\mathcal{D}(A)$ is disconnected. □

Corollary 3.6.3. *Algebras A and A/R^2 are either both decomposable or both indecomposable.*

In addition to the diagram $\mathcal{D}(A)$, an algebra A has a number of important invariants. First of all, such are the division algebras $D_i = E_A(U_i)$ and the multiplicities n_i of the P_i in the decomposition of the regular module. For a basic algebra, all $n_i = 1$, but the division algebras can be arbitrary. If the field K is algebraically closed then the situation simplifies significantly: All D_i coincide with the ground field K.

An algebra A over the field K is called *split* if $A/R \simeq M_{n_1}(K) \times M_{n_2}(K) \times \ldots \times M_{n_s}(K)$. All algebras over algebraically closed fields are split.

For split algebras, Lemma 3.6.1 can be strengthened.

Lemma 3.6.4. *Let A be a basic split algebra. Then every element $r \in R_{ij}$ can be represented in the form $r = \sum r_{\sigma_1} r_{\sigma_2} \ldots r_{\sigma_k} c_{\sigma_1\sigma_2\ldots\sigma_k}$, where $c_{\sigma_1\sigma_2\ldots\sigma_k} \in K$*

and the summation runs through all paths $\sigma_1\sigma_2\ldots\sigma_k : i \to j$ (here, possibly, $i = j$).

Proof. The proof of Lemma 3.6.1 can be repeated word by word. Observe that in this case $A_{jj}/R_{jj} = K$, i.e. every element of the algebra A_{jj} is of the form $c + x$, where $c \in K$, $x \in R_{jj}$. □

A *cycle* of the diagram $\mathcal{D}$ is a path whose tail coincides with its head.

Corollary 3.6.5. *If there is no cycle in the diagram of a basic split algebra A, then $E_A(P) = K$ for each principal A-module P.*

Lemma 3.6.4 enables us to construct for every diagram $\mathcal{D}$ a K-algebra $K(\mathcal{D})$, in general infinite dimensional, such that every basic split algebra with the given diagram $\mathcal{D}$ is its quotient algebra.

A basis of the space $K(\mathcal{D})$ is formed by all possible paths of the diagram and by the symbols $\{\varepsilon_i\}$ (indexed by the points of $\mathcal{D}$). In this way, every element of $K(\mathcal{D})$ can be uniquely written in the form $\sum_{i=1}^{s} c_i\varepsilon_i + \sum c_{\sigma_1\sigma_2\ldots\sigma_k}\sigma_1\sigma_2\ldots\sigma_k$ (the second sum runs over all paths of the diagram $\mathcal{D}$), where $c_i \in K$, $c_{\sigma_1\sigma_2\ldots\sigma_k} \in K$. It will be convenient to interpret the symbol ε_i as the path of length 0 with its head and tail at the point i.

Define the product of the paths α and β as the path $\alpha\beta$ if the head of α coincides with the tail of β, and as 0 otherwise. In other words,

$$(\sigma_1\sigma_2\ldots\sigma_k)(\tau_1\tau_2\ldots\tau_\ell) = \begin{cases} \sigma_1\sigma_2\ldots\sigma_k\tau_1\tau_2\ldots\tau_\ell & \text{if the head of } \sigma_k \text{ coincides with the tail of } \tau_1, \\ 0 & \text{otherwise;} \end{cases}$$

$$\varepsilon_i\sigma_1\sigma_2\ldots\sigma_k = \begin{cases} \sigma_1\sigma_2\ldots\sigma_k & \text{if } i \text{ is the tail of } \sigma_1, \\ 0 & \text{otherwise;} \end{cases}$$

$$\sigma_1\sigma_2\ldots\sigma_k\varepsilon_i = \begin{cases} \sigma_1\sigma_2\ldots\sigma_k & \text{if } i \text{ is the head of } \sigma_k, \\ 0 & \text{otherwise;} \end{cases}$$

$$\varepsilon_i\varepsilon_j = \begin{cases} \varepsilon_i & \text{if } i = j, \\ 0 & \text{if } i \neq j. \end{cases}$$

Extend this definition to the whole space $K(\mathcal{D})$ "by linearity" putting $\big(\sum_\alpha c_\alpha\alpha\big)\big(\sum_\beta c'_\beta\beta\big) = \sum_{\alpha,\beta} c_\alpha c'_\beta(\alpha\beta)$, where α, β are paths of the diagram $\mathcal{D}$ and c_α, c'_β elements of the field K. A trivial verification shows that in this way $K(\mathcal{D})$ becomes an algebra over the field K with the identity $1 = \varepsilon_1 + \varepsilon_2 + \ldots + \varepsilon_s$.

Denote by J the set of those elements of the algebra $K(\mathcal{D})$ whose coefficients of ε_i are equal to 0 for all i. Evidently, J is an ideal of $K(\mathcal{D})$. An ideal $I \subset K(\mathcal{D})$ is called *admissible* if $J^2 \supset I \supset J^n$ for some $n \geq 2$.

Theorem 3.6.6. *For any admissible ideal $I \subset K(\mathcal{D})$, the quotient algebra $K(\mathcal{D})/I$ is a split basic algebra with the diagram $\mathcal{D}$. Conversely, every split* *basic algebra with a diagram $\mathcal{D}$ is isomorphic to a quotient algebra of the algebra $K(\mathcal{D})$ by an admissible ideal I.*

Proof. Let $A_n = K(\mathcal{D})/J^{n+1}$ $(n \geq 1)$. The classes $\bar{\alpha} = \alpha + J^{n+1}$, where α is an arbitrary path of the diagram $\mathcal{D}$ of length smaller or equal than n, form a basis of the algebra A_n. The ideal $\bar{J} = J/J^{n+1}$ of A_n is nilpotent and $A_n/\bar{J} \simeq K(\mathcal{D})/J \simeq K^s$ (a basis of this algebra is formed by the classes $\bar{\varepsilon}_i = \varepsilon_i + J$; moreover $\varepsilon_i\varepsilon_j = \delta_{ij}\varepsilon_i$). By Proposition 3.1.13, $\bar{J} = \operatorname{rad} A_n$. In this way, A_n is a basic split algebra and $A_n/J^2 \simeq A_1$, i.e. $\mathcal{D}(A_n) = \mathcal{D}(A_1)$. In the algebra A_1, $\bar{\varepsilon}_i\bar{J}\bar{\varepsilon}_j$ is a vector space over K with a basis $\{\bar{\sigma}\}$, where σ are the arrows from i to j. Therefore, if $[\mathcal{D}] = (t_{ij})$ is the incidence matrix of the diagram $\mathcal{D}$, then $\bar{\varepsilon}_i\bar{J}\bar{\varepsilon}_j \simeq t_{ij}U_j$, where $U_j = \bar{\varepsilon}_jA_1/\bar{\varepsilon}_j\bar{J}$ and $\mathcal{D}(A_1) = \mathcal{D}$. From here we get the first statement of the theorem (taking into account Corollary 3.1.14).

Now, let A be an arbitrary basic split algebra with the diagram $\mathcal{D}$, $1 = e_1 + e_2 + \ldots + e_s$ be a decomposition of the identity into minimal idempotents and $\{r_\sigma\}$ a generating set of the radical constructed earlier (before Lemma 3.6.1).

For every path $\alpha = \sigma_1\sigma_2\ldots\sigma_k$ of the diagram $\mathcal{D}$, we write $r_\alpha = r_{\sigma_1}r_{\sigma_2}\ldots r_{\sigma_k}$, $r_{\varepsilon_i} = e_i$ and put $f(\sum\limits_\alpha c_\alpha\alpha) = \sum\limits_\alpha c_\alpha r_\alpha$. The relations between r_σ and e_i imply that f is a homomorphism of the algebra $K(\mathcal{D})$ into the algebra A, and that, in view of Lemma 3.6.4, it is an epimorphism. Therefore, $A \simeq K(\mathcal{D})/I$ with $I = \operatorname{Ker} f$. One can see easily that $f(J) = R$, where $R = \operatorname{rad} A$. Since $R^n = 0$ for some n, $J^n \subset I$. Finally, the elements $v_\sigma = r_\sigma + R^2$ are linearly independent in R/R^2 and therefore the homomorphism $A_1 \to A/R^2$ is an isomorphism; hence $I \subset J^2$. The proof of the theorem is completed. □

The algebra $K(\mathcal{D})$ is called the *path algebra* or the *universal algebra* of the diagram $\mathcal{D}$.

Of course, a similar construction can be performed for the left diagrams. However, it turns out that the following proposition holds.

Proposition 3.6.7. *If A is a split algebra, then the (left) diagram $\mathcal{D}'(A)$ can be obtained from the diagram $\mathcal{D}(A)$ by reversing all arrows, or by transposition of the incidence matrix.*

Proof. The algebra A can be assumed to be basic. Then, as we have already seen, $[\mathcal{D}(A)] = (t_{ij})$, where t_{ij} is the dimension of the space e_iVe_j (here, $V = R/R^2$ and $1 = e_1 + e_2 + \ldots + e_s$ is a decomposition of the identity into minimal idempotents). Similarly, $[\mathcal{D}'(A)] = (t'_{ij})$, where t'_{ij} is the dimension of e_jVe_i, i.e. $t'_{ij} = t_{ji}$, as required. □

3.7 Hereditary Algebras

The construction of the universal algebra allows us to give a description of an interesting class of algebras, viz. of hereditary algebras.

An algebra A is called *hereditary* if every right ideal of A is projective.[8]

Theorem 3.7.1. *The following conditions for an algebra A are equivalent:*

1) *A is hereditary;*
2) *every submodule of a principal A-module is projective;*
3) *every submodule of a projective A-module is projective;*
4) *$\operatorname{rad} A$ is projective (as a right module).*

Proof. 1) $\Rightarrow$ 2) and 3) $\Rightarrow$ 1) trivially.

2) $\Rightarrow$ 4). If $A = P_1 \oplus P_2 \oplus \ldots \oplus P_n$, where P_i are principal modules, then $\operatorname{rad} A = R = R_1 \oplus R_2 \oplus \ldots \oplus R_n$, where $R_i = \operatorname{rad} P_i$. Thus, since all R_i are projective, R is projective as well.

4) $\Rightarrow$ 3). Let M be a submodule of a projective A-module P. The fact that M is projective will be proved by induction on $\ell(P) = \ell$. For $\ell = 1$, the assertion is trivial and thus, we assume that the assertion holds for modules P' with $\ell(P') < \ell$.

The module P has a principal direct summand P_1, i.e. $P = P_1 \oplus P_2$ (possibly with $P_2 = 0$). Denote by π the projection of P onto P_1. If $\pi(M) = P_1$, then, by Theorem 3.3.5, $M \simeq P_1 \oplus N$, where $N = M \cap P_2$ is a submodule of P_2. Since $\ell(P_2) < \ell$, N (and thus also M) is projective.

If $\pi(M) \neq P_1$, then $M \subset R_1 \oplus P_2$, where $R_1 = P_1 R$ is a direct summand of the radical and thus a projective module. Again, $\ell(R_1 \oplus P_2) < \ell$, and therefore M is projective. The theorem is proved. □

Lemma 3.7.2. *If an algebra A is hereditary, then every non-zero homomorphism $f : P_i \to P_j$ between principal A-modules is a monomorphism.*

Proof. This follows from the fact that $\operatorname{Im} f$ is a projective module and, by Theorem 3.3.5, $P_i \simeq \operatorname{Im} f \oplus \operatorname{Ker} f$. Thus, if $\operatorname{Im} f \neq 0$, necessarily $\operatorname{Ker} f = 0$. □

Corollary 3.7.3. *If an algebra A is hereditary, then the diagram $\mathcal{D}(A)$ has no cycles.*

Proof. If there is an arrow σ in $\mathcal{D}(A)$ with the tail at i and the head at j, then there is a non-zero homomorphism $f_\sigma : P_j \to P_i$ and $\operatorname{Im} f_\sigma \subset \operatorname{rad} P_i$. Let A be a hereditary algebra and $\sigma_1 \sigma_2 \ldots \sigma_k$ a path with both tail and head at the point i. Then $f = f_{\sigma_1} f_{\sigma_2} \ldots f_{\sigma_k}$ is a monomorphism $P_i \to P_i$ because all

[8] In a similar manner, one can define left hereditary algebras. However, in Chapter 8, we shall see that, for finite dimensional algebras, these two concepts coincide.

$f_{\sigma_1}, f_{\sigma_2}, \ldots, f_{\sigma_k}$ are monomorphisms, and $\operatorname{Im} f \subset \operatorname{rad} P_i$. But by a dimension argument, this is impossible. □

Now we are going to present a description of the hereditary basic split algebras.

Theorem 3.7.4. *If $\mathcal{D}$ is a diagram without cycles, then $K(\mathcal{D})$ is a hereditary algebra. Conversely, a hereditary basic split algebra A is isomorphic to the algebra $K(\mathcal{D})$, where $\mathcal{D} = \mathcal{D}(A)$. In this way, there is a bijective correspondence between diagrams without cycles and basic split hereditary algebras.*

Proof. It is clear that, in a diagram without cycles, the lengths of paths are bounded and therefore $A = K(\mathcal{D})$ is a finite dimensional algebra. Write $P_i = \varepsilon_i A$. The elements of P_i are linear combinations $\sum\limits_{\alpha} c_\alpha \alpha$, where α are the paths with tails at i (including ε_i). Then $\operatorname{rad} P_i = P_i J$ consists of all linear combinations $\sum\limits_{\alpha} c_\alpha \alpha$, where α are the paths of non-zero lengths with tails at i. Therefore $P_i J = \bigoplus\limits_{\sigma} \sigma A$, where σ runs through all arrows with tails at the point i. Now, if $\sigma : i \to j$, then assigning to every linear combination $\sum\limits_{\beta} c_\beta \beta$, where β is a path with tail at j (including ε_j), the element $\sum\limits_{\beta} c_\beta \sigma \beta$, we obtain, as one can easily see, an isomorphism $P_j \simeq \sigma A$. Consequently, $P_i J$ are projective modules, and therefore $J = \bigoplus\limits_{i=1}^{s} P_i J$ is a projective module and the algebra A is hereditary by Theorem 3.7.1.

It remains to verify that if $I \subset J^2$ and $\tilde{A} = A/I$ is a hereditary algebra, then $I = 0$. Write $R = J/I$, $\tilde{P}_i = P_i/P_i I$. Then $R = \operatorname{rad} \tilde{A}$ and $\tilde{P}_i$ are the principal $\tilde{A}$-modules. Besides, $\tilde{A}/R \simeq A/J$ and $R/R^2 \simeq J/J^2$. We shall show by induction on k that also $R^k/R^{k+1} \simeq J^k/J^{k+1}$. Thus, assume that $R^{k-1}/R^k \simeq J^{k-1}/J^k$ and let $\tilde{P} = P(R^{k-1}/R^k) = \bigoplus\limits_{i=1}^{s} m_i \tilde{P}_i$. Then $\tilde{P} = P(R^{k-1})$ by Theorem 3.3.7, and since R^{k-1} is projective, $\tilde{P} \simeq R^{k-1}$. Therefore $R^k \simeq \tilde{P}R$ and $R^k/R^{k+1} \simeq \tilde{P}R/\tilde{P}R^2$ and since $\tilde{P}_i R/\tilde{P}_i R^2 \simeq P_i J/P_i J^2$, we get that $R^k/R^{k+1} \simeq J^k/J^{k+1}$. Now it is clear that, in view of the latter isomorphism, $I \subset J^k$ for all k, and therefore $I = 0$, as required. □

Exercises to Chapter 3

1. Find the radical of the algebra $T_n(K)$.

2. Find the radical of the monogenic algebra $K[x]/f(x)K[x]$.

3. Prove that every algebra which is not simple contains a maximal ideal whose annihilator is non-zero.

4. Prove that a minimal right ideal is either a principal module or is contained in the radical of the algebra.

5. Let M be an A-module, $R = \operatorname{rad} A$, $\bar{A} = A/R$, $\bar{A} = n_1U_1 \oplus n_2U_2 \oplus \ldots \oplus n_sU_s$ a decomposition of $\bar{A}$ into a direct sum of principal modules and $\bar{M} = M/MR \simeq t_1U_1 \oplus t_2U_2 \oplus \ldots \oplus t_sU_s$. Prove that $\mu_A(M) = \mu_A(\bar{M})$. Show that $\mu_A(M) = \max_i \left\{ \left[\frac{t_i - 1}{n_i}\right] + 1 \right\}$ where $[t/n]$ is the integral part of the number t/n.

6. Denote by $I(M)$ the set of all endomorphisms of M whose image is in $\operatorname{rad} M$.
   a) Verify that $I(M)$ is a nilpotent ideal of $E_A(M)$.
   b) Prove that, for a projective module P, $I(P) = \operatorname{rad} E_A(P)$.
   c) Construct a module M which satisfies $I(M) \neq \operatorname{rad} E_A(M)$.

7. Prove that an algebra is basic if and only if its nilpotent elements form an ideal.

8. Let $1 = e_1 + e_2 + \ldots + e_n$ be a decomposition of the identity of an algebra A in which all the idempotents are minimal.
   a) Prove that the K-dimension of a faithful representation of the algebra A is at least n. If A has a faithful representation of K-dimension n then A is called a *minimal algebra of degree n*.
   b) Let A be a minimal algebra. Let us write $i \to j$ if $e_iAe_j \neq 0$. Prove that $\to$ is a *quasi-order relation* on the set $S = \{1, 2, \ldots, n\}$, i.e. $i \to i$ and $i \to j$, $j \to k$ implies $i \to k$.
   c) Prove that a minimal algebra is always a split algebra, and that it is basic if and only if $\to$ is an order relation, i.e. $i \to j$, $j \to i$ implies $i = j$.

9. Assume that there is a quasi-order relation $\to$ defined on the set $S = \{1, 2, \ldots, n\}$. Construct a minimal algebra A of degree n so that $i \to j$ if and only if $e_iAe_j \neq 0$. (Hint: Consider the subalgebra of $M_n(K)$ with the basis consisting of the matrix units e_{ij} with $i \to j$.)

10. Prove that two minimal algebras of degree n defining on S the same quasi-order relation are isomorphic.

11. Find the radical, the diagram and the basic algebra of a minimal algebra. Prove that an algebra is minimal if and only if its basic algebra is minimal.

12. Let $(\operatorname{rad} A)^2 = 0$ and $P_1, P_2, \ldots, P_s$ be the principal A-modules. Prove that $\operatorname{Hom}_A(P_j, P_i) \neq 0$ for $i \neq j$ if and only if there is an arrow from i to j in the diagram $\mathcal{D}(A)$.

13. Prove that a split basic algebra A with $(\operatorname{rad} A)^2 = 0$ is (up to an isomorphism) uniquely determined by its diagram.

14. Let A be an algebra over the reals $\mathbb{R}$ consisting of 2×2 complex matrices of the form
$$\begin{pmatrix} a & b \\ 0 & c \end{pmatrix}$$
where $a \in \mathbb{R}$, $b, c \in \mathbb{C}$. Find $\mathcal{D}(A)$ and $\mathcal{D}'(A)$.

15. Describe the three-dimensional algebras over an algebraically closed field.

16. Let φ be an endomorphism of an algebra A, $T : A \to M_n(K)$ a representation of this algebra and M the corresponding module. Denote by $M\varphi$ the module corresponding to the representation $T\varphi : A \to M_n(K)$.
    a) Give an intrinsic description of the module $M\varphi$ and verify that $M(\varphi\psi) \simeq (M\varphi)\psi$ and $(M \oplus N)\varphi \simeq M\varphi \oplus N\varphi$.
    b) If φ is an automorphism, prove that $P\varphi$ is projective if and only if the module P is projective. (Hint: Verify that $A\varphi \simeq A$.)
    c) Show that if A is a basic algebra and P and Q are projective A-modules, then $E_A(P) \simeq E_A(Q)$ if and only if $Q \simeq P\varphi$ for a suitable automorphism φ of the algebra A. (Hint: Make use of matrix notation for endomorphisms.)
    d) Give an example of a projective module P over a basic algebra A and an automorphism φ of the algebra A such that $P \not\simeq P\varphi$.

17. A *diagram with multiplicities* is a diagram $\mathcal{D}$ in which a natural number n_i is assigned to every vertex i. Write $A = K(\mathcal{D})$, $P_i = \varepsilon_i A$, $P = n_1 P_1 \oplus n_2 P_2 \oplus \ldots \oplus n_s P_s$, $\tilde{A} = E_A(P)$. For the algebra $\tilde{A}$, prove an analogue of Theorem 3.6.6, considering arbitrary split algebras and diagrams with multiplicities.

18. Prove that for diagrams without cycles, the algebra of Exercise 17 is hereditary and that every split hereditary (finite dimensional) algebra has such a form.

19. The *socle* soc M of a module M is the sum of all its minimal submodules.
    a) Show that $\operatorname{soc} M = M$ if and only if M is semisimple.
    b) Prove that $f(\operatorname{soc} M) \subset \operatorname{soc} N$ for every homomorphism $f : M \to N$, and that f is a monomorphism if and only if the induced map $\operatorname{soc} M \to \operatorname{soc} N$ is a monomorphism.
    c) Prove that the socles of the right and left regular A-modules are ideals of A. They are called the *right* and *left socles* of the algebra A and are denoted by $r.\operatorname{soc} A$ and $l.\operatorname{soc} A$, respectively.
    d) Verify that $r.\operatorname{soc} A = \{a \in A \mid aR = 0\}$ and $l.\operatorname{soc} A = \{a \in A \mid Ra = 0\}$, where $R = \operatorname{rad} A$.
    e) Compute the right and left socle of the algebra $A = T_n(K)$ and show that $r.\operatorname{soc} A \neq l.\operatorname{soc} A$.

20. Let $P_1, P_2, \ldots, P_s$ be pairwise non-isomorphic principal A-modules. Denote by c_{ij} the multiplicity of the simple module $U_j = P_j/\operatorname{rad} P_j$ in a composition series of P_i. The integers c_{ij} are called *Cartan numbers* and the matrix $c(A) = (c_{ij})$ the *Cartan matrix* of the algebra A.
    a) Prove that $c(A) = c(B)$ where B is the basic algebra of the algebra A.
    b) If $(\operatorname{rad} A)^2 = 0$, verify that $c(A) = E + [\mathcal{D}]$, where E is the identity matrix and $[\mathcal{D}]$ the incidence matrix of the diagram $\mathcal{D} = \mathcal{D}(A)$.
    c) Write $A_n = K(\mathcal{D})/J^{n+1}$. Prove that $c(A_n) = E + [\mathcal{D}] + [\mathcal{D}]^2 + \ldots + [\mathcal{D}]^n$.
    d) Prove that, for a hereditary algebra A, the matrix $[\mathcal{D}]$, where $\mathcal{D} = \mathcal{D}(A)$, is always nilpotent, $c(A) = \sum_{m=0}^{\infty} [\mathcal{D}]^m$, the Cartan number c_{ij} is equal to the number of paths of the diagram $\mathcal{D}$ starting at i and ending at j (for $i \neq j$) and $c_{ii} = 1$. (Hint: In this case $c_{ij} = \sum_k t_{ik} c_{kj}$ for $i \neq j$, i.e. $c(A) = [\mathcal{D}]c(A) + E$, where $[\mathcal{D}] = (t_{ij})$.)

4. Central Simple Algebras

The Wedderburn-Artin theorem reduces the study of semisimple algebras to the description of division algebras over a field K. If D is a finite dimensional division algebra over K and C its center, then C is a field (an extension of the field K) and D can be considered as an algebra over the field C. In this way, the investigation is divided into two steps: the study of the extensions of the field K and the study of *central division algebras*, i. e. of division algebras whose center coincides with the ground field. It turns out that these are two separate problems. However, one can conveniently apply common methods of investigation based on the concept of a bimodule and tensor product of algebras.

The present chapter is devoted to a description of these methods and to their application to the study of central division algebras. Here, the fundamental role is played obviously by Theorem 4.3.1, from which the main theorems on division algebras (the Skolem-Noether theorem and the centralizer theorem) follow relatively easily.

4.1 Bimodules

Let A and B be two algebras over a field K. A vector space M endowed with a left A-module and a right B-module structure which are connected by the associative law is called an *A-B-bimodule*; thus

$$(am)b = a(mb)$$

for all $a \in A$, $b \in B$, $m \in M$. If $A = B$, then M is called simply an *A-bimodule.*

For bimodules, one can introduce all concepts which were introduced for modules in Chapter 1: the concept of a homomorphism, isomorphism, subbimodule, factor bimodule, direct sum, etc. Moreover, it is easy to see that the main results of Chapter 1 such as the homomorphism theorem, the parallelogram law and the Jordan-Hölder theorem can be translated word by word to bimodules. In fact, in the next section, we shall see that the study of A-B-bimodules is in fact equivalent to the study of modules over a new algebra, namely over the tensor product of the algebras A and B^o.

Let us consider some examples which will play an important role in what follows and which will illustrate the importance of the concept of a bimodule for the structure theory of algebras.

Examples. 1. Obviously, every algebra A can be considered as a bimodule over itself. This bimodule is called *regular*. Subbimodules of a regular bimodule are the subspaces $I \subset A$ which are closed with respect to multiplication by arbitrary elements $a \in A$ both from the left and the right, i. e. the ideals of the algebra A. From this point of view, simple algebras are those whose regular bimodule is simple.

Let us clarify the form of the endomorphisms of a regular bimodule. If $f : A \to A$ is such an endomorphism, then f is, in particular, an endomorphism of the regular module and thus, by Theorem 1.7.1, it has the form $f(x) = ax$, where a is a fixed element of A. However, f is also an endomorphism of the left regular module and this means that $f(bx) = bf(x)$ for any $b \in A$; consequently, $abx = bax$ for all $b, x \in A$ and therefore $a \in C(A)$. Conversely, if a belongs to the center, then the same relations show that the multiplication by a is an endomorphism of the regular bimodule. Hence, we have proved the following proposition.

Proposition 4.1.1. *The submodules of a regular bimodule are the ideals of the algebra. The endomorphism algebra of a regular bimodule is isomorphic to the center of the algebra.*

2. Let $f : B \to A$ be an algebra homomorphism. We shall attach to it a B-A-bimodule which will be denoted by ${}_fA$. To construct it, we consider the regular A-module and define the left B-module structure by $ba = f(b)a$. It is clear that the associative law is satisfied and thus A becomes a B-A-module ${}_fA$. The previous example can be obtained if $B = A$ and f is the identity endomorphism.

3. Consider D-bimodules, where D is a finite dimensional division algebra over K. If M is such a bimodule, then it is, in particular, a vector space over the division algebra D. The map $m \mapsto md$ with a fixed $d \in D$ is an endomorphism of the vector space M, and if we fix an isomorphism $M \simeq nD$, where $n = [M : D]$, there is a matrix $T(d) \in M_n(D)$ corresponding to it. One can verify easily that $T(d + d') = T(d) + T(d')$; $T(\alpha d) = \alpha T(d)$ for $\alpha \in K$; $T(dd') = T(d)T(d')$ and $T(1) = 1$. The correspondence $d \mapsto T(d)$ is called a *self-representation of the division algebra D of dimension n*.

In particular, the case $n = 1$ is of interest. Then $T(d) \in D$ and $T : D \to D$ is an *automorphism* of the division algebra D.

Conversely, for any self-representation of dimension n, one can define a D-bimodule by considering the vector space nD and putting $xd = xT(d)$, $x \in nD$. In particular, to every automorphism, there corresponds a D-bimodule M such that $[M : D] = 1$.

4.2 Tensor Products

In Chapter 1, we have attached to the concept of a module one of a representation, i. e. of a linear map compatible with multiplication. We shall try to establish a similar construction for bimodules.

Let M be an A-B-bimodule. To any pair a, b of elements $a \in A$, $b \in B$, we attach the endomorphism of the vector space M which sends m into amb. In this way, we obtain a map $A \times B \to E(M)$ which is easily seen to be bilinear (i. e. linear in b for a fixed a and vice versa). Thus, leaving multiplication aside for a moment, we face the problem of classifying bilinear maps $U \times V \to W$, where U, V and W are vector spaces over a field K.

We choose bases $\{u_1, u_2, \ldots, u_n\}$ and $\{v_1, v_2, \ldots, v_n\}$ in the spaces U and V. Then a bilinear map $F : U \times V \to W$ is uniquely determined by the values $F(u_i, v_j) = w_{ij}$; here, w_{ij} can be arbitrary. This leads to the following definition.

The *tensor product* of a space U with a basis $\{u_1, u_2, \ldots, u_n\}$ and a space V with a basis $\{v_1, v_2, \ldots, v_m\}$ is a vector space $U \otimes V$ with the basis $\{u_i \otimes v_j\}$, $i = 1, 2, \ldots, n$, $j = 1, 2, \ldots, m$ and a fixed bilinear map $\otimes : U \times V \to U \otimes V$ defined by $\otimes(u_i, v_j) = u_i \otimes v_j$. The image of the pair (u, v) under the map $\otimes$ is denoted by $u \otimes v$.

The above considerations establish a universal property of the tensor product.

Theorem 4.2.1. *For every bilinear map $F : U \times V \to W$, there is a unique linear map $\tilde{F} : U \otimes V \to W$ such that $F = \tilde{F}\otimes$ (i. e. $F(u, v) = \tilde{F}(u \otimes v)$).*

Corollary 4.2.2. *If $\phi : U \times V \to W_0$ is a bilinear map such that for every bilinear map $F : U \times V \to W$ there is a unique map $\tilde{F} : W_0 \to W$ such that $\tilde{F}\phi = F$, then there is a unique isomorphism $\varphi : W_0 \xrightarrow{\sim} U \otimes V$ such that $u \otimes v = \varphi\phi(u, v)$ for all elements $u \in U$, $v \in V$.*

Proof. From the conditions of the theorem, there exists a unique homomorphism $\varphi : W_0 \to U \otimes V$ for which $u \otimes v = \varphi\phi(u, v)$. On the other hand, it follows from Theorem 4.2.1 that there is a homomorphism $\tilde{\phi} : U \otimes V \to W_0$ satisfying $\tilde{\phi}(u \otimes v) = \phi(u, v)$. But then $\varphi\tilde{\phi}(u \otimes v) = \varphi\phi(u, v) = u \otimes v$, and by Theorem 4.2.1, $\varphi\tilde{\phi} = 1$. Similarly, $\tilde{\phi}\varphi\phi(u, v) = \tilde{\phi}(u \otimes v) = \phi(u, v)$ and thus $\tilde{\phi}\varphi = 1$, i. e. $\tilde{\phi}$ is the inverse of φ. □

The existence and uniqueness of the isomorphism φ allows us to identify every space W_0 satisfying the conditions of Corollary 4.2.2 with the tensor product $U \otimes V$. In particular, one identifies all the tensor products obtained for various choices of bases in U and V.

Besides, Theorem 4.2.1 helps to establish basic properties of the tensor product.

Proposition 4.2.3. *For any spaces U, V, W, there is a unique isomorphism $f : U \otimes V \tilde{\to} V \otimes U$ such that $f(u \otimes v) = v \otimes u$ and a unique isomorphism $g : (U \otimes V) \otimes W \tilde{\to} U \otimes (V \otimes W)$ such that $g((u \otimes v) \otimes w) = u \otimes (v \otimes w)$.*

Proof. Since $(u, v) \to v \otimes u$ is evidently a bilinear map, there is a unique homomorphism $f : U \otimes V \to V \otimes U$ for which $f(u \otimes v) = v \otimes u$. Similarly, there is a homomorphism $f' : V \otimes U \to U \otimes V$ for which $f'(v \otimes u) = u \otimes v$ and, by Theorem 4.2.1, it follows immediately that f and f' are inverse isomorphisms.

Now, let F be a bilinear map $(U \otimes V) \times W \to Z$. For a fixed $w \in W$, it becomes a linear map $F_w : U \otimes V \to Z$, i.e. a bilinear map $U \times V \to Z$ depending on w linearly. Hence for a fixed $u \in U$, the map $V \times W \to Z$, assigning to a pair (v, w) the vector $F_w(u, v)$, is bilinear and consequently defines a linear map $V \otimes W \to Z$ depending on u linearly. Therefore, a bilinear map $U \times (V \otimes W) \to Z$ is defined. Passing to the tensor products, we associate every linear map $(U \otimes V) \otimes W \to Z$ with an (obviously unique) map $U \otimes (V \otimes W) \to Z$. Conversely, for every map $U \otimes (V \otimes W) \to Z$, there corresponds a unique map $(U \otimes V) \otimes W \to Z$. Setting subsequently $Z = (U \otimes V) \otimes W$ and $Z = U \otimes (V \otimes W)$, we get, respectively, a required isomorphism g and its inverse. □

In what follows, we shall identify $U \otimes V$ with $V \otimes U$, and also $(U \otimes V) \otimes W$ with $U \otimes (V \otimes W)$ (and write simply $U \otimes V \otimes W$ without brackets).

Every element of $U \otimes V$ can be evidently represented uniquely in the form $\sum_{i,j} \alpha_{ij}(u_i \otimes v_j)$. If we put $\sum_{i=1}^{n} \alpha_{ij} u_i = x_j$, such an element can be written in the form $\sum_{j=1}^{m} x_j \otimes v_j$, and it is easy to verify that this expression is unique. Similarly, every element from $U \otimes V$ can be written uniquely in the form $\sum_{i=1}^{n} u_i \otimes y_i$, $y_i \in V$.

If U' is a subspace of U, then $(u', v) \to u' \otimes v$ is a bilinear map $U' \otimes V \to U \otimes V$ which defines a homomorphism $f : U' \otimes V \to U \otimes V$. It is clear that if a basis of U is chosen so that $\{u_1, u_2, \ldots, u_k\}$ $(k \leq n)$ is a basis of U', then the image of f consists of elements of the form $\sum_{i=1}^{k} u_i \otimes y_i$, and thus f is a monomorphism. Consequently, we may consider $U' \otimes V$ to be a subspace of $U \otimes V$. Moreover, one can see easily that if $U = U' \oplus U''$, then $U \otimes V = (U' \otimes V) \oplus (U'' \otimes V)$. Similar statements hold for subspaces of the space V.

Now, let A and B be algebras over a field K. Then, for a fixed $a_0 \in A$ and $b_0 \in B$, the map $A \times B \to A \otimes B$ assigning to a pair (a, b) the element $aa_0 \otimes bb_0$ is bilinear and thus defines a linear map $F : A \otimes B \to A \otimes B$ such that $a \otimes b \mapsto aa_0 \otimes bb_0$. On the other hand, F depends bilinearly on (a_0, b_0), and thus we can, to every element $x \in A \otimes B$, assign a linear map F_x which depends on x linearly and satisfies $F_{a_0 \otimes b_0}(a \otimes b) = aa_0 \otimes bb_0$.

Writing $F_x(y) = yx$, we define a bilinear multiplication in $A \otimes B$ such that $(a \otimes b)(a_0 \otimes b_0) = aa_0 \otimes bb_0$ for every $a, a_0 \in A$ and $b, b_0 \in B$. Since $[(a \otimes b)(a' \otimes b')](a'' \otimes b'') = (aa' \otimes bb')(a'' \otimes b'') = aa'a'' \otimes bb'b'' = (a \otimes b)[(a' \otimes b')(a'' \otimes b'')]$ and $(a \otimes b)(1 \otimes 1) = (1 \otimes 1)(a \otimes b) = a \otimes b$, the space $A \otimes B$ becomes an algebra over the field K which is called the *tensor product of the algebras A and B.*

It follows from Proposition 4.2.3 that $A \otimes B \simeq B \otimes A$ and $(A \otimes B) \otimes C \simeq A \otimes (B \otimes C)$ as algebras.

The tensor products allow us to reduce the study of bimodules to the study of modules.

Let M be an A-B-bimodule. Denote by A^o the *algebra opposite to* A, i. e. the algebra consisting of the same elements with multiplication $a^o b^o = (ba)^o$ (here, a^o denotes the element $a \in A$, considered as an element of the algebra A^o). We introduce a module structure over the algebra $B \otimes A^o$ on M by setting $m(b \otimes a^o) = amb$. The fact that the structure is well-defined and that all module axioms are satisfied can be verified easily.

Conversely, every $B \otimes A^o$-module N can be considered as an A-B-bimodule by setting $an = n(1 \otimes a^o)$ and $nb = n(b \otimes 1)$. In this way, the concepts of an A-B-bimodule and of a $B \otimes A^o$-module, in fact, coincide.

In conclusion, we determine the center of the tensor product of algebras.

Theorem 4.2.4 $C(A \otimes B) = C(A) \otimes C(B)$.

Proof. Let A' be a complement of $C(A)$ in A, i. e. $A = A' \oplus C(A)$ as a vector space. Then $A \otimes B = (A' \otimes B) \oplus (C(A) \otimes B)$. We shall show that $C(A \otimes B) \subset C(A) \otimes B$. Indeed, let $c \in C(A \otimes B)$, $c = x + y$ with $x \in A' \otimes B$ and $y \in C(A) \otimes B$. Then $c(a \otimes 1) = (a \otimes 1)c$, and thus $x(a \otimes 1) = (a \otimes 1)x$ because $y(a \otimes 1) = (a \otimes 1)y$ is satisfied trivially.

Choose a basis $\{b_1, b_2, \ldots, b_m\}$ of B and write $x = \sum_{i=1}^{m} x_i \otimes b_i$. Since this form is unique, we deduce that $x_i a = a x_i$ for all a, and hence $x_i \in C(A)$. But $x_i \in A'$ and thus $x_i = 0$ and $x = 0$.

Similarly, choosing a basis in $C(A)$ and decomposing $C(A) \otimes B = (C(A) \otimes C(B)) \oplus (C(A) \otimes B')$, where B' is a complement of $C(B)$ in B, we can show that $C(A \otimes B) \subset C(A) \otimes C(B)$. Since the converse inclusion is trivial, the theorem follows. □

4.3 Central Simple Algebras

An algebra A over a field K is called *central* if $C(A) = K$.

Our aim in this chapter is to study central division algebras (finite dimensional over K). However, as we shall see from what follows, it is convenient to consider at the same time all central simple algebras, i. e. all algebras of the form $M_n(D)$, where D is a central division algebra.

From Proposition 4.1.1, it follows that an algebra is central simple if and only if its regular bimodule is simple and its endomorphism algebra coincides with K. This yields a characterization of the algebra $A \otimes A^o$.

Define the homomorphism T of the algebra $A \otimes A^o$ into the endomorphism algebra $E(A)$ of the vector space A by putting $T(a \otimes b^o)$ to be the linear operator mapping $x \in A$ into bxa.

Theorem 4.3.1. *An algebra A is a central simple algebra over a field K if and only if the homomorphism $T : A \otimes A^o \to E(A)$ defined above is an isomorphism.*

Proof. If A is a central simple algebra, then A can be considered as a simple module over the algebra $A \otimes A^o$ with the endomorphism algebra K. But then, by Theorem 2.6.7, the homomorphism $T : A \otimes A^o \to E(A)$ is an epimorphism and, since $[A \otimes A^o : K] = n^2 = [E(A) : K]$, where $n = [A : K]$, T is an isomorphism.

Conversely, if T is an isomorphism, then identifying $A \otimes A^o$ with $E(A)$, we see that A is a simple $A \otimes A^o$-module, i. e. a simple bimodule, and that its endomorphism algebra is K. In other words, A is a central simple algebra. □

We shall apply Theorem 4.3.1 to investigate the structure of the algebra $A \times B$, where A is a central simple and B is an arbitrary K-algebra.

Theorem 4.3.2. *Every ideal of the algebra $A \otimes B$, where A is a central simple algebra, is of the form $A \otimes I$, where I is an ideal of the algebra B.*

Proof. Evidently, if I is an ideal of B, then $A \otimes I$ is an ideal of $A \otimes B$. Conversely, let J be an ideal of the algebra $A \otimes B$. Selecting a basis $\{a_1, a_2, \ldots, a_n\}$ in A, we can express every element of $A \otimes B$ uniquely in the form $\sum_{i=1}^{n} a_i \otimes b_i$, where $b_i \in B$. Consider the linear operator T_k of the space A mapping a_k into 1 and all the other elements of the basis into 0. By Theorem 4.3.1, $T_k = T(y_k)$, where $y_k \in A \otimes A^o$ and thus $y_k = \sum_{j=1}^{n} a_j \otimes x_j^o$ with $x_j \in A$. But then

$$\sum_{j=1}^{n} (x_j \otimes 1)\Big(\sum_{i=1}^{n} a_i \otimes b_i\Big)(a_j \otimes 1) = \sum_{i=1}^{n}\Big(\sum_{j=1}^{n} x_j a_i a_j \otimes b_i\Big) = \sum_{i=1}^{n} (a_i T_k) \otimes b_i = 1 \otimes b_k .$$

Thus, if $\sum_{i=1}^{n} a_i \otimes b_i$ belongs to the ideal J, then also $1 \otimes b_k \in J$ for every k. Let $I = \{b \in B \mid 1 \otimes b \in J\}$. Clearly, I is an ideal of the algebra B, and as we have just seen, every element of J has the form $\sum_{i=1}^{n} a_i \otimes b_i$, where $b_i \in I$, i. e. $J = A \otimes I$, as required. □

Corollary 4.3.3. *If A is a central simple algebra, then the algebra $A \otimes B$ is simple if and only if B is simple.*

Corollary 4.3.4. *If A is a central simple algebra, then* $\operatorname{rad}(A \otimes B) = A \otimes \operatorname{rad} B$ *for any algebra B.*

The proof follows from the fact that the radical is the intersection of maximal two-sided ideals.

Corollary 4.3.5. *If A is a central simple algebra, then the algebra $A \otimes B$ is semisimple if and only if the algebra B is semisimple.*

Theorem 4.3.2, in combination with Theorem 4.2.4 describing the center of a tensor product, implies also the following corollary.

Corollary 4.3.6. *If A is a central simple algebra, then $A \otimes B$ is a central simple algebra if and only if B is a central simple algebra.*

4.4 Fundamental Theorems of the Theory of Division Algebras

Theorem 4.4.1 (Skolem-Noether). *If f and g are two homomorphisms of a simple algebra B into a central simple algebra A, then there is an invertible element a in A such that $g(b) = af(b)a^{-1}$ for all $b \in B$.*

Proof. Consider the B-A-modules ${}_fA$ and ${}_gA$ (see Sect. 4.1, Example 2). These are modules over the algebra $A \otimes B^o$ which is simple by Corollary 4.3.4. Since the dimensions of these modules coincide (they are equal to $[A : K]$), we get, by Corollary 2.3.5, ${}_fA \simeq {}_gA$.

Let φ be an isomorphism of ${}_fA$ onto ${}_gA$. Then φ is an automorphism of the regular A-module and therefore $\varphi(x) = ax$, where a is a fixed invertible element of A. Moreover, φ is a homomorphism of the left B-modules, i.e. $\varphi(bx) = b\varphi(x)$. By the definition of ${}_fA$ and ${}_gA$, $\varphi(f(b)x) = g(b)\varphi(x)$. Taking $x = 1$, we obtain $af(b) = \varphi(f(b)) = g(b)\varphi(1) = g(b)a$ for every $b \in B$, i.e. $g(b) = af(b)a^{-1}$. □

The map $x \mapsto axa^{-1}$ is obviously an automorphism of the algebra A. Such an automorphism is called *inner*. If B is a subalgebra of A, then $aBa^{-1} = \{aba^{-1} \mid b \in B\}$ is also a subalgebra. We say that B and aBa^{-1} are *conjugate* in A.

Corollary 4.4.2. *Isomorphic simple subalgebras B and B' of a central simple algebra A are conjugate. Moreover, every isomorphism $g : B \xrightarrow{\sim} B'$ can be extended to an inner automorphism of the algebra A, i.e. it is of the form $g(b) = aba^{-1}$ for some invertible element a of A.*

The proof follows from the Skolem-Noether theorem if we consider, together with g, the embedding f of the algebra B in the algebra A.

Corollary 4.4.3. *Every automorphism of a central simple algebra is inner. In particular, every automorphism of the algebra $M_n(K)$ is inner.*

Let us remark that the algebras A and B appear in the proof of the Skolem-Noether theorem symmetrically: we have used only the simplicity of the tensor product $A \otimes B^o$ and for that, it is sufficient that one of these algebras is central simple and the other simple. Therefore, we can formulate a "dual" theorem whose proof is left to the reader.

Theorem 4.4.4. *If f and g are two homomorphisms of a central simple algebra B into a simple algebra A, then there is an invertible element a in A such that $g(b) = af(b)a^{-1}$ for all $b \in B$.*

Corollary 4.4.5. *Isomorphic central simple subalgebras B and B' of a simple algebra A are conjugate. Moreover, every isomorphism $g : B \xrightarrow{\sim} B'$ can be extended to an inner automorphism of the algebra A, i. e. it is of the form $g(b) = aba^{-1}$ for some invertible element a of A.*

Of course, a counterpart of Corollary 4.4.3 does not hold for non-central algebras. The most simple example is the complex conjugation which is a non-inner automorphism of the field of complex numbers considered as an algebra over the field of real numbers.

The Skolem-Noether theorem is often called the first fundamental theorem of the theory of division algebras. The second theorem is related to the concept of a centralizer.

A *centralizer* of a subset X of an algebra A is the subset of all elements $a \in A$ such that $ax = xa$ for every $x \in X$. A centralizer of a subset X is a subalgebra of A which will be denoted by $C_A(X)$. In the particular case when $X = A$, $C_A(A) = C(A)$ is the center of the algebra.

Theorem 4.4.6. *Let A be a central simple algebra, B its simple subalgebra and $B' = C_A(B)$. Then*

1) *B' is a simple algebra;*
2) *$C_A(B') = B$;*
3) *$[B : K][B' : K] = [A : K]$;*
4) *if $B' \simeq M_m(D)$, then $A \otimes B^o \simeq M_n(D)$ where m divides n.*

Proof. Consider the B-A-bimodule ${}_fA$, where f is the embedding of B into A. By Corollary 4.3.4, the algebra $A \otimes B^o$ is simple. Hence, $A \otimes B^o \simeq M_n(D)$ and ${}_fA \simeq mU$, where U is a simple module over $A \otimes B^o$. Therefore, $E_{A\otimes B^o}({}_fA) \simeq M_m(D)$.

Let φ be an endomorphism of ${}_fA$. Then, φ is an endomorphism of the regular module, i. e. $\varphi(x) = ax$ for some $a \in A$. Besides, $\varphi(bx) = b\varphi(x)$, which implies (for $x = 1$) that $ab = ba$ for all $b \in B$ and thus $a \in B'$. Conversely, if $a \in B'$, then the map $x \mapsto ax$ is evidently an endomorphism of ${}_fA$. Therefore $B' \simeq E(A) \simeq M_m(D)$. The statements 1) and 4) (except the statement on the divisibility) follow.

Write $d = [D : K]$. Then $U \simeq nD$, thus $[U : K] = nd$ and $[A : K] = mnd$. On the other hand, $[A : K][B : K] = [A \otimes B^o : K] = n^2d$ and $[B' : K] = m^2d$. It turns out that $[B : K] = \dfrac{n^2d}{mnd} = \dfrac{n}{m}$, i. e. m divides n and $[A : K] = [B : K][B' : K]$. Hence, 3) and 4) follow.

Finally, let $B'' = C_A(B')$. Clearly, $B \subset B''$. However, the algebra B' is simple and therefore assertion 3) holds also for B'. This means that $[B'' : K] = [A : K]/[B' : K] = [B : K]$. Consequently, $B'' = B$, completing the proof of the theorem. □

4.5 Subfields of Division Algebras. Splitting Fields

We shall apply the results of the preceding section to an investigation of subfields of central division algebras. We shall be interested in maximal subfields, i. e. in subfields which are not contained in any larger subfield.

Theorem 4.5.1. *A subfield L of a division algebra D is maximal if and only if $L = C_D(L)$. If the division algebra D is central, then $[D : K] = [L : K]^2$ and $D \otimes L \simeq M_n(L)$, where $n = [L : K]$.*

Proof. Clearly, every subfield containing L is contained in $C_D(L)$. Therefore, if $L = C_D(L)$, then L is maximal. On the other hand, if a belongs to $C_D(L)$ and does not belong to L, then the set of elements of the form $f(a)$, where $f(x)$ is a polynomial over the field L, forms a commutative subalgebra in D, i. e. a subfield properly larger than L. This proves the first statement.

Now, let D be a central division algebra and L its maximal subfield. Then $L = C_D(L)$ and thus, by Theorem 4.4.6, $[D : K] = [L : K]^2$ and $D \otimes L \simeq M_n(L)$ ($L^o \simeq L$, since L is commutative). Finally, a dimension argument yields immediately that $n = [L : K]$. □

Corollary 4.5.2. *The dimension of a central simple algebra is always the square of an integer.*

For a K-algebra A and an extension L of the field K, the algebra $A \otimes L$ can be considered as an L-algebra by setting $\alpha x = x(1 \otimes \alpha)$, where $x \in A \otimes L$, $\alpha \in L$. We shall denote this L-algebra by A_L. We say that A_L is obtained from A *by extension of the field of scalars.* Clearly, $[A_L : L] = [A : K]$.

Theorem 4.5.1 shows that if L is a maximal subfield of a central division algebra D, then $D_L \simeq M_n(L)$. Obviously, $M_k(D) \otimes L \simeq M_{nk}(L)$.

If A is a central K-algebra, then Theorem 4.2.4 shows that the L-algebra A_L is also central. In particular, if A is a central simple K-algebra, then A_L is a central simple L-algebra.

A field L is called a *splitting field* for a central simple algebra A if $A_L \simeq M_n(L)$. It follows from Theorem 4.5.1 that a splitting field always exists.

However, the splitting fields are not unique; after all, every extension of a splitting field is also a splitting field. In fact, even minimal splitting fields (containing no other splitting field) are not uniquely determined.

The following theorem establishes a characterization of splitting fields which will be useful in the sequel.

Theorem 4.5.3. *A field L is a splitting field for a central division algebra D of dimension d^2 if and only if $[L : K] = md$ and L is isomorphic to a subalgebra of the algebra $M_m(D)$.*

Proof. Let $D \otimes L \simeq M_d(L)$. Consider a simple D_L-module, i. e. a simple L-D-bimodule U. Thus U is a right vector space over D and multiplication by an element $\alpha \in L$ defines an endomorphism of this vector space. Denoting by $T(\alpha)$ the matrix of this endomorphism, we obtain a homomorphism (in fact, since L is a field, a monomorphism) $T : L \to M_m(D)$, where $m = [U : D]$.

On the other hand, since $D_L \simeq M_d(L)$, $U \simeq dL$ and $[U : K] = d[L : K]$. Taking into account that $[U : K] = md^2$, we get $[L : K] = md$.

Conversely, let L be a subfield of the algebra $A = M_m(D)$ of dimension md and L' the centralizer of L. Then $L' \supset L$ and, by Theorem 4.4.6, $[L : K][L' : K] = [A : K] = m^2d^2$. From here, $[L' : K] = md = [L : K]$; thus $L = L'$ and therefore $A \otimes L \simeq M_n(L)$. We conclude that L is a splitting field for A, and thus also for D. □

4.6 Brauer Group. The Frobenius Theorem

In Sect. 4.3, we have observed that the class of central simple algebras is closed with respect to the tensor multiplication. The ground field K plays the role of identity since $A \otimes K \simeq A$ for any algebra A. Finally, Theorem 4.3.1 shows that the opposite algebra A^o is an inverse of the algebra A in the sense of this operation. All this allows us to define a group structure on the set of isomorphism classes of the central division algebras in the following way.

We fix a representative in each isomorphism class of the central division algebras. If D_1 and D_2 are such representatives, then $D_1 \otimes D_2$ is a central simple algebra, and therefore isomorphic to $M_n(D)$, where D is a central division algebra. Put $D = D_1D_2$. It follows from Sect. 4.2 that $D_1D_2 = D_2D_1$ and $D_1(D_2D_3) = (D_1D_2)D_3$. Furthermore $DK = KD = D$, and by Theorem 4.3.1, $DD^o = D^oD = K$. In this way, our set of central division algebras forms a commutative group. This group is called the *Brauer group* of the field K and is denoted by $\mathrm{Br}(K)$.

If L is an extension of the field K, then for every central division algebra D also the L-algebra D_L is central simple and thus isomorphic to $M_n(D')$, where D' is a central division algebra over L. It is easy to verify that assigning to D the division algebra D', we obtain a group homomorphism $\mathrm{Br}(K) \to \mathrm{Br}(L)$. The kernel of this homomorphism consists of those division algebras for which L is a splitting field. This subgroup of the Brauer group is denoted by $\mathrm{Br}(L/K)$. Theorem 5.1 shows that every element of the Brauer group belongs to some subgroup of the form $\mathrm{Br}(L/K)$ and thus $\mathrm{Br}(K) = \cup \mathrm{Br}(L/K)$.

A concrete calculation of the Brauer group is, as a rule, rather complex, and the structure of this group is known only for some fields K. We shall limit ourselves to the most simple cases: the field of real numbers and the finite fields (see Chap. 5).

Of course, if K is an algebraically closed field, then there are no central division algebras (in fact, no division algebras) different from K, and thus the Brauer group is trivial.

Over the field $\mathbb{R}$ of real numbers, there is at least one proper central field, viz. the quaternion algebra $\mathbb{H}$. A remarkable result asserts that this is the only central division algebra over the field $\mathbb{R}$.

Theorem 4.6.1 (Frobenius). *The only finite dimensional division algebras over the field $\mathbb{R}$ of real numbers are the field $\mathbb{R}$ itself, the field $\mathbb{C}$ of complex numbers and the quaternion algebra $\mathbb{H}$.*

Proof. First, let L be a finite extension of the field $\mathbb{R}$, a an element of L and $m_a(x)$ the minimal polynomial of the element a over the field $\mathbb{R}$ (see Sect. 1.2). Since $m_a(x)$ is irreducible, it is either linear (and then $a \in \mathbb{R}$) or quadratic of the form $x^2 + 2px + q$, where $p^2 < q$. In the second case, the element $a + p$ is a root of the polynomial $x^2 + q - p^2$. Thus, $\dfrac{a+p}{\sqrt{q-p^2}}$ is a root of the polynomial $x^2 + 1$. Therefore the subfield $\mathbb{R}[a]$ is isomorphic to $\mathbb{C}$. Since $\mathbb{C}$ is algebraically closed, $L \simeq \mathbb{C}$.

Thus, the finite extensions of the field $\mathbb{R}$ are either $\mathbb{R}$ itself or $\mathbb{C}$. Therefore $\mathbb{C}$ is a splitting field of any central division algebra D over $\mathbb{R}$. Let $D \neq \mathbb{R}$, $d^2 = [D : \mathbb{R}]$ and L a maximal subfield of D. Since $L \neq \mathbb{R}$, necessarily $L \simeq \mathbb{C}$, and by Theorem 4.5.1, $d = [\mathbb{C} : \mathbb{R}] = 2$, i.e. $[D : \mathbb{R}] = 4$.

Denote by i an element of the subfield L such that $i^2 = -1$ (the image of the element $i \in \mathbb{C}$ in the isomorphism $\mathbb{C} \simeq L$). The complex conjugation determines an automorphism of the field L in which i is mapped to $-i$. By Corollary 4.4.2, there is a non-zero element j in D such that $jij^{-1} = -i$, i.e. $ji = -ij$.

Since j and i do not commute, $j \notin L$ and thus 1, i, j are linearly independent. Besides, $j^2 i = -jij = ij^2$, i.e. $j^2 \in C_D(L) = L$. Thus $j^2 = \alpha + \beta i$ with $\alpha, \beta \in \mathbb{R}$. But j^2 must commute with j, and thus $j(\alpha + \beta i) = \alpha j + \beta ji = (\alpha - \beta i)j = (\alpha + \beta i)j$ and therefore $\beta = 0$. It follows that $j^2 = \alpha \in \mathbb{R}$. Clearly, $\alpha < 0$ (otherwise $\alpha = \gamma^2$ and $(j-\gamma)(j+\gamma) = 0$ which is impossible). Replacing j by $j/\sqrt{-\alpha}$, we may assume that $j^2 = -1$.

Thus, we have already identified two elements i and j such that $i^2 = j^2 = -1$ and $ji = -ij$. Write $k = ij$. Then $k^2 = ijij = -i^2j^2 = -1$; $ik = i^2j = -j$; $ki = iji = -i^2j = j$. Similarly, $jk = -kj = i$. In other words, the elements i, j, k have the same multiplication table as the canonic basis of the quaternions $\mathbb{H}$. Therefore, there is a homomorphism $f : \mathbb{H} \to D$ which is a monomorphism because $\mathbb{H}$ is a division algebra. But $[\mathbb{H} : \mathbb{R}] = [D : \mathbb{R}]$, and therefore f is an isomorphism, as required. □

Corollary 4.6.2. $\mathrm{Br}(\mathbb{R}) = \mathrm{Br}(\mathbb{C}/\mathbb{R})$ *is the cyclic group of order two.*

Exercises to Chapter 4

1. Let D be a finite dimensional division algebra over K. Prove that two D-bimodules are isomorphic if and only if the corresponding self-representations are similar (i. e. differ by an inner automorphism of the algebra $M_n(D)$).

2. Let S and T be finite sets with quasi-order relations $\to$, A and B the corresponding minimal algebras over a field K (see Exercises 8–10 to Chap. 3). Introduce a quasi-order on the Cartesian product $S \times T$ by $(s,t) \to (s',t')$ if $s \to s'$ and $t \to t'$. Prove that $A \otimes B$ is the minimal algebra corresponding to this quasi-order relation on the Cartesian product $S \times T$.

3. For the $\mathbb{R}$-algebra $\mathbb{C}$, prove that $\mathbb{C} \otimes \mathbb{C} \simeq \mathbb{C} \oplus \mathbb{C}$. This example shows that the tensor product of simple algebras need not be a simple algebra.

4. A linear transformation $\partial : A \to A$ is said to be a *derivation* on the algebra A if, for arbitrary elements $a, b \in A$, $\partial(ab) = a\partial(b) + \partial(a)b$.
   a) Show that the map ∂_x, where x is a fixed element of A, given by the formula $\partial_x a = ax - xa$ is a derivation on the algebra A. This derivation is called *inner*.
   b) If ∂ is a derivation on the algebra A, prove that the map $T : A \to M_2(A)$ given by the formula
   $$T(a) = \begin{pmatrix} a & \partial a \\ 0 & a \end{pmatrix}$$
   is an algebra homomorphism.
   c) Prove that every derivation on a central simple algebra is inner. (Hint: Use Exercise b) and the Skolem-Noether theorem.)

5. Let A be a simple algebra, B its central simple subalgebra and $B' = C_A(B)$. Prove that:
   a) B' is a simple algebra;
   b) $[B : K][B' : K] = [A : K]$;
   c) if $B' \simeq M_m(D)$, then $A \otimes B^o \simeq M_n(D)$, where m divides n.
   Give an example in which $C_A(B') \neq B$.

6. Consider the algebra D over the field $\mathbb{Q}$ of rational numbers with basis $\{1, i, j, k\}$ and multiplication table

	i	j	k
i	-1	k	$-j$
j	$-k$	-2	$2i$
k	j	$-2i$	-2

(verify that, indeed, it is an algebra).
 a) Prove that D is a central division algebra.
 b) Verify that $L_1 = \mathbb{Q}[i]$ and $L_2 = \mathbb{Q}[j]$ are non-isomorphic maximal subfields of D.

7. Prove that if D_1 and D_2 are two central division algebras over K such that $[D_1 : K]$ and $[D_2 : K]$ are relatively prime, then $D_1 \otimes D_2$ is a division algebra. (Hint: Assuming that $D_1 \otimes D_2 \simeq M_n(D)$, calculate $D_1 \otimes D_2 \otimes D_2^\circ$ in two different ways and deduce that n divides $[D_2 : K]$.)

8. (Dickson's theorem) Prove that two elements of a central division algebra are conjugate if and only if they have the same minimal polynomials.

9. (Hilbert division algebra) Let L be a field and φ an automorphism of L. Consider the "power series" of the form $\sum_{i \gg -\infty}^{\infty} a_i t^i$, where $a_i \in L$, t a symbol (variable) and $i \gg -\infty$ indicates, as usual, that there is only a finite number of powers with negative exponents. Addition of the series is given simply by $\sum_{i \gg -\infty}^{\infty} a_i t^i + \sum_{i \gg -\infty}^{\infty} b_i t^i = \sum_{i \gg -\infty}^{\infty} (a_i + b_i) t^i$ and multiplication by the rules $ta = \varphi(a)t$ $(a \in L)$ and $a\left(\sum_{i \gg -\infty}^{\infty} a_i t^i\right) = \sum_{i \gg -\infty}^{\infty} (aa_i) t^i$.
 a) Verify that the set of the series with the above operations of addition and multiplication forms a division algebra $L[\![t, \varphi]\!]$. This division algebra is called the *Hilbert division algebra*.
 b) Let $K = \{a \in L \mid \varphi(a) = a\}$ and n be the order of the automorphism φ, i.e. the least natural number for which φ^n is the identity automorphism, or ∞ if such a natural number does not exist. Prove that the center of the Hilbert division algebra is $K[\![t^n]\!]$ if $n \neq \infty$ and K otherwise.
 c) Construct an example of a central infinite dimensional division algebra. (Hint: Take for L the field of rational functions $K(x)$.)

5. Galois Theory

In this chapter, we shall apply the machinery of bimodules and tensor products to the study of extensions of a field K, i. e. to Galois theory.

5.1 Elements of Field Theory

In what follows, we shall need some well-known results on the structure of fields and their extensions.

Let K be any field. Recall that the *characteristic* of the field K is the least natural number p such that

$$p1 = \underbrace{1 + 1 + \ldots + 1}_{p \text{ times}} = 0 \quad \text{(if such a number exists).}$$

If there is no number with the above property, i. e. if $m1 \neq 0$ for all m, then K is said to have characteristic 0. Since any decomposition $p = mn$ obviously implies $p1 = (m1)(n1)$, the characteristic of a field is a prime number or zero.

Assume that K is a field of characteristic 0. Then the map $n \mapsto n1$ is a ring embedding of $\mathbb{Z}$ into K. In fact, considering the ratios $n1/m1$, we can embed the entire field $\mathbb{Q}$ of rational numbers into K.

If K is of prime characteristic p, then the situation is still simpler: It is easy to see that in this case the elements of the form $n1$, where $0 \leq n < p$, constitute a subfield isomorphic to the field $\mathbb{F}(p)$ of the congruence classes of the integers modulo p.

The fields $\mathbb{Q}$ and $\mathbb{F}(p)$ contain no proper subfields. Fields with this property are called *prime*. A result of our consideration is the following theorem.

Theorem 5.1.1. *Every field K contains a prime field which is isomorphic either to $\mathbb{Q}$ if K is of characteristic 0, or $\mathbb{F}(p)$ if K is of characteristic $p > 0$.*

Corollary 5.1.2. *If a field K is finite then the number of its elements is p^n, where p is a prime.*

Proof. The characteristic of a finite field must be non-zero and thus K contains an isomorphic copy of $\mathbb{F}(p)$ for some prime p. Hence, K is a finite extension of $\mathbb{F}(p)$, so it is a vector space of finite dimension $n = [K : \mathbb{F}(p)]$ over $\mathbb{F}(p)$. It is now clear that K contains p^n elements. □

We are primarily interested in finite extensions of a fixed field K. The following result plays an important role in the construction and investigation of such extensions.

Theorem 5.1.3 (Kronecker). *Let $p(x)$ be an irreducible polynomial over a field K and $(p(x))$ be the ideal of $K[x]$ consisting of all polynomials which are multiples of $p(x)$: $(p(x)) = \{p(x)g(x) \mid g(x) \in K[x]\}$. Then $K[x]/(p(x))$ is an extension field of K which contains a root of $p(x)$. Conversely, if L is an extension of a field K in which $p(x)$ has a root a, then $K[a] \simeq K[x]/(p(x))$.*

Proof. Write $I = (p(x))$ and consider a class $\bar{x} = x + I$ in the quotient algebra $K[x]/I$. In view of the definition of the operations in a quotient algebra, it follows that $p(\bar{x}) = p(x) + I = 0$, and thus $\bar{x}$ is a root of $p(x)$. It remains to verify that $K[x]/I$ is a field.

Let $\bar{f} = f(x) + I$ be a non-zero class of $K[x]/I$, i.e. $f(x) \notin I$. Then the polynomials $f(x)$ and $p(x)$ are relatively prime. Therefore $1 = f(x)h(x) + p(x)g(x)$ for some polynomials $h(x)$ and $g(x)$. Denoting $h(x) + I$ by $\bar{h}$, we get $\bar{f}\bar{h} = 1$ in the quotient algebra $K[x]/I$.

Conversely, let L be an extension of a field K and $a \in L$ be a root of $p(x)$. Then $p(x)$ is the minimal polynomial of a over K. Defining the homomorphism $\varphi : K[x] \to L$ by the formula $\varphi(f(x)) = f(a)$, we get, by the homomorphism theorem, that $K[a] \simeq K[x]/I$.

Let us remark that the extension $K[x]/I$ is finite over K. Indeed, if the degree of $p(x)$ is n, then it is easy to verify that $1, \bar{x}, \bar{x}^2, \ldots, \bar{x}^{n-1}$ is a basis of $K[x]/I$ over K. □

The Kronecker theorem has the following important consequence. Let $f(x)$ be an arbitrary polynomial over the field K and let $p(x)$ be an irreducible factor. Then the polynomial $p(x)$, and thus also the polynomial $f(x)$, has a root a_1 in the field $K_1 = K[x]/(p(x))$; consequently, by the Bézout theorem, $f(x) = (x - a_1)f_1(x)$ for some polynomial $f_1(x)$ over K_1. Continuing this process, we can construct a chain of finite extensions $K \subset K_1 \subset K_2 \subset \ldots$[9] such that $f(x)$ has i roots (counting multiplicities) in K_i. It follows that if $f(x)$ is of degree n, then it decomposes into linear factors over K_n.

A field $L \supset K$ is called a *splitting field of the polynomial $f(x)$ over K* if $f(x)$ decomposes into linear factors over L, and does not decompose into linear factors over any proper subfield of L containing K.

Theorem 5.1.4 *For any polynomial $f(x) \in K[x]$, there is a splitting field L of $f(x)$ over K and any two such splitting fields are K-isomorphic, i. e. there is an isomorphism which coincides with the identity on K.*

Proof. The existence of a splitting field follows from the argument above; we can take L as the subfield of K_n generated by K and the roots of $f(x)$. The

[9] Observe that the inclusion in this chain may not be proper. For example, if $p(x)$ is a linear polynomial, then already $K_1 = K$.

fact that the splitting field is unique will be proved by induction on the degree n of the polynomial $f(x)$. The result is trivial for $n = 1$: here $L = K$.

Let $f(x)$ be a polynomial of degree n, L and L' be splitting fields of $f(x)$ over K and $p(x)$ an irreducible factor of $f(x)$ (over K). Then $p(x)$ has a root a in the field L and a root a' in the field L'. By the Kronecker theorem, the fields $K[a]$ and $K[a']$ are isomorphic. Identifying them, we can assume that both L and L' contain a common subfield $K_1 = K[a]$.

However, then L and L' are extensions of the field K_1. Moreover, they are splitting fields of the polynomial $f_1(x) = f(x)/(x-a)$ of degree $n-1$ over K_1. By the induction hypothesis, $L \simeq L'$, as required. □

We shall prove that the splitting field L is a finite extension of the ground field K. Since L is a subfield of a field which can be obtained from K by constructing a chain of finite extensions, our statement is a particular instance of the following result.

Theorem 5.1.5. *If $K = K_0 \subset K_1 \subset \ldots \subset K_{n-1} \subset K_n$ is a chain of fields in which, for every i, K_{i+1} is a finite extension of K_i, then K_n is a finite extension of K and $[K_n : K] = \prod_{i=1}^{n} [K_i : K_{i-1}]$.*

Proof. Obviously, it is sufficient to prove the result for $n = 2$; in general, it follows by induction.

Hence, let $K \subset F \subset L$, with $[F : K] = n$ and $[L : F] = m$. Choose a basis $\{a_1, a_2, \ldots, a_n\}$ of the field F over K and a basis $\{b_1, b_2, \ldots, b_m\}$ of the field L over F. Then every element of F has the form $\sum_{i=1}^{n} \alpha_i a_i$, where $\alpha_i \in K$, and every element of the field L has the form $\sum_{j=1}^{m} \beta_j b_j$, where $\beta_j \in F$. Consequently, writing $\beta_j = \sum_{i=1}^{n} \alpha_{ij} a_i$, $\alpha_{ij} \in K$, we obtain $\sum_{j=1}^{m} \beta_j b_j = \sum_{i,j} \alpha_{ij} a_i b_j$, $\alpha_{ij} \in K$ and thus $\{a_i b_j\}$ is a generating system of the vector space L over the field K.

On the other hand, if $\sum_{i,j} \alpha_{ij} a_i b_j = 0$, then, in view of the linear independence of $\{b_j\}$ over F, we get that $\sum_{i=1}^{n} \alpha_{ij} a_i = 0$ for all j. This shows that $\alpha_{ij} = 0$ for all i, j, since $\{a_i\}$ is linearly independent over K. Thus, we have constructed a basis $\{a_i b_j\}$ of the field L over K consisting of nm elements, and the theorem follows. □

5.2 Finite Fields. The Wedderburn Theorem

We shall apply the previous results to a description of finite fields (and also all finite division rings!). First, let us prove a lemma on commutative groups.

Lemma 5.2.1. *If a commutative group G contains elements of order m and n, then there is an element of order k in G, where k is the least common multiple of m and n.*

Proof. Let x be an element of order m and y an element of order n. First consider the case where m and n are relatively prime. Then $k = mn$ and $(xy)^k = x^k y^k = 1$. Conversely, if $(xy)^\ell = 1$, then $x^\ell = y^{-\ell}$ and the elements x^ℓ and y^ℓ have the same order. But the order of the element x^ℓ is a divisor of m and the order of y^ℓ is a divisor of n, and thus $x^\ell = y^\ell = 1$. This implies that ℓ is divisible by m and n, and therefore also by k.

In the general case decompose m and n into prime factors and, for each prime number p, choose either in m or in n the factor p^t, where t is the exponent with which p appears in k. In this way we can write $m = m_0 m'$, $n = n_0 n'$ such that $k = m_0 n_0$ and m_0 and n_0 are relatively prime.

Now, the elements $x' = x^{m'}$ and $y' = y^{n'}$ have orders m_0 and n_0, respectively, and therefore the order of the element $x'y'$ is $m_0 n_0 = k$, as required. □

An immediate consequence of this lemma is the following theorem.

Theorem 5.2.2. *A finite subgroup of the multiplicative group of a field is cyclic. In particular, the multiplicative group of a finite field is always cyclic.*

Proof. Let G be a subgroup of order n of the multiplicative group of a field K. It follows from Lemma 5.2.1 that there is an element g in G whose order m is such that the order of any other element of G is a divisor of m. Therefore $a^m = 1$ for every $a \in G$, and thus all the elements of G are roots of the equation $x^m - 1 = 0$. Consequently, $n \leq m$. Hence $n = m$ and g generates the group G. □

Theorem 5.2.3. *For any prime p and any natural number n there is, up to an isomorphism, a unique field of p^n elements.*

Proof. Put $K = \mathbb{F}(p)$ and consider the polynomial $f(x) = x^{p^n} - x$ over K. Let L be its splitting field and S be the set of the roots of $f(x)$ in L. Since $f'(x) = -1$, $f(x)$ has no multiple roots, and thus S consists of p^n elements. Now, $a \in S$ if and only if $a^{p^n} = a$. Applying the Newton binomial formula, we see that $(a+b)^{p^n} = a^{p^n} + b^{p^n}$ in any field of characteristic p. Since also $(ab)^{p^n} = a^{p^n} b^{p^n}$ and $(a^{-1})^{p^n} = \left(a^{p^n}\right)^{-1}$, it follows that S is a subfield of L containing K. Therefore $S = L$ and L consists of p^n elements.

Now, let L' be any field consisting of p^n elements and let G be its multiplicative group. Then G consists of $p^n - 1$ elements and therefore $a^{p^n - 1} = 1$

for all $a \in G$. Hence $a^{p^n} = a$. Since the last equality holds also for $a = 0$, we conclude that the elements of L' are just the roots of the equation $x^{p^n} - x = 0$. Their number is p^n, and thus L' is the splitting field of $f(x)$ and $L' \simeq L$, by Theorem 5.1.4. □

Theorem 5.2.3 implies, in particular, that two finite fields with the same number of elements are isomorphic. In combination with the Skolem-Noether theorem, this leads to the following remarkable result.

Theorem 5.2.4 (Wedderburn). *Every finite division ring is commutative, i. e. it is a field.*

Proof. If D is a finite division ring, then its center K is a finite field. Let $[D : K] = d^2$. For any maximal subfield L of D, $[L : K] = d$. This means that all maximal subfields are isomorphic. By the Skolem-Noether theorem, they are all conjugate. On the other hand, every element $a \in D$ belongs, obviously, to some maximal field. Thus, if G is the multiplicative group of the division ring D and H the multiplicative group of a maximal field L, then $G = \bigcup_g gHg^{-1}$, where g runs through all G. We shall show that, for $H \neq G$, this is impossible.

Let n be the order of G and m be the order of H with $m < n$. Then $G = \bigcup_g gHg^{-1}$ implies that $n < mk$, where k is the number of distinct subgroups of the form gHg^{-1} (since all of them contain the identity). On the other hand, if $g_1 = gh$ with $h \in H$, then $g_1 H g_1^{-1} = gHg^{-1}$ and we conclude that k does not exceed the index i of H in G. This contradicts the Lagrange theorem asserting that $n = mi$.

Thus, we conclude $H = G$ and $L = D$. The proof of the theorem is completed. □

5.3 Separable Extensions

We return to the study of finite extensions of arbitrary fields. As in the previous chapter, an important role is played here by the algebra $L \otimes L$ (note that $L^o = L$). Therefore, we shall need some information on the structure of the tensor products of fields.

We shall consider tensor products over various fields, including the ground field K and its extensions. The tensor product of vector spaces (or algebras) over a field $L \supset K$ will be denoted by the symbol $\otimes_L$. It is not difficult to verify that the "associativity formula" of Proposition 4.2.3 also holds in this more general situation. Namely, if $L \subset M$ are two extensions of the field K, U is a vector space over L, and V and W are vector spaces over M, then $(U \otimes_L V) \otimes_M W \simeq U \otimes_L (V \otimes_M W)$. The proof of the formula is left to the reader. In place of $\otimes_K$ we shall write simply $\otimes$.

Consider the simplest situation when one of the factors is monogenic, i. e. of the form $K[a]$.

Proposition 5.3.1. *Let L and F be finite extensions of the field K and suppose that $F = K[a]$, where the minimal polynomial of a over K is $p(x)$. Then $L \otimes F \simeq L[x]/(p(x))$.*

Proof. Evidently, $L \otimes F$ is a monogenic L-algebra: $L \otimes F = L[1 \otimes a]$. Therefore, we get an epimorphism of $L[x]$ onto $L \otimes F$ by mapping $f(x) \in L[x]$ onto the element $f(1 \otimes a) \in L \otimes F$.

It follows that $L \otimes F \simeq L[x]/(m(x))$, where $m(x)$ is the minimal polynomial of the element $1 \otimes a$ over L. However, clearly $p(1 \otimes a) = 1 \otimes p(a) = 0$. At the same time, if n is the degree of $p(x)$, then $1, 1 \otimes a, \dots, 1 \otimes a^{n-1}$ are linearly independent over L. We conclude that $m(x) = p(x)$. □

We now clarify the structure of the quotient algebra $K[x]/(f(x))$ for an arbitrary polynomial $f(x)$.

Lemma 5.3.2. *If $f(x) = f_1(x) \dots f_t(x)$, where the polynomials $f_1(x), \dots, f_t(x)$ are pairwise relatively prime, then $K[x]/(f(x)) \simeq \prod_{i=1}^{t} K[x]/(f_i(x))$.*

Proof. Write $I = (f(x))$, $I_i = (f_i(x))$. Consider the map which sends each class $g(x) + I$ of the quotient algebra $K[x]/I$ onto the t-tuple $(g(x) + I_1, g(x) + I_2, \dots, g(x) + I_t) \in \prod_{i=1}^{t} K[x]/I_i$. This is clearly an algebra homomorphism, and also a monomorphism, since whenever $g(x)$ is divisible by each of the $f_i(x)$, it is also divisible by their product $f(x)$. However, the dimension of $K[x]/I$ equals the degree n of the polynomial $f(x)$ and the dimension of each $K[x]/I_i$ equals the degree n_i of $f_i(x)$. Consequently, the dimensions of $K[x]/I$ and of $\prod_{i=1}^{t} K[x]/I_i$ are equal, and thus the above monomorphism is necessarily an isomorphism. □

Corollary 5.3.3. *The algebra $K[x]/(f(x))$ is semisimple if and only if the polynomial $f(x)$ has no multiple irreducible factors.*

Proof. If $f(x) = p_1(x)p_2(x) \dots p_s(x)$, where $p_1(x), p_2(x), \dots, p_s(x)$ are pairwise different irreducible polynomials, then $K[x]/(f(x)) \simeq \prod_{i=1}^{s} K[x]/(p_i(x))$ and all $K[x]/(p_i(x))$ are fields (by the Kronecker theorem). Hence the algebra is semisimple. On the other hand, if $f(x) = p^2(x)g(x)$, then the class of the polynomial $p(x)g(x)$ is easily seen to be a non-zero nilpotent element of $K[x]/(f(x))$. □

Corollary 5.3.4. *Under the assumptions of Proposition 5.3.1, the algebra $L \otimes F$ is semisimple if and only if the polynomial $p(x)$ has no multiple irreducible factors in the field L.*

Corollary 5.3.5. *Let $F = K[a]$ be a monogenic extension and let a be a simple root of its minimal polynomial $p(x)$. Then, for any commutative semisimple algebra A, the algebra $A \otimes F$ is semisimple.*

Proof. Decomposing A, in accordance with the Weierstrass-Dedekind theorem, into a direct product of fields, we see that it is sufficient to prove the result for the algebra $L \otimes F$, where L is a field. By Corollary 5.3.4, we have to show that $p(x)$ does not possess any multiple irreducible factors over L.

Assume that $g(x)$ is a multiple irreducible factor of $p(x)$ over the field L and that L' is an extension in which $g(x)$ has a root b. Then b is a multiple root of $p(x)$. However, by the Kronecker theorem, $K[b] \simeq K[a]$ and thus a is a multiple root, contrary to our assumption. □

An irreducible polynomial is called *separable* if $p(x)$ has no multiple roots in any extension of the field K. The argument given in the proof of Corollary 5.3.5 shows that $p(x)$ is separable whenever it has a simple root in some extension.

An element of a finite dimensional division algebra is called *separable* if its minimal polynomial is separable.

Theorem 5.3.6. *The following conditions are equivalent for a finite extension L of the field K:*

1) *$L \otimes L$ is a semisimple algebra;*
2) *$A \otimes L$ is semisimple for any commutative semisimple algebra A;*
3) *every element of L is separable;*
4) *$L = K[a_1, a_2, \ldots, a_t]$, where all a_i are separable.*

Proof. Trivially, 2) ⇒ 1) and 3) ⇒ 4).

1) ⇒ 3). Let a be a non-separable element of L and $F = K[a]$. Then a is a multiple root of its minimal polynomial and, by Corollary 5.3.4, $L \otimes F$ is not a semisimple algebra. Thus the algebra $L \otimes L \supset L \otimes F$ is also not semisimple.

4) ⇒ 2) can be proved by induction on t. For $t = 1$, the assertion is Corollary 5.3.5. Denote $K[a_t]$ by F. Then the F-algebra $A_F = A \otimes F$ is semisimple. But $A \otimes L \simeq A_F \otimes_F L$ and $L = F[a_1, a_2, \ldots, a_{t-1}]$; hence, by induction, $A \otimes L$ is semisimple.

The proof of the theorem is completed. □

An extension satisfying the equivalent conditions of Theorem 5.3.6 is called *separable.*

Corollary 5.3.7. *Consider a chain of fields $K = K_0 \subset K_1 \subset \ldots \subset K_n = L$, where K_i is a finite extension of K_{i-1}. Then L is separable over K if and only if each K_i is separable over K_{i-1}.*

Proof. It is sufficient to prove the statement for $n = 2$. If K_1 is separable over K and L is separable over K_1, then $L \otimes L \simeq L \otimes (K_1 \otimes_{K_1} L) \simeq A \otimes_{K_1} L$, where $A = L \otimes K_1$ is a semisimple algebra. Therefore, $L \otimes L$ is a semisimple algebra, and thus L is separable over K.

On the other hand, $K_1 \otimes K_1$ is a subalgebra of $L \otimes L$ and therefore, since $L \otimes L$ is semisimple, $K_1 \otimes K_1$ is semisimple as well. Consider the algebra $L \otimes_{K_1} L$ and the map $f : L \otimes L \to L \otimes_{K_1} L$ assigning to $a \otimes b$ the element $a \otimes_{K_1} b$. It is easy to verify that f is an algebra epimorphism. Thus $L \otimes_{K_1} L$ is isomorphic to a quotient algebra of $L \otimes L$. Since a quotient algebra of a semisimple algebra is semisimple, the separability of L over K implies the separability of L over K_1. □

A field K is called *perfect* if every finite extension of K is separable or, in other words, if every irreducible polynomial from $K[x]$ is separable. It is not difficult to establish the following criteria.

Theorem 5.3.8. *Every field of characteristic 0 is perfect. A field K of characteristic p is perfect if and only if the equation $x^p = \alpha$ has a solution in K for each $\alpha \in K$.*

Proof. If $f(x)$ is an irreducible polynomial and $f'(x)$ is its derivative, then either f and f' are relatively prime, or f divides f'. The latter is impossible unless $f'(x) = 0$. Let $f(x) = \alpha_0 x^n + \alpha_1 x^{n-1} + \ldots + \alpha_n$, $\alpha_0 \neq 0$; then $f'(x) = n\alpha_0 x^{n-1} + (n-1)\alpha_1 x^{n-2} + \ldots + \alpha_{n-1}$. If K is a field of characteristic 0, $n\alpha_0 \neq 0$ and thus $f'(x) \neq 0$. As a consequence, $f(x)$ has no multiple roots in any extension of the field K, i. e. $f(x)$ is separable.

If K is a field of characteristic p, then $f'(x) = 0$ if and only if $f(x)$ is of the form $\beta_0 x^{pk} + \beta_1 x^{p(k-1)} + \ldots + \beta_k$. Assume that the equation $x^p = \beta_i$ has a solution $\gamma_i \in K$. Then the binomial theorem shows that $f(x) = (\gamma_0 x^k + \gamma_1 x^{k-1} + \ldots + \gamma_k)^p$, and f is not irreducible. If, on the other hand, the equation $x^p = \alpha$ has no solution for some $\alpha \in K$, then $f(x) = (x - \beta)^p$, where $\beta^p = \alpha$, over the splitting field of the polynomial $f(x) = x^p - \alpha$. Thus $f(x)$ is not separable in this case. The theorem follows. □

Corollary 5.3.9. *Every finite field is perfect.*

Proof. Let K be a finite field of characteristic p. Then the map $\varphi : K \to K$ which sends α into α^p is injective; indeed, if $\alpha^p = \beta^p$, then $\alpha = \beta$ because $\alpha^p - \beta^p = (\alpha - \beta)^p$. Since K is finite, φ is a bijection, and so each element in K is a pth power. □

5.4 Normal Extensions. The Galois Group

We proceed to the main topic of this chapter, namely the study of the automorphisms of finite extensions. Recall the connection between the automorphisms and bimodules established in Chapter 4 (Sect. 4.1, Examples 2, 3). Let L be a finite extension of the field K and σ a K-automorphism of L. Then ${}_\sigma L$ is an L-bimodule on which the operators from L act from the right as on the regular L-module and from the left according to the rule $ax = x\sigma(a)$. Conversely, if M is an L-bimodule such that $[M : L] = 1$ [10], then $M \simeq L$ as a right L-module. Assigning to each $a \in L$ the endomorphism $x \mapsto ax$ of the right L-module M, we get a homomorphism $L \to E_L(M) \simeq L$, i. e. an automorphism of the field L. Therefore $M \simeq {}_\sigma L$ for some automorphism σ. It is easy to check that ${}_\sigma L \simeq {}_\tau L$ if and only if $\sigma = \tau$.

In this way, we have established a bijective correspondence between the set of automorphisms of the field L and the isomorphism classes of the one-dimensional (over L) L-bimodules. Since a one-dimensional bimodule is obviously simple, we obtain, by considering it as an $L \otimes L$-module, the following result.

Theorem 5.4.1. *There is a bijective correspondence between the K-automorphisms of the field L and the one-dimensional simple components of the algebra $(L \otimes L)/\mathrm{rad}(L \otimes L)$.*

Counting the dimensions, we get from here the following corollary.

Corollary 5.4.2. *The number of distinct K-automorphisms of the field L does not exceed $[L : K]$ and can equal $[L : K]$ only in the case when L is separable.*

If an extension L has precisely $[L : K]$ distinct automorphisms, it is called *normal*. In view of Theorem 5.4.1, this is equivalent to the isomorphism $L \otimes L \simeq L^n$. A normal extension is always separable.

Corollary 5.4.3. *If the extension L of the field K is normal and $K \subset K_1 \subset L$, then the extension L of the field K_1 is also normal.*

Proof. The statement follows from the fact that $L \otimes_{K_1} L$ is a quotient algebra of $L \otimes_K L$ (see the proof of Corollary 5.3.7) and every quotient algebra of L^n has again the form L^m for some $m \leq n$. □

The K-automorphisms of an extension L of the field K evidently form a group; it will be denoted by $G(L/K)$. If the extension is normal, $G(L/K)$ is called its *Galois group*.

[10] A priori, one has to distinguish the left and right dimensions of M over L, however, due to finite dimensionality, they coincide: both are equal to $[M : K]/[L : K]$.

An element a of the field L is called an *invariant* of the automorphism σ if $\sigma(a) = a$. This is equivalent to saying that, in the bimodule ${}_\sigma L$, $ax = xa$ for every $x \in {}_\sigma L$. If H is a subset of $G(L/K)$, then an *invariant of* H is every element $a \in L$ which is an invariant of every element $\sigma \in H$. The invariants of a subset H form a subfield of L. This subfield is called the *field of the invariants* of H and is denoted by $\operatorname{Inv} H$.

Theorem 5.4.4. *The following conditions are equivalent:*

1) *the extension* L *of the field* K *is normal;*
2) $\operatorname{Inv} G(L/K) = K$*;*
3) L *is a splitting field of a separable polynomial* $f(x) \in K[x]$.

(An arbitrary polynomial is called separable if all its irreducible factors are separable.)

Proof. 1) $\Rightarrow$ 2). Write $K_1 = \operatorname{Inv} G(L/K)$. Then L is an extension of K_1 and $G(L/K) = G(L/K_1)$, which shows that $[L : K] = |G(L/K)| = |G(L/K_1)| \leq [L : K_1]$. Since $K \subset K_1$, we conclude $K = K_1$.

2) $\Rightarrow$ 3). Let a be an element of the field L. Apply all automorphisms of $G(L/K)$ to a and consider the distinct elements which we obtain this way: $a = a_1, a_2, \ldots, a_k$. Consider the polynomial $f(x) = (x-a_1)(x-a_2)\ldots(x-a_k)$. It is invariant under the action of any $\sigma \in G(L/K)$. Therefore $f(x) \in K[x]$. Hence, a is a root of a separable polynomial from $K[x]$ which decomposes over L into linear factors.

Now choose a generating system (for instance, a basis) of L, say $L = K[w_1, w_2, \ldots, w_t]$, and for each w_i construct a separable polynomial $f_i(x)$ which decomposes into linear factors over L and has w_i as a root. Then L is the splitting field of the polynomial $f(x) = f_1(x)f_2(x)\ldots f_t(x)$ which is separable over the field K.

3) $\Rightarrow$ 1) will be proved by induction on the degree d of the polynomial $f(x)$. For $d = 1$ (i.e. $L = K$), the assertion is trivial. We assume that the implication holds for all polynomials of degree $d-1$.

Let $L = K[w_1, w_2, \ldots, w_t]$ where w_i are the roots of $f(x)$. Write $K_1 = K[w_1]$. Now, L is a splitting field (over K_1) of the separable polynomial $f(x)/(x - w_1)$ of degree $d - 1$, and thus normal over K_1: $L \otimes_{K_1} L \simeq L^m$. But $L \otimes L \simeq (L \otimes K_1) \otimes_{K_1} L$. Consider the structure of $L \otimes K_1$. Let $p(x)$ be the minimal polynomial of the element w_1. It is separable and decomposes over L into linear factors: $p(x) = (x - a_1)(x - a_2)\ldots(x - a_s)$, where $a_1 = w_1$, $a_2, \ldots$, a_s are distinct and $s = [K_1 : K]$. Consequently, $K[a_i] \simeq K_1$ and we get s distinct homomorphisms $\sigma_i : K_1 \to L$. Accordingly, we can construct s K_1-L-bimodules ${}_{\sigma_i}L$ which are one-dimensional over L. Therefore, $L \otimes K_1$ has a quotient algebra which is isomorphic to L^s. A simple dimension count now shows that $L \otimes K_1 \simeq L^s$ and $L \otimes L \simeq L^s \otimes_{K_1} L \simeq L^{ms}$. The theorem follows. □

Corollary 5.4.5. *For any separable extension L of the field K, there is a normal extension L' containing L.*

Proof. Let $L = K[w_1, w_2, \dots, w_t]$, where w_i is a root of a separable polynomial $f_i(x)$. Then we can take L' as the splitting field of the polynomial $f(x) = f_1(x)f_2(x)\dots f_t(x)$. □

5.5 The Fundamental Theorem of Galois Theory

In this section, we shall prove the central theorems of the theory of fields: the normal basis theorem and the fundamental theorem of Galois theory. We begin with the following useful remark.

Lemma 5.5.1. *Let A be an algebra over the field K, let M and N be A-modules, and let L be a finite extension of the field K. Put $A_L = L \otimes A$, $M_L = L \otimes M$ and $N_L = L \otimes N$ (obviously, M_L and N_L can be considered as A_L-modules). If $M_L \simeq N_L$ as A_L-modules, then $M \simeq N$ as A-modules.*

Proof. Consider M_L and N_L as A-modules. Since L is an n-dimensional vector space over K, $M_L \simeq nM$ and $N_L \simeq nN$. In view of the Krull-Schmidt theorem, the isomorphism $nM \simeq nN$ implies that $M \simeq N$. □

Now let L be an extension of K, let $G = G(L/K)$, and let KG be the group algebra of G over the field K (see Sect. 1.1, Example 6). Then L can be considered as a left KG-module by setting $\left(\sum_{g\in G} \alpha_g g\right)a = \sum_{g\in G} \alpha_g g(a)$ for each $a \in L$.

Theorem 5.5.2. *An extension L is normal if and only if L is isomorphic, as a left KG-module, to the regular left KG-module.*

Proof. If $L \simeq KG$ as left KG-modules, then $[KG : K] = [L : K]$. But $[KG : K] = (G : 1)$ (the order of the group G), and thus L is normal.

Conversely, let L be a normal extension and consider $L \otimes L$ as a left module over the algebra $L \otimes KG \simeq LG$. In view of Lemma 5.5.1, it is sufficient to show that $L \otimes L \simeq LG$.

Now $L \otimes L$, as a module over itself (or a bimodule over L), decomposes into a direct sum $L \otimes L \simeq \bigotimes_{\sigma\in G} {}_\sigma L$. Let $1 = \sum_{\sigma\in G} e_\sigma$ be the corresponding decomposition of the identity. It is easy to verify that the map $x \mapsto gx$ is an automorphism of the algebra $L \otimes L$. Therefore ge_σ is a minimal idempotent of that algebra, and so Theorem 2.5.1 implies that $ge_\sigma = e_\tau$ for some $\tau \in G$. In accordance with the definition of ${}_\tau L$, this means that $age_\sigma = ge_\sigma\tau(a)$ for every $a \in L$.

Now, for each $x = \sum a_i \otimes b_i \in L \otimes L$, we have $gx = \sum a_i \otimes gb_i$ and $ax = \sum aa_i \otimes b_i$, $agx = gax$ for all $a \in L$. Consequently, $age_\sigma = gae_\sigma = g(e_\sigma \sigma(a)) = ge_\sigma(g\sigma(a))$, and thus $\tau = g\sigma$.

Thus $ge_\sigma = e_{g\sigma}$. This shows that the map which sends $\sigma \in LG$ onto e_σ gives an isomorphism of LG-modules $LG \simeq L \otimes L$, as required. □

Observe that the isomorphism of the KG-modules $L \simeq KG$ implies that there exists an element $w \in L$ such that the elements $\sigma(w)$ form a basis of L as σ runs through the group G. A basis of this kind is called *normal* and Theorem 5.5.2 is the *normal basis theorem.*

Corollary 5.5.3. *A normal extension is monogenic.*

Proof. If w is an element such that the $\sigma(w)$ form a basis as σ runs through G, then the $\sigma(w)$ are the roots of the polynomial $m_w(x)$ and the degree of $m_w(x)$ is equal to the dimension of the extension L. It follows that $L = K[w]$. □

The *fundamental theorem of Galois theory* follows easily from the normal basis theorem.

Theorem 5.5.4. *Let L be a normal extension of the field K and $G = G(L/K)$. For each subfield F of the field L containing K, let* Inv F *denote the subgroup of G consisting of all those elements σ for which $\sigma(a) = a$ for all $a \in F$. Then*

1) *for any subgroup $H \subset G$,* $\operatorname{Inv}(\operatorname{Inv} H) = H$, *and for any field F with $K \subset F \subset L$,* $\operatorname{Inv}(\operatorname{Inv} F) = F$;
2) *the map $H \to \operatorname{Inv} H$ is a bijective correspondence between the set of subgroups of the Galois group and the set of intermediate fields between K and L; and $H \supset H_1$ if and only if* $\operatorname{Inv} H \subset \operatorname{Inv} H_1$;
3) $\operatorname{Inv} F \simeq G(L/F)$ *for any intermediate field F;*
4) *the intermediate field F is normal if and only if the subgroup* $\operatorname{Inv} F$ *is normal, and then* $G(F/K) \simeq G/\operatorname{Inv} F$.

Proof. Evidently, $\operatorname{Inv}(\operatorname{Inv} H) \supset H$ and $\operatorname{Inv}(\operatorname{Inv} F) \supset F$. We can determine the degree of the field $\operatorname{Inv} H$ over K by using the isomorphism of the left KG-modules $L \simeq KG$. In this isomorphism, $\operatorname{Inv} H$ maps into the subspace $V \subset KG$ consisting of the elements x such that $\sigma x = x$ for all $\sigma \in H$. If we write x in the form $x = \sum_{g\in G} \alpha_g g$, then $\sigma x = \sum_{g\in G} \alpha_g(\sigma g)$, and thus $\alpha_g = \alpha_{\sigma g}$ for any $\sigma \in H$ and the elements of the form $\sum_{\sigma\in H} \sigma g$, where g is a fixed element of G, constitute a basis of V. The number of distinct elements of this form is just the number of cosets of G by H, i.e. $(G : H)$. Hence, $[\operatorname{Inv} H : K] = (G : H)$.

On the other hand, the field L is normal over every subfield (by Corollary 5.4.3). Consequently, there are $[L : F]$ distinct automorphisms of L leaving the elements of F invariant and thus the order of the group $\operatorname{Inv} F$ is equal to $[L : F]$.

In particular, the order of the group $\operatorname{Inv}(\operatorname{Inv} H)$ equals

$$[L : \operatorname{Inv} H] = [L : K]/[\operatorname{Inv} H : K] = (G : 1)/(G : H) = (H : 1);$$

so it follows that $\operatorname{Inv}(\operatorname{Inv} H) = H$. In a similar way, the degree of the field $\operatorname{Inv}(\operatorname{Inv} F)$ over K is equal to

$$(G : \operatorname{Inv} F) = (G : 1)/(\operatorname{Inv} F : 1) = [L : K]/[L : F] = [F : K],$$

and this proves that $\operatorname{Inv}(\operatorname{Inv} F) = F$. This yields both assertions 1) and 2), and we also note that along the way we established the isomorphism $G(L/F) \simeq \operatorname{Inv} F$.

Next, we determine $\operatorname{Inv}(g(F))$, where g is an automorphism from G. If $\sigma \in \operatorname{Inv}(g(F))$ then $\sigma g(a) = g(a)$ for all $a \in F$, i.e. $g^{-1}\sigma g(a) = a$ and thus $g^{-1}\sigma g \in \operatorname{Inv} F$. Consequently, $\operatorname{Inv}(g(F)) = g(\operatorname{Inv} F)g^{-1}$. If $\operatorname{Inv} F$ is a normal subgroup, then $\operatorname{Inv}(g(F)) = \operatorname{Inv} F$, which shows that $g(F) = F$ for any $g \in G$. In this way, every K-automorphism of the field L induces a K-automorphism of the field F; moreover, it is easy to see that g and h induce the same automorphism of the field F if and only if they are in the same coset of G by $\operatorname{Inv} F$. Altogether, we obtain $(G : \operatorname{Inv} F)$ automorphisms of F over K. Since $(G : \operatorname{Inv} F) = [F : K]$, it follows that the field F is normal and that $G(F/K) \simeq G/\operatorname{Inv} F$.

Conversely, if F is normal, then Theorem 5.4.4 shows that F is the splitting field of a separable polynomial $f(x) \in K[x]$. Since every automorphism g of the field L maps a root of $f(x)$ again to a root of $f(x)$, therefore $g(F) = F$ and $g(\operatorname{Inv} F)g^{-1} = \operatorname{Inv} F$, and thus $\operatorname{Inv} F$ is a normal subgroup of G. The proof of the theorem is completed. □

Corollary 5.5.5. *A separable extension L of the field K contains only a finite number of subfields.*

Proof. If L is normal, the corollary is an immediate consequence of the fundamental theorem of Galois theory, because a finite group possesses a finite number of subgroups. In the general case, it is sufficient to embed L into a normal extension; this is always possible by Corollary 5.4.5. □

Corollary 5.5.6. *A separable extension is monogenic.*

Proof. Every extension of a finite field is normal by Theorems 5.2.3 and 5.4.4; consequently, by Corollary 5.5.3, it is monogenic. Therefore, we can assume that the ground field K is infinite.

A separable extension L of the field K contains a finite number of subfields. If a is an element of L which does not belong to any of these subfields, then clearly $L = K[a]$. Therefore, the proof is reduced to the following fact from linear algebra.

Lemma 5.5.7. *A vector space over an infinite field cannot be expressed as a finite union of its proper subspaces.*

Proof. Evidently, a space V cannot be expressed as a union of its two proper subspaces V_1 and V_2; for if $x_1 \in V_1 \setminus V_2$ and $x_2 \in V_2 \setminus V_1$ then $x_1 + x_2$ does not belong to either V_1 or V_2. Now, let $V = \bigcup_{i=1}^{m} V_i$, where V_i are proper subspaces. We are going to show that one of them is contained in the union of the remaining ones.

Indeed, otherwise there exist x_1 and x_2 such that x_1 belongs only to V_1 and x_2 belongs only to V_2. Then for every non-zero $\alpha \in K$, $x_1 + \alpha x_2$ does not belong to $V_1 \cup V_2$, and therefore $x_1 + \alpha x_2 \in V_i$ for some $i > 2$. Since K is infinite, there are two distinct elements α and β in the field K such that $x_1 + \alpha x_2 \in V_i$ and $x_1 + \beta x_2 \in V_i$ (for the same $i > 2$). But then $x_2 \in V_i$, in contradiction to the assumption.

As a consequence, we can omit one of the subspaces and obtain V as a union of $m - 1$ subspaces. Continuing this process, we arrive finally to two subspaces, which is impossible. □

5.6 Crossed Products

The Galois theory allows a new approach to the study of central simple algebras. In this section we shall present a construction allowing us to describe the Brauer group in terms of normal extensions and to construct an algebra of the form $M_m(D)$ for every central division algebra D.

We begin with the following important result.

Lemma 5.6.1 (Noether). *Let D be a finite dimensional central division algebra over a field K. Then there is a maximal subfield $L \subset D$ which is separable over K.*

Proof. If the characteristic of K equals 0, then every subfield of D which contains K is separable over K. Hence, we may assume that K is a field of characteristic $p > 0$. In this case, we shall show that there is an element in D which is separable over K and does not belong to K.

Take an arbitrary element $a \in D \setminus K$. Let $f(x) = m_a(x)$. If a is not separable over K, then the irreducible polynomial $f(x)$ has a multiple root. It follows that $f'(x) = 0$, and thus $f(x) = g(x^p)$ for some $g(x) \in K[x]$. The element a^p is a root of the polynomial $g(x)$. If $g(x)$ is not separable over K then again $g(x) = h(x^p)$ for some $h(x) \in K[x]$. Continuing this process we reach an element b which is not separable over K such that b^p is separable.

Assume that $b^p \in K$ and consider the map $\delta : D \to D$ which maps $d \in D$ to $db - bd$. Since $b \notin K$, there is a $d_0 \in D$ such that $\delta(d_0) \neq 0$, and $\delta^p(d_0) = d_0 b^p - b^p d_0 = 0$ because $b^p \in K$. Let m be the least natural number such that $\delta^m(d_0) = 0$, and let $\delta^{m-1}(d_0) = t$, $\delta^{m-2}(d_0) = w$ and $u = b^{-1}t$. Then $t = \delta(w) = wb - bw$, and $ub = bu$ because $\delta(t) = 0$. But then $b = tu^{-1} = (wb - bw)u^{-1} = wbu^{-1} - bwu^{-1} = (wu^{-1})b - b(wu^{-1})$, and consequently, $b = cb - bc$, where $c = wu^{-1}$. Multiplying by b^{-1}, we get $c = 1 + bcb^{-1}$.

The argument used above for a shows that there is an exponent $p^n = q$ such that c^q is separable over K. Now $c^q = 1 + bc^q b^{-1}$ and so b does not commute with c^q. Hence $c^q \notin K$, as required.

The lemma now follows easily by induction on the dimension $[D : K]$. For $[D : K] = 1$, the statement is trivial, so suppose $[D : K] > 1$. Assume that the lemma holds for all division algebras whose dimension over the center is smaller than $[D : K]$.

Choose an element $a \in D \setminus K$ which is separable over K and put $F = K[a]$ and $D_1 = C_D(F)$. According to Theorem 4.4.6, $F = C_D(D)$ and thus $F = C(D_1)$, because $F \subset D_1$. However, $[D_1 : F] < [D : K]$, and so there is a maximal subfield L in D_1 which is separable over F. Then $[D_1 : F] = [L : F]^2$ and, using Theorem 4.4.6 again, $[D : K] = [D_1 : K][F : K] = [D_1 : F][F : K]^2 = [L : F]^2[F : K]^2 = [L : K]^2$; thus L is a maximal subfield of D in view of Theorem 4.5.1. Since F is separable over K, L is also separable over K by Corollary 5.3.7. This completes the induction and proves the lemma. □

Corollary 5.6.2 *Every simple central algebra has a normal splitting field.*

Proof. The statement follows immediately, using Theorem 4.5.1 and Corollary 5.4.5. □

In terms of the Brauer group, Corollary 5.6.2 reads

$$\operatorname{Br} K = \bigcup_L \operatorname{Br}(L/K),$$

where L runs through all normal extensions of the field K.

Now, let D be a central division algebra with a normal splitting field L. By Theorem 4.5.3, L is a subalgebra of the algebra $A = M_m(D)$ for some m with $[A : K] = [L : K]^2$. If $\sigma \in G(L/K)$ then σ extends to an inner automorphism of the algebra A by the Skolem-Noether theorem (see Corollary 4.4.2). In other words, there is an invertible element a_σ in A such that $a_\sigma x = \sigma(x) a_\sigma$ for all $x \in L$. The element a_σ is obviously determined up to a factor which commutes with all $x \in L$; since $L = C_A(L)$, this means up to a factor from L.

If τ is another element of the group $G = G(L/K)$, then $\sigma\tau(x) = a_{\sigma\tau} x a_{\sigma\tau}^{-1}$. Since

$$\sigma\tau(x) = \sigma(a_\tau x a_\tau^{-1}) = a_\sigma a_\tau x a_\tau^{-1} a_\sigma^{-1} = (a_\sigma a_\tau) x (a_\sigma a_\tau)^{-1},$$

we conclude that $a_\sigma a_\tau = \gamma_{\sigma,\tau} a_{\sigma\tau}$ for some $\gamma_{\sigma,\tau} \in L$. In this way, we get a function $\gamma_{\sigma,\tau}$ of two variables $\sigma, \tau \in G$ whose values lie in the multiplicative group L^* of the field L. Calculating $a_\sigma a_\tau a_\rho$ in two ways, we obtain

$$(a_\sigma a_\tau) a_\rho = \gamma_{\sigma,\tau} a_{\sigma\tau} a_\rho = \gamma_{\sigma,\tau} \gamma_{\sigma\tau,\rho} a_{\sigma\tau\rho},$$

$$a_\sigma (a_\tau a_\rho) = a_\sigma \gamma_{\tau,\rho} a_{\tau\rho} = \sigma(\gamma_{\tau,\rho}) a_\sigma a_{\tau\rho} = \sigma(\gamma_{\tau,\rho}) \gamma_{\sigma,\tau\rho} a_{\sigma\tau\rho},$$

which shows that

$$\gamma_{\sigma,\tau}\gamma_{\sigma\tau,\rho} = \sigma(\gamma_{\tau,\rho})\gamma_{\sigma,\tau\rho} \,. \tag{5.6.1}$$

A function satisfying the equation (5.6.1) is called a *cocycle* of the group G with values in L^* (more precisely, a two-dimensional cocycle). Hence, one can speak about the *cocycle group* $Z(G, L^*)$.

Conversely, let $\gamma \in Z(G, L^*)$. We shall construct an algebra $A = A(G, L, \gamma)$ which will be called the *crossed product* of the group G and the field L relative to γ.

The elements of the algebra $A(G, L, \gamma)$ are the formal linear combinations $\sum_{\sigma \in G} a_\sigma e_\sigma$, where the a_σ are elements of the field L and the e_σ certain symbols indexed by G.

The vector space structure of A is defined in the usual "coordinatewise manner", and the multiplication is determined by the rules

$$e_\sigma x = \sigma(x)e_\sigma \text{ for } x \in L \quad\text{and}\quad e_\sigma e_\tau = \gamma_{\sigma,\tau}e_{\sigma\tau}$$

(the elements of L are multiplied in the usual way). The associativity of this multiplication follows immediately from the condition (5.6.1) for the cocycle (the verification of this fact is left to the reader).

Theorem 5.6.3. *$A = A(G, L, \gamma)$ is a simple central algebra over the field K and L is a splitting field of K.*

Proof. If an element $\sum_\sigma a_\sigma e_\sigma$ belongs to the center of A, then it commutes with all elements of L, and therefore

$$\sum_\sigma (aa_\sigma)e_\sigma = a\sum_\sigma a_\sigma e_\sigma = \Big(\sum_\sigma a_\sigma e_\sigma\Big)a = \sum_\sigma a_\sigma e_\sigma a = \sum_\sigma a_\sigma \sigma(a)e_\sigma \,.$$

This means that whenever $a_\sigma \neq 0$, then $a = \sigma(a)$ for all $a \in L$, and consequently $\sigma = 1$. Thus $C_A(L) = L$, and so $C(A) \subset L$. However, if an element $a \in L$ belongs to $C(A)$, then $ae_\sigma = e_\sigma a = \sigma(a)e_\sigma$ for all $\sigma \in G$, and consequently $a \in \operatorname{Inv} G = K$. Thus $C(A) = K$ and the algebra A is central.

Now, let I be an ideal of A. Choose a non-zero element $x = \sum_\sigma a_\sigma e_\sigma$ of I with the least number of non-zero coefficients a_σ. Multiplying x by e_σ for a suitable σ, we can assume that $a_1 \neq 0$. Let a be an arbitrary element of L. Then $ax - xa \in I$. But

$$\begin{aligned} ax - xa &= \sum_\sigma aa_\sigma e_\sigma - \sum_\sigma a_\sigma e_\sigma a = \sum_\sigma aa_\sigma e_\sigma - \sum_\sigma a_\sigma\sigma(a)e_\sigma = \\ &= \sum_\sigma (aa_\sigma - \sigma(a)a_\sigma)e_\sigma \,, \end{aligned}$$

and so $aa_1 - 1(a)a_1 = 0$. Thus, the number of non-zero coordinates of $ax - xa$ is smaller than the number in x. We conclude that $ax - xa = 0$ and $x \in C_A(L) = L$. This shows that x is invertible and so $I = A$; hence A is simple.

The fact that L is a splitting field of A now follows immediately from Theorem 4.4.6 because $C_A(L) = L$. □

Corollary 5.6.4. *Let D be a central division algebra and L be its normal splitting field. Then, for suitable m and a cocycle $\gamma \in Z(G, L^*)$, $M_m(D) \simeq A(G, L, \gamma)$.*

Proof. Earlier we have constructed in $M_m(D)$ the elements e_σ whose products with the elements of L are the same as those of the corresponding elements e_σ of $A = A(G, L, \gamma)$. This allows us to define an algebra homomorphism $f : A \to M_m(D)$. Since the algebra A is simple, f is a monomorphism. However, $(G : 1) = [L : K]$, $[A : K] = [L : K]^2 = [M_m(D) : K]$ (see Theorem 4.5.3), and thus f is an isomorphism, as required. □

A cocycle $\delta \in Z(G, L^*)$ is called a *coboundary* if there is a function μ_σ defined on G with values in L^* such that

$$\delta_{\sigma,\tau} = \mu_\sigma \sigma(\mu_\tau)\mu_{\sigma\tau}^{-1}$$

for any $\sigma, \tau \in G$.

Theorem 5.6.5. *Let γ and η be cocycles from $Z(G, L^*)$. The algebras $A = A(G, L, \gamma)$ and $B = A(G, L, \eta)$ are isomorphic if and only if $\gamma = \delta\eta$ for some coboundary δ.*

Proof. Assume that $f : A \xrightarrow{\sim} B$. Then $f(L)$ is a subfield of B which is isomorphic to L. Applying Corollary 4.4.2 of the Skolem-Noether theorem, we can assume that $f(a) = a$ for all $a \in L$. Write $f_\sigma = f(e_\sigma) \in B$. Then

$$f(e_\sigma a) = f_\sigma a = f(\sigma(a)e_\sigma) = \sigma(a)f_\sigma\,,$$

from which it follows that $f_\sigma = \mu_\sigma e_\sigma$ for some $\mu_\sigma \in C_B(L) = L$. But then

$$\begin{aligned} f(e_\sigma e_\tau) = f_\sigma f_\tau &= \mu_\sigma e_\sigma \mu_\tau e_\tau = \mu_\sigma \sigma(\mu_\tau) e_\sigma e_\tau = \mu_\sigma\sigma(\mu_\tau)\eta_{\sigma,\tau} e_{\sigma\tau} \\ &= f(\gamma_{\sigma,\tau} e_{\sigma\tau}) = \gamma_{\sigma,\tau} f_{\sigma\tau} = \gamma_{\sigma,\tau}\mu_{\sigma\tau} e_{\sigma\tau}\,, \end{aligned}$$

and thus $\gamma_{\sigma,\tau} = \mu_\sigma\sigma(\mu_\tau)\mu_{\sigma\tau}^{-1}\eta_{\sigma,\tau}$.

Conversely, let $\gamma = \delta\eta$ where $\delta_{\sigma,\tau} = \mu_\sigma\sigma(\mu_\tau)\mu_{\sigma\tau}^{-1}$. Define the map $f : A \to B$ by putting $f(\sum_\sigma a_\sigma e_\sigma) = \sum_\sigma a_\sigma \mu_\sigma e_\sigma$. It is not difficult to see that f is an algebra homomorphism. Since A is simple and the dimensions of A and B are equal, therefore f is an isomorphism, as required. □

The coboundaries form a subgroup $B(G, L^*)$ of the group $Z(G, L^*)$ and Theorem 5.6.5 together with Corollary 5.6.4 implies that the elements of the group $\mathrm{Br}\,(L/K)$ correspond bijectively to the respective elements of the factor group

$$H(G, L^*) = Z(G, L^*)/B(G, L^*)\,.$$

Theorem 5.6.6. $\mathrm{Br}\,(L/K) \simeq H(G, L^*)$.

Proof. The proof is based on the following lemma.

Lemma 5.6.7. *$A(G, L, \gamma) \otimes A(G, L, \eta) \simeq M_n(A(G, L, \gamma\eta))$, where $n = [L : K]$.*

Proof. Write $A = A(G, L, \gamma)$ and $B = A(G, L, \eta)$. Since L is embedded both in A and in B, the algebra $A \otimes B$ contains the subalgebra $L \otimes L$. As the proof of Theorem 5.5.2 shows, there is a unique non-zero idempotent f in $L \otimes L$ such that $(x \otimes 1)f = f(1 \otimes x)$ for all $x \in L$. We want to show that for every $\sigma \in G$, $f(e_\sigma \otimes e_\sigma) = (e_\sigma \otimes e_\sigma)f$. Indeed, $(e_\sigma \otimes e_\sigma)^{-1} f(e_\sigma \otimes e_\sigma)$ is also an idempotent in $L \otimes L$, moreover,

$$(x \otimes 1)(e_\sigma \otimes e_\sigma)^{-1} f(e_\sigma \otimes e_\sigma) = (e_\sigma \otimes e_\sigma)^{-1}(\sigma(x) \otimes 1) f(e_\sigma \otimes e_\sigma) =$$
$$= (e_\sigma \otimes e_\sigma)^{-1} f(1 \otimes \sigma(x))(e_\sigma \otimes e_\sigma) = (e_\sigma \otimes e_\sigma)^{-1} f(e_\sigma \otimes e_\sigma)(1 \otimes x)$$

because $e_\sigma x = \sigma(x) e_\sigma$, and thus $x e_\sigma^{-1} = e_\sigma^{-1}\sigma(x)$. This means that we have $(e_\sigma \otimes e_\sigma)^{-1} f(e_\sigma \otimes e_\sigma) = f$, as required.

Now consider the algebra $T = f(A \otimes B)f$. The map which sends $a \in L$ into the element $\bar{a} = f(1 \otimes a) = (a \otimes 1)f$ is an embedding of the field L into the algebra T. Write $\bar{e}_\sigma = f(e_\sigma \otimes e_\sigma) = (e_\sigma \otimes e_\sigma)f$. Then

$$\bar{a}\bar{b} = \overline{ab},$$
$$\bar{e}_\sigma \bar{a} = (e_\sigma \otimes e_\sigma) f(1 \otimes a) = (e_\sigma \otimes e_\sigma)(a \otimes 1)f = (\sigma(a) \otimes 1)(e_\sigma \otimes e_\sigma)f =$$
$$= \overline{\sigma(a)}\bar{e}_\sigma\,,$$
$$\bar{e}_\sigma \bar{e}_\tau = f(e_\sigma \otimes e_\sigma)(e_\tau \otimes e_\tau) = f(e_\sigma e_\tau \otimes e_\sigma e_\tau) = f(\gamma_{\sigma,\tau} e_{\sigma\tau} \otimes \eta_{\sigma,\tau} e_{\sigma\tau}) =$$
$$= f(\gamma_{\sigma,\tau} \otimes 1)(1 \otimes \eta_{\sigma,\tau})(e_{\sigma\tau} \otimes e_{\sigma\tau}) = (\gamma_{\sigma,\tau}\eta_{\sigma,\tau} \otimes 1) f(e_{\sigma\tau} \otimes e_{\sigma\tau}) =$$
$$= \bar{\gamma}_{\sigma,\tau}\bar{\eta}_{\sigma,\tau}\bar{e}_{\sigma\tau}\,.$$

Therefore, the map $\sum_\sigma a_\sigma e_\sigma \mapsto \sum_\sigma \bar{a}_\sigma \bar{e}_\sigma$ is a homomorphism (and thus, a monomorphism) from $A(G, L, \gamma\eta)$ to T.

On the other hand, $T \simeq E_{A \otimes B}(f(A \otimes B))$. However, the decomposition of the identity in $L \otimes L$ has the form $1 = \sum_{\sigma \in G} f_\sigma$, where f_σ is a (unique) idempotent such that $x f_\sigma = f_\sigma \sigma(x)$ for all $x \in L$. Besides, $(1 \otimes e_\tau) f_\sigma (1 \otimes e_\tau)^{-1} = f_{\tau\sigma}$ (see the proof of Theorem 5.5.2). Therefore all the modules $f_\sigma(A \otimes B)$ are mutually isomorphic and in particular, isomorphic to $f(A \otimes B)$, $f = f_1$. Consequently, $A \otimes B \simeq M_n(T)$, and so $[T : K] = n^2 = [A(G, L, \gamma\eta) : K]$. Hence $T \simeq A(G, L, \gamma\eta)$, which completes the proof. □

The proof of Theorem 5.6.6 now follows from the fact that the map $Z(G, L^*) \to \mathrm{Br}\,(L/K)$ is a homomorphism (by Lemma 5.6.7), that its kernel is $B(G, L^*)$ (by Theorem 5.6.5), and that it is an epimorphism (by Corollary 5.6.4). □

Corollary 5.6.8. *The algebra $A(G, L, \gamma)$ is isomorphic to $M_n(K)$ if and only if $\gamma \in B(G, L^*)$.*

Exercises to Chapter 5

1. Let K be a field of characteristic $p > 0$. Prove that the map φ defined by $\varphi(a) = a^p$ is an endomorphism of the field K. This map is called the *Frobenius endomorphism*. Prove that if K is finite, then φ is an automorphism. Evaluate $\operatorname{Im}\varphi$ if $K = F(t)$ is the field of rational functions over a field F.

2. Prove that the automorphism group of a finite field is cyclic and that the Frobenius automorphism is its generator.

3. Assume that the characteristic of the field K is not 2. Show that every quadratic extension L of the field K, i. e. a field satisfying $[L : K] = 2$, is a splitting field for a polynomial of the form $x^2 - a$, where $a \in K$ is not a square. One writes usually $L = K[\sqrt{a}]$. Verify that $K[\sqrt{a}] \simeq K[\sqrt{b}]$ if and only if ab^{-1} is a square in K.

4. Find a splitting field for the polynomial $x^3 - 2$ over the field of rational numbers.

5. a) Let K be a finite field of q elements and $f(x)$ an irreducible polynomial of degree d over the field K. Prove that $f(x)$ divides $x^{q^d} - x$ and conversely, if $f(x)$ divides $x^{q^n} - x$, then d divides n.
   b) Denote by $\psi(d)$ the number of irreducible polynomials over the field K of degree d with the leading coefficient equal to 1. Prove that $q^n = \sum_{d \mid n} d\psi(d)$.
   c) Using the Möbius inversion formula (see e.g. I.M. Vinogradov: An Introduction to the Theory of Numbers, Pergamon Press, London, 1955) prove that
   $$\psi(n) = \frac{1}{n}\sum_{d \mid n} \mu(\frac{n}{d})q^d,$$
   where μ is the Möbius function.

6. Let K be a field of characteristic $p > 0$, $F = K[a]$ a finite monogenic extension of K and $m(x)$ the minimal polynomial of the element a over K. Prove that the algebra $F \otimes F$ is local if and only if $m(x) = x^{p^k} - \alpha$ for an integer k and some $\alpha \in K$. In this case, the irreducible polynomial $m(x)$ and the element a are called *purely inseparable* and the exponent k is called the *height* of the element a.

7. Let L be a finite extension of a field K. Prove that the following statements are equivalent:
   1) $L \otimes L$ is a local algebra;
   2) for every local commutative algebra A, the algebra $A \otimes L$ is local;
   3) every element of the field L is purely inseparable;
   4) $L = K[a_1, a_2, \ldots, a_t]$, where all elements a_i are purely inseparable.
   In the last case, show that the height of any element of the field L is bounded by the largest of the heights of the elements $a_1, a_2, \ldots, a_t$.
   If these conditions are satisfied, L is called a *purely inseparable extension* of K.

8. Let $K = K_0 \subset K_1 \subset \ldots \subset K_n = L$ be a chain of finite extensions. Prove that L is purely inseparable over K if and only if every K_i is purely inseparable over K_{i-1}.

9. Given a finite extension L of a field K, denote by L_s the set of all separable elements and by L_i the set of all purely inseparable elements of the field L. Prove that:
   a) L_s and L_i are subfields of L and $L_s \cap L_i = K$;
   b) L is purely inseparable over L_s.

Now, assume that L_s is normal. Prove that:
c) L is separable over L_i;
d) if $\{a_1, a_2, \ldots, a_n\}$ is a basis of L_s and $\{b_1, b_2, \ldots, b_m\}$ a basis of L_i, then $\{a_k b_j \mid 1 \le k \le n,\ 1 \le j \le m\}$ is a basis of L (over K).
In particular, $[L : K] = [L_s : K][L_i : K]$ and L is the least field containing L_s and L_i. The subfields L_s and L_i are called *separable* and *purely inseparable extensions*, respectively.

10. For a finite extension L of a field K, prove that the following conditions are equivalent:
1) L is monogenic;
2) L contains only a finite number of subfields.
(Hint: To prove 1) $\Rightarrow$ 2), show that if $L = K[a]$, F a subfield of L and $x^m + b_1 x^{m-1} + \ldots + b_m$ the minimal polynomial of a over F, then $F = K[b_1, b_2, \ldots, b_m]$.)
Construct an example of a non-monogenic extension. (Hint: Let $K = F(x,y)$ be the field of rational functions in two variables over a field of characteristic $p > 0$.)

11. Let F be a subfield of a finite extension L of a field K and $d = [F : K]$. Prove that the number of the homomorphisms $F \to L$ (including the identical one) is not greater than d and it is equal to d if and only if F is separable and L contains a subfield $F' \supset F$ which is normal over K.

12. Prove that the least normal extension containing a given separable extension is uniquely determined, up to an isomorphism.

13. Assuming that the finite extension L of a field K is a *join* of its subfields L_1 and L_2 (i.e. the least field containing L_1 and L_2), and that L_2 is normal over K, prove that L is normal over L_1 and that $G(L/L_1) \simeq \mathrm{Inv}(L_1 \cap L_2) \subset G(L_2/K)$.

The following set of exercises (14 to 22) deals with solving of equations by radicals. For simplicity, the characteristic of the ground field K is assumed in these exercises to be 0; an exception is Exercise 22 which indicates the changes necessary for fields of positive characteristic.

14. Let L be a splitting field for the polynomial $x^n - 1$. Prove that L contains a *primitive n-th root of unity*, i.e. a root ξ of the given polynomial such that all other roots are powers of ξ. The number of such roots is $\varphi(n)$, where φ is the Euler function. Using this fact, show that $G(L/K)$ is a cyclic group whose order divides $\varphi(n)$.

15. Assuming that K contains a primitive nth root of unity, prove that $L = K[a]$, where a is a root of the polynomial $x^n - \alpha$, $\alpha \in K$, is a normal extension, $G(L/K)$ is cyclic and its order divides n. In particular, if p is a prime, then the polynomial $x^p - \alpha$ is either irreducible or a product of linear factors.

16. Conversely, let, as before, K contain a primitive nth root of unity ξ and let L be a normal extension of K with a cyclic Galois group of order n. Prove that then $L = K[a]$, and the minimal polynomial of the element a is of the form $x^n - \alpha$. (Hint: Take $a = \omega + \xi\sigma(\omega) + \xi^2\sigma^2(\omega) + \ldots + \xi^{n-1}\sigma^{n-1}(\omega)$, where σ is a generator of $G(L/K)$ and ω a generator of the KG-module L.)

17. The field L is said to be a *radical extension* of the field K if there is a chain of subfields

$$K = L_0 \subset L_1 \subset L_2 \subset \ldots \subset L_m = L$$

such that $L_i = L_{i-1}[a_i]$, where the minimal polynomial of the element a_i over the field L_{i-1} is of the form $x^{n_i} - \alpha_i$; it is obvious that, refining the chain, we may achieve that all exponents n_i be primes. Assuming that K contains primitive n_ith roots of unity, show that every radical extension is contained in a normal radical extension.

18. An extension L of K is said to be ***solvable*** if it is contained in a radical extension of K. Prove that the splitting field for the polynomial $x^n - 1$ is solvable for every n. (Hint: Prove by induction, using the results of Exercises 14–16.)

19. Prove that an extension L of K is solvable if and only if the least normal extension of K containing L (see Exercise 12) is solvable. (Hint: Use the results of Exercises 17 and 18.)

20. Prove that a normal extension L of K is solvable if and only if its Galois group G is ***solvable***, i. e. there is a chain of subgroups $G = G_0 \supset G_1 \supset G_2 \supset \ldots \supset G_m = \{1\}$ such that G_i is normal in G_{i-1} and the factor group G_{i-1}/G_i is abelian for all i. (Hint: We may assume that G_{i-1}/G_i is cyclic and L a radical extension, and use the results of Exercises 15 and 16; Exercise 18 allows to adjoin roots of unity and Exercise 13 controls the behaviour of the Galois groups.)

21. An irreducible equation $f(x) = 0$, where $f(x) \in K[x]$, is said to be ***solvable by radicals*** if it has a root in a radical extension of K. Prove that this is equivalent to the fact that the splitting field for the polynomial $f(x)$ over K is solvable.

22. Develop a theory producing results similar to those in Exercises 14–21 for fields of characteristic $p > 0$. Note that in the definition of a radical extension, one has to allow also the polynomials of the form $x^p - x - \alpha$. Accordingly, it is necessary to modify Exercise 16 (for $n = p$, $\alpha \in K$). Note also that every purely inseparable extension is radical.

The Exercises 23–27 deal with the important example of crossed products, viz. cyclic algebras. A well-known result of Brauer, Noether and Hasse states that if K is a field of algebraic numbers (i. e. a finite extension of the field $\mathbb{Q}$), then this construction yields all central division algebras over K. All these exercises assume that L is a normal extension of a field K with a cyclic Galois group of order n; a generator of this group is denoted by σ.

23. Let $\gamma \in Z(G, L^*)$ be a cocycle of the group G with values in L^*. Prove that it belongs to the same coset of the subgroup of coboundaries $B(G, L^*)$ as a cocycle η of the form
$$\eta_{\sigma^i,\sigma^j} = \begin{cases} 1 & \text{if } i+j < n\,, \\ \alpha & \text{if } i+j \geq n\,, \end{cases}$$
where $\alpha \in K^*$. Denote by $A(L, \sigma, \alpha)$ the corresponding algebra $A(G, L, \gamma)$ (in general, it depends on the choice of the generator σ). (Hint: In the algebra $A(G, L, \gamma)$, the element e_{σ^i} can be changed to e_σ^i.)

24. Prove that $A(L, \sigma, \alpha) \simeq A(L, \sigma, \beta)$ if and only if there is an element $\lambda \in L^*$ such that $\beta = \alpha N(\lambda)$, where $N(\lambda) = \lambda\sigma(\lambda)\sigma^2(\lambda)\ldots\sigma^{n-1}(\lambda)$. Verify that N is a homomorphism from the group L^* to the group K^* and deduce that $\mathrm{Br}\,(L/K) \simeq K^*/\mathrm{Im}\,N$. The homomorphism N is called the *norm*.

25. Let K be a finite field and L a finite extension. Prove that the norm $N : L^* \to K^*$ is an epimorphism in this case. (Hint: Apply the Wedderburn theorem on finite division rings.)

26. Denote by $K = F(t)$ the field of rational functions over the field F of two elements. Prove that the polynomial $x^2 + x + 1$ is irreducible over the field K. (It is possible that the exercise may be easier with K being the field $F((t))$ of formal power series over the field F.)

27. Using the notation of the preceding exercise, let $L = K[a]$, where a is a root of the polynomial $x^2 + x + 1$. Prove that if $N : L^* \to K^*$ is the norm (see Exercise 24), then $t \notin \operatorname{Im} N$. Making use of the results of Exercise 24, construct a four-dimensional division algebra D over K which contains a subfield isomorphic to $K[\sqrt{t}]$; this is an example of a purely inseparable splitting field of a central division algebra.

28. Prove the theorem on "independence of automorphisms": If L is a normal extension of a field K and $G = G(L/K)$, then for every function $f : G \to L$ there is an element $a \in L$ such that $\sum_{\sigma \in G} f(\sigma)\sigma(a) \neq 0$. (Hint: Use Theorem 5.6.3 by considering L as a module over the algebra $A(G, L, 1)$, where 1 is the identity cocycle.)

6. Separable Algebras

Those algebras which remain semisimple under arbitrary ground field extensions play a particular role among the semisimple algebras. They are called separable. Examples of separable algebras are, on the one hand, central simple algebras, and on the other, separable fields. It turns out that a general case represents, in a certain sense, a combination of these two examples. Furthermore, we shall establish the following fundamental properties of separable algebras: semisimplicity of all bimodules, the Wedderburn-Malcev theorem on lifting separable quotient algebras (which will be used in Chapter 8 for a generalization of the "universal algebra" construction of Sect. 3.6) and non-degeneracy of the principal trace form (which plays an important role in the study of arithmetical properties of separable algebras).

6.1 Bimodules over Separable Algebras

An algebra A over a field K is called *separable* if, for every extension L of the field K, $A_L = A \otimes L$ is semisimple.

In particular, every separable algebra is semisimple; however, the converse is, in general, false: If L is an inseparable extension of a field K, then L is a semisimple K-algebra, but the algebra $L \otimes L$ is no longer semisimple (cf. Theorem 5.3.6).

Generalizing Theorem 5.3.6, we shall give a description of separable algebras and a criterion of separability.

First of all, we are going to establish the following simple result.

Lemma 6.1.1. *For every K-algebra A, there is an extension L of the field K such that the L-algebra A_L splits, i. e.*

$$A_L/\operatorname{rad} A_L \simeq M_{n_1}(L) \times M_{n_2}(L) \times \ldots \times M_{n_s}(L).$$

Proof. Let $\bar{A} = A/\operatorname{rad} A \simeq \prod_{i=1}^{t} M_{k_i}(D_i)$, where D_i are division algebras and $[D_1 : K] = d > 1$.

Since $\operatorname{rad} A \otimes L$ is obviously a nilpotent ideal, it is contained in $\operatorname{rad} A_L$ for any field L. Consequently, $A_L/\operatorname{rad} A_L$ is a quotient algebra of the algebra $\bar{A}_L$.

Choose an element $a \in D_1$, $a \notin K$. Let $p(x)$ be its minimal polynomial and $F = K[a]$. Since $p(x)$ has a root in F, $D_1 \otimes F$ is not a division algebra. Thus, if we write $B = M_{k_1}(D_1)$, then the division algebras which appear in the decomposition of the algebra $B_F/\operatorname{rad} B_F$, have F-dimensions smaller than the dimension $[D_1 : K]$. Continuing in this "reduction of dimension", we obtain finally a field L such that A_L is a split algebra. □

A field L whose existence is asserted by Lemma 6.1.1 is called a *splitting field* for the algebra A. Let us remark that it is far from being unique. Even a minimal splitting field, i. e. such that its subfields are no longer splitting fields, is not determined uniquely (see Exercise 6 of Chap. 4).

Theorem 6.1.2. *The following conditions for an algebra A are equivalent:*

1) *A is a separable algebra;*
2) *$A \otimes A^o$ is a semisimple algebra;*
3) *$A \simeq A_1 \times A_2 \times \ldots \times A_s$, where A_i are simple algebras with separable centers.*

Proof. 1) $\Rightarrow$ 2). Let L be a splitting field for the algebra A. Since A_L is a semisimple algebra, $A_L \simeq M_{n_1}(L) \times M_{n_2}(L) \times \ldots \times M_{n_t}(L)$. Then $(A_L) \otimes_L (A_L^o)$ is a direct product of algebras of the form $M_k(L) \otimes_L M_m(L) \simeq M_{km}(L)$ and thus a semisimple algebra. It remains to observe that $(A_L) \otimes_L (A_L^o) = (A \otimes L) \otimes_L (L \otimes A^o) \simeq A \otimes (L \otimes_L L) \otimes A^o \simeq A \otimes L \otimes A^o \simeq (A \otimes A^o)_L$. Finally, the semisimplicity of $(A \otimes A^o)_L$ implies that $A \otimes A^o$ is semisimple.

2) $\Rightarrow$ 3). If $A \otimes A^o$ is semisimple, then A is semisimple, as well. Hence $A = A_1 \times A_2 \times \ldots \times A_s$, where A_i are simple algebras and $A_i \otimes A_i^o$ are semisimple. Consequently, the center of $A_i \otimes A_i^o$ is semisimple. However, $C(A_i \otimes A_i^o) = C(A_i) \otimes C(A_i^o) = C(A_i) \otimes C(A_i)$ and therefore, in view of Theorem 5.3.6, $C(A_i)$ is separable.

3) $\Rightarrow$ 1). Let $A = A_1 \times A_2 \times \ldots \times A_s$, where A_i are simple algebras and all centers $C_i = C(A_i)$ are separable. According to Corollary 5.6.2, there is a separable extension F of the field C_i such that $A_i \otimes_{C_i} F \simeq M_k(F)$. If L is an arbitrary extension of the field K, then $(A_{i\,L}) \otimes_{C_i} F \simeq L \otimes (A_i \otimes_{C_i} F) \simeq M_k(L \otimes F)$. It follows, by Corollary 5.3.7, that F is separable over K. This means that $L \otimes F$ and therefore also $A_i \otimes_{C_i} F$ are semisimple algebras. From here it follows immediately that every $A_{i\,L}$ and thus also A_L is semisimple for an arbitrary field L, i. e. that A is separable. □

Observe that we have established, in fact, the following result.

Corollary 6.1.3. *An algebra A is separable if and only if $A_L \simeq M_{n_1}(L) \times M_{n_2}(L) \times \ldots \times M_{n_k}(L)$ for some extension L of the field K. In addition, the field L may be assumed to be separable.*

Corollary 6.1.4. *If a field K is perfect (for example, of characteristic 0 or finite), then every semisimple K-algebra is separable.*

Moreover, the semisimplicity of $A \otimes A^o$ implies the following result.

Corollary 6.1.5. *An algebra A is separable if and only if every A bimodule is semisimple.*

Obviously, the last assertion can be reformulated in the following way: An algebra A is separable if and only if every A-bimodule (or equivalently, every $A \otimes A^o$-module) is projective. It is remarkable that, in fact, it is sufficient to verify the projectivity of the regular bimodule, i. e. of the algebra A considered as an $A \otimes A^o$-module, alone.

Theorem 6.1.6. *An algebra A is separable if and only if the regular A-bimodule is projective.*

Proof. Assume that the regular A-bimodule is projective. Choose a splitting field L for the algebra A; thus, $A_L/\operatorname{rad} A_L \simeq M_{n_1}(L) \times M_{n_2}(L) \times \ldots \times M_{n_s}(L)$. In view of Theorem 3.3.5, A is a direct summand of a free $A \otimes A^o$-module F. But then A_L is a direct summand of the free module F_L over the algebra $(A \otimes A^o)_L \simeq (A_L) \otimes_L (A^o_L)$, i. e. A_L is a projective A_L-bimodule. In view of Corollary 3.1.8, the radical of a regular bimodule coincides with the radical of the algebra. Moreover, a decomposition of the quotient algebra $A_L/\operatorname{rad} A_L$ into a direct product of simple algebras yields a decomposition of the A_L-bimodule $A_L/\operatorname{rad} A_L$ into a direct sum of simple bimodules (minimal ideals of that quotient algebra).

By virtue of the relationship between projective and semisimple modules (Theorem 3.3.6), we obtain a decomposition of A_L into a direct sum of ideals, i. e. into a direct product of algebras A_i: $A_L = A_1 \times A_2 \times \ldots \times A_s$, where $A_i/\operatorname{rad} A_i \simeq M_n(L)$ with $n = n_i$.

Then, by Theorem 3.3.4, $A_i \simeq M_n(B)$, where $B/\operatorname{rad} B \simeq L$ (here B depends, in general, on the index i). Besides, since A_i is a projective A_L-bimodule and the components A_j, $j \neq i$, operate on A_i trivially, A_i is a projective A_i-bimodule. Now, note that if $R = \operatorname{rad} B$, then $I = (R \otimes_L B^o) \oplus (B \otimes_L R^o)$ is a nil ideal of $B \otimes_L B^o$ and $(B \otimes_L B^o)/I \simeq L \otimes_L L \simeq L$. Consequently, by Proposition 3.1.3, $I = \operatorname{rad}(B \otimes_L B^o)$, $B \otimes_L B^o$ is a local algebra, and $A_i \otimes A_i^o \simeq M_{n^2}(B \otimes_L B^o)$ is a primary algebra. According to Theorem 3.3.10, it has precisely one principal module (obviously, equal to A_i), while $A_i \otimes_L A_i^o \simeq n^2 A_i$ as an A_i-bimodule. But $[A_i : L] = n^2 b$, where $b = [B : L]$, $[A_i \otimes_L A_i^o : L] = n^4 b^2$ and thus $b = 1$. It follows that $B \simeq L$, $A_i \simeq M_n(L)$ and A is separable by Corollary 6.1.3. □

Finally, observe that if $(A_L) \otimes_L (A^o_L)$ is semisimple, then the algebra $A \otimes A^o$ is semisimple as well, and hence we get the following corollary.

Corollary 6.1.7. *Let A be a K-algebra. If the L-algebra A_L is separable for some extension L of the field K, then A is separable.*

6.2 The Wedderburn-Malcev Theorem

Let A be an arbitrary, in general not semisimple, algebra and R its radical. Let $\bar{A} = A/R$ and π be the (canonical) projection of the algebra A onto the quotient algebra $\bar{A}$. In many problems of the theory of algebras, one requires to "lift" the quotient algebra $\bar{A}$ to an isomorphic subalgebra of A. Let us give the following definition.

An algebra homomorphism $\varepsilon : \bar{A} \to A$ such that $\pi\varepsilon = 1$ will be called a *lifting of the quotient algebra* $\bar{A}$. Evidently, a lifting ε is always a monomorphism and $\operatorname{Im}\varepsilon = A_0$ is a subalgebra of A which is isomorphic to $\bar{A}$; moreover, $A = A_0 \oplus R$ (as a direct sum of vector spaces).

Conversely, if A_0 is a subalgebra of A which is isomorphic to $\bar{A}$, then $A_0 \cap R = 0$ (because A_0 is semisimple). Consequently, $A = A_0 \oplus R$ (because $[A : K] = [A_0 : K] + [R : K]$). Then the restriction of the projection π to the subalgebra A_0 results in an isomorphism $\bar{\pi} : A_0 \xrightarrow{\sim} \bar{A}$. Taking $\varepsilon = \pi^{-1}$, we obtain a lifting of the quotient algebra. As a result, the existence of a lifting is equivalent to the existence of a complement to the radical.

Two liftings $\varepsilon : \bar{A} \to A$ and $\eta : \bar{A} \to A$ are said to be *conjugate* if there is an invertible element a of the algebra A such that $\eta(x) = a\varepsilon(x)a^{-1}$ for all $x \in \bar{A}$. If, in addition, $a = 1 + r$, where $r \in R$ (such elements are called *unipotent*), we say that ε and η are *unipotently conjugate*.

This section will be devoted to a proof of the following fundamental result.

Theorem 6.2.1 (Wedderburn-Malcev). *If the quotient algebra $\bar{A}$ is separable, then a lifting always exists and any two liftings are unipotently conjugate.*

Let us remark that without assumption of separability, the statements no longer hold: a lifting may not exist (see Exercise 4), and two liftings may not be conjugate (see Exercise 5).

Proof. We shall prove the existence of a lifting in several stages, gradually extending the class of algebras for which the results hold.

1) First, we assume that A is a split algebra, i. e. that $\bar{A} \simeq M_{n_1}(K) \times M_{n_2}(K) \times \ldots \times M_{n_s}(K)$. Denote by U_i a simple A_i-module corresponding to the i-th component of the algebra $\bar{A}$ and by P_i the respective principal A-module (see Corollary 3.2.9). Then $\bar{A} = n_1U_1 \oplus n_2U_2 \oplus \ldots \oplus n_sU_s$. Consequently, $A \simeq n_1P_1 \oplus n_2P_2 \oplus \ldots \oplus n_sP_s$ (Theorem 3.3.6).

Utilizing the isomorphism $A \simeq E_A(A)$ and matrix notation for the endomorphisms of a direct sum (see Sect. 1.7), we find that A is isomorphic to the algebra of matrices of the form

$$\begin{pmatrix} x_{11} & x_{12} & \ldots & x_{1s} \\ x_{21} & x_{22} & \ldots & x_{2s} \\ \ldots & \ldots & \ldots & \ldots \\ x_{s1} & x_{s2} & \ldots & x_{ss} \end{pmatrix},$$

where $x_{ij} \in \mathrm{Hom}_A(n_jP_j, n_iP_i)$. In particular, A contains the subalgebra of all "diagonal matrices" which is isomorphic to $E_A(n_1P_1) \times E_A(n_2P_2) \times \ldots \times E_A(n_sP_s) \simeq M_{n_1}(A_1) \times M_{n_2}(A_2) \times \ldots \times M_{n_s}(A_s)$ with $A_i = E_A(P_i)$, and therefore also a subalgebra isomorphic to $\bar{A}$. We conclude that a lifting exists.

2) Now, let $\bar{A}$ be an arbitrary separable algebra with $R^2 = 0$. Choose a basis $\{a_1, a_2, \ldots, a_n\}$ of the algebra A such that $\{\bar{a}_1, \bar{a}_2, \ldots, \bar{a}_m\}$, where $\bar{a}_i = \pi(a_i)$, forms a basis of the quotient algebra $\bar{A}$, and $\{a_{m+1}, a_{m+2}, \ldots, a_n\}$ a basis of the radical of A. Denote by $\{\alpha_{ij}^k\}$ the structure constants of the algebra A. In other words, $a_ia_j = \sum\limits_{k=1}^{n} \alpha_{ij}^k a_k$, and thus $\bar{a}_i\bar{a}_j = \sum\limits_{k=1}^{m} \alpha_{ij}^k \bar{a}_k$. Being a linear transformation, a lifting ε is determined by the images of basis elements. The condition $\pi\varepsilon = 1$ means that $\varepsilon(\bar{a}_i)$ has the form $a_i + \sum\limits_{j=m+1}^{n} x_{ij}a_j$, where $x_{ij} \in K$. Furthermore, ε is a homomorphism if and only if $\varepsilon(\bar{a}_i\bar{a}_j) = \varepsilon(\bar{a}_i)\varepsilon(\bar{a}_j)$. However,

$$\begin{aligned}
\varepsilon(\bar{a}_i\bar{a}_j) &= \varepsilon\Big(\sum_{k=1}^{m} \alpha_{ij}^k \bar{a}_k\Big) = \sum_{k=1}^{m} \alpha_{ij}^k \Big(a_k + \sum_{l=m+1}^{n} x_{k\ell}a_\ell\Big) = \\
&= \sum_{l=1}^{m} \alpha_{ij}^\ell a_\ell + \sum_{k=1}^{m} \sum_{\ell=m+1}^{n} \alpha_{ij}^k x_{k\ell} a_\ell\,; \\
\varepsilon(\bar{a}_i)\varepsilon(\bar{a}_j) &= \Big(a_i + \sum_{k=m+1}^{n} x_{ik}a_k\Big)\Big(a_j + \sum_{k=m+1}^{n} x_{jk}a_k\Big) = \\
&= \sum_{\ell=1}^{n} \alpha_{ij}^\ell a_\ell + \sum_{k,\ell=m+1}^{n} x_{jk}\alpha_{ik}^\ell a_\ell + \sum_{k,\ell=m+1}^{n} x_{ik}\alpha_{kj}^\ell a_\ell\,.
\end{aligned}$$

Here we have used the fact that products of elements from the radical are zero and that a product of an arbitrary element and an element from the radical belongs to the radical. Comparing the coefficients of the basis vectors a_ℓ, we obtain a system of linear equations in $x_{k\ell}$:

$$\sum_{k=1}^{m} \alpha_{ij}^k x_{k\ell} = \alpha_{ij}^\ell + \sum_{k=m+1}^{n} \alpha_{ik}^\ell x_{jk} + \sum_{k=m+1}^{n} \alpha_{kj}^\ell x_{ik}\,;\ i,j = 1,2,\ldots,n;$$
$$\ell = m+1, m+2, \ldots, n.$$

Hence, in this case a lifting exists if and only if the system of linear equations has a solution. Since the coefficients of the equations are the structure constants, they do not change under ground field extensions. However, if L is a splitting field for the algebra A, then $\bar{A}_L$ is a split semisimple algebra and thus, in view of 1), there is a lifting $\bar{A}_L \to A_L$. Therefore our system of linear equations (with coefficients from the field K) has a solution in the field L. Here we may apply the following simple lemma whose proof follows immediately from the Kronecker-Capelli theorem.

Lemma 6.2.2. *If a system of linear equations with coefficients from a field K has a solution in an extension of the field, then it has a solution in K.*

In this way, the fact that there is a lifting over L implies that there is a lifting for our original algebra (in case that $R^2 = 0$).

3) Now, the general case can easily be handled by induction on the dimension of the radical.

Let $R^2 \neq 0$ and $B = A/R^2$. According to the previous result, there is a lifting $\bar{\varepsilon} : \bar{A} \to B$. Denote by A' the inverse image of $\operatorname{Im}\bar{\varepsilon}$ in the algebra A. Then $A' \supset R^2$ and $A'/R^2 \simeq \operatorname{Im}\bar{\varepsilon} \simeq \bar{A}$ is a semisimple algebra. By Proposition 3.1.13, it follows that $R^2 = \operatorname{rad} A'$. Since $\dim R^2 < \dim R$, we apply the induction hypothesis to the algebra A' and obtain a lifting $\varepsilon : \bar{A} \to A'$. But then ε is also a lifting of $\bar{A}$ into A.

We precede the proof of conjugacy by the following theorem which generalizes Theorem 4.4.4 ("dual" to the Skolem-Noether theorem). This result is of an independent interest.

Theorem 6.2.3. *If f and g are homomorphisms of a central simple algebra B into an algebra A, then there is an invertible element a in A such that $g(b) = af(b)a^{-1}$ for all $b \in B$.*

Corollary 6.2.4. *Two isomorphic central simple subalgebras B and B' of an algebra A are conjugate. Moreover, every isomorphism $g : B \xrightarrow{\sim} B'$ extends to an inner automorphism of the algebra A, i. e. it satisfies $g(b) = aba^{-1}$, where a is an invertible element of A.*

In order to give a proof, it is sufficient, as in the Skolem-Noether theorem, to establish an isomorphism of B-A-bimodules ${}_fA$ and ${}_gA$ (see Sect. 4.1, Example 2). Both of them, as right A-modules, coincide with the regular module. The statement of the theorem is thus reduced to the following lemma.

Lemma 6.2.5. *Let A be an algebra, B a central simple algebra and M and N two B-A-bimodules. If M and N are isomorphic as A-modules, then they are isomorphic as B-A-bimodules.*

Proof. Let L be a splitting field for the algebra B. Since $(A_L)\otimes_L(B_L^o) \simeq (A \otimes B^o)_L$, it is sufficient, in view of Lemma 5.5.1, to prove that if M_L and N_L are isomorphic as A_L-modules, then they are isomorphic as B_L-A_L-bimodules. Hence, we may assume from the beginning that $B = M_n(K)$. Then $A \otimes B^o \simeq M_n(A)$ and we need to establish that two $M_n(A)$-modules M and N which are isomorphic as A-modules are also isomorphic as $M_n(A)$-modules.

Write $M_i = Me_{ii}$. Clearly, M_i is an A-submodule of M and $M = \bigoplus_{i=1}^{n} M_i$. Moreover, $M_i e_{ij} \subset M_j$ and multiplications by e_{ij} and e_{ji} are mutually inverse A-homomorphisms. Consequently, $M_i \simeq M_j$ and $M \simeq nM_1$ as A-modules. If M and N are isomorphic as A-modules, then by the Krull-Schmidt theorem, so are M_1 and N_1. Let φ be such an isomorphism. Note that every element $x \in M$ has a unique form $x = x_1 + x_2 e_{12} + \ldots + x_n e_{1n}$, where $x_i \in M_1$.

Define the map $\psi : M \to N$ by

$$\psi(x) = \varphi(x_1) + \varphi(x_2)e_{12} + \ldots + \varphi(x_n)e_{1n}.$$

It is easy to show that ψ is a homomorphism of $M_n(A)$-modules and that, since φ is one-to-one, ψ is also a one-to-one correspondence, as was to be shown. □

Now we return to the proof of the Wedderburn-Malcev theorem. Let ε and η be two liftings of $\bar{A}$ into A. We need to find an element $a = 1 + r$, $r \in R$, such that $a\varepsilon(x)a^{-1} = \eta(x)$, i.e. $a\varepsilon(x) = \eta(x)a$, for all $x \in \bar{A}$. Again, choosing bases in $\bar{A}$ and $\operatorname{rad} A$, we can write r with "indeterminate coefficients" $r = \sum_{i=m+1}^{n} x_i a_i$ and turn the equality into a system of linear equations with respect to x_i. In view of Lemma 6.2.2, it suffices to find a solution of this system, or equivalently to prove unipotent conjugacy of ε and η, in an extension L of the field K. Of course, we should take a splitting field L and reduce the problem to the case of split algebras.

Hence, we can assume that $\bar{A} = M_{n_1}(K) \times M_{n_2}(K) \times \ldots \times M_{n_s}(K)$.

Denote by $\bar{e}_{ij}^k$ the matrix units of the kth component of the algebra $\bar{A}$, $k = 1, 2, \ldots, s$; $i, j = 1, 2, \ldots, n_k$, and put $e_{ij}^k = \varepsilon(\bar{e}_{ij}^k)$, $f_{ij}^k = \eta(\bar{e}_{ij}^k)$. Then $1 = \sum_{k=1}^{s}\sum_{i=1}^{n_k} e_{ii}^k = \sum_{k=1}^{s}\sum_{i=1}^{n_k} f_{ii}^k$ are two decompositions of the identity of the algebra A with $e_{ii}^k A \simeq f_{ii}^k A$. By Theorem 3.5.1, there is an invertible element a in the algebra A such that $f_{ii}^k = a e_{ii}^k a^{-1}$ for all $k = 1, 2, \ldots, s$; $i = 1, 2, \ldots, n_k$.

Applying the projection π, we obtain $\bar{e}_{ii}^k = \bar{a}\bar{e}_{ii}^k\bar{a}^{-1}$, when $\bar{a} = \pi(a)$, and thus $\bar{a} = \sum_{i,k} \alpha_{ik}\bar{e}_{ii}^k$ with $\alpha_{ik} \neq 0$ for all i, k. Put $b = \sum_{i,k} \alpha_{ik} e_{ii}^k$. Then b is an invertible element commuting with all e_{ii}^k and thus ab^{-1} is a unipotent element with $f_{ii}^k = (ab^{-1})e_{ii}^k(ab^{-1})^{-1}$. Hence, in what follows, we may assume that $e_{ii}^k = f_{ii}^k$ for all i, k.

Write $e_k = \sum_{i=1}^{n_k} e_{ii}^k$. Then $\bar{e}_k = \sum_{i=1}^{n_k} \bar{e}_{ii}^k$ is a central idempotent of the quotient algebra $\bar{A}$. Taking into account that $e_{ij}^k = e_{ii}^k e_{ij}^k e_{jj}^k$ and $f_{ij}^k = e_{ii}^k f_{ij}^k e_{jj}^k$, i.e. that e_{ij}^k and f_{ij}^k lie in $A_k = e_k A e_k$, we can see that by restricting ε and η to $\bar{A}_k = \bar{e}_k\bar{A}\bar{e}_k$, we obtain homomorphisms $\varepsilon_k : \bar{A}_k \to A_k$ and $\eta_k : \bar{A}_k \to A_k$.

Since $\bar{A}_k$ is a central simple algebra, then by Theorem 6.2.3, there is an invertible element $a_k \in A_k$ such that $\eta_k(x) = a_k\varepsilon_k(x)a_k^{-1}$ for all $x \in \bar{A}_k$. Applying the projection π again, we get $x = \bar{a}_k x \bar{a}_k^{-1}$ for all $x \in \bar{A}_k$, where $\bar{a}_k = \pi(a_k)$. Thus $\bar{a}_k = \alpha_k e_k$ for some $\alpha_k \in K$. Replacing a_k by $\alpha_k^{-1} a_k$, we may assume that $a_k = r_k + e_k$, where $r_k \in R$. But then, taking $a = \sum_{k=1}^{s} a_k$, we get that $\eta(x) = a\varepsilon(x)a^{-1}$, where $a = 1 + r$ with $r \in R$. The proof is completed. □

6.3 Trace, Norm, Discriminant

Let T be a representation of an algebra A. Consider the characteristic polynomial $\det\big(xE - T(a)\big)$ of a matrix $T(a)$. Since characteristic polynomials of similar matrices coincide, it is determined by the corresponding A-module M. This polynomial is called the *characteristic polynomial* of the element a with respect to the module M (or representation T) and is denoted by $\chi_{M,a}(x)$.

Similarly, the *trace* and the *norm* of an element a with respect to a module M are, respectively, the trace and the determinant of a matrix $T(a)$. The trace and norm are denoted, respectively, by $\mathrm{Tr}_M(a)$ and $\mathrm{N}_M(a)$.

It follows immediately from the definition that the trace is a linear map $A \to K$ such that

$$\mathrm{Tr}_M(a+b) = \mathrm{Tr}_M(a) + \mathrm{Tr}_M(b); \quad \mathrm{Tr}_M(\alpha a) = \alpha \mathrm{Tr}_M(a), \ \alpha \in K.$$

In addition, $\mathrm{Tr}_M(ab) = \mathrm{Tr}_M(ba)$, $\mathrm{N}_M(ab) = \mathrm{N}_M(a)\mathrm{N}_M(b)$ and, for arbitrary $\alpha \in K$,

$$\chi_{M,\alpha}(x) = (x-\alpha)^d; \quad \mathrm{Tr}_M(\alpha) = d\alpha; \quad \mathrm{N}_M(\alpha) = \alpha^d,$$

where $d = [M : K]$.

The following simple statement reduces computation of characteristic polynomials, traces and norms to the case of simple modules.

Proposition 6.3.1 *Let $M = M_0 \supset M_1 \supset \ldots \supset M_s = 0$ be a composition series of the module M and $U_i = M_i/M_{i+1}$ its simple factors. Then, for an element $a \in A$,*

$$\chi_{M,a}(x) = \prod_{i=1}^{s} \chi_{U_i,a}(x); \quad \mathrm{Tr}_M(a) = \sum_{i=1}^{s} \mathrm{Tr}_{U_i}(a); \quad \mathrm{N}_M(a) = \prod_{i=1}^{s} \mathrm{N}_{U_i}(a).$$

Proof. This follows immediately from the fact that a representation T corresponding to a module M can be brought to the form

$$T(a) = \begin{pmatrix} T_1(a) & & & 0 \\ & T_2(a) & & \\ & & \ddots & \\ * & & & T_s(a) \end{pmatrix},$$

where T_i is a representation corresponding to the module U_i. □

Corollary 6.3.2. *If an element a belongs to $\operatorname{rad} A$, then $\chi_{M,a}(x) = x^d$, $\mathrm{Tr}_M(a) = \mathrm{N}_M(a) = 0$, where $d = [M : K]$.*

Indeed, if U is a simple module, then $ua = 0$ for all $u \in U$, i. e. a is mapped in the corresponding representation to the zero matrix.

Proposition 6.3.3. *For every extension L of a field K,*

$$\chi_{M_L,a\otimes 1}(x) = \chi_{M,a}(x), \quad \mathrm{Tr}_{M_L}(a\otimes 1) = \mathrm{Tr}_M(a), \quad \mathrm{N}_{M_L}(a\otimes 1) = \mathrm{N}_M(a).$$

Proof. If $\{m_1, m_2, \dots, m_d\}$ is a K-basis of M, then $\{m_1\otimes 1, m_2\otimes 1, \dots, m_d\otimes 1\}$ is an L-basis of M_L and the matrix of the elements a and $a\otimes 1$ with respect to these bases is the same. □

As we shall see in the sequel, sometimes it is convenient to consider together with representations of an algebra A over a ground field K also its representations over an extension L of the field K, or, equivalently, representations of the L-algebra A_L, i. e. A_L-modules. We shall identify an element $a \in A$ with the element $a\otimes 1 \in A_L$ and write $\chi_{M,a}(x)$, $\mathrm{Tr}_M(a)$, $\mathrm{N}_M(a)$ instead of $\chi_{M,a\otimes 1}(x)$, $\mathrm{Tr}_M(a\otimes 1)$, $\mathrm{N}_M(a\otimes 1)$ for an A_L-module M.

In general, the coefficients of $\chi_{M,a}(x)$, and in particular, $\mathrm{Tr}_M(a)$, and $\mathrm{N}_M(a)$ are elements of the field L. If they belong to K for every element $a \in A$, we call the module M *proper* (using this ad hoc term only in the present section).

A *trace form* on an algebra A corresponding to an A-module M is the function $B_M(a,b) = \mathrm{Tr}_M(ab)$ with $a, b \in A$. In view of the properties of the trace, B_M is a symmetric bilinear form on the space A. The discriminant of the form B_M, i. e. the element

$$\Delta_M = \begin{vmatrix} \mathrm{Tr}_M(a_1a_1) & \mathrm{Tr}_M(a_1a_2) & \dots & \mathrm{Tr}_M(a_1a_n) \\ \mathrm{Tr}_M(a_2a_1) & \mathrm{Tr}_M(a_2a_2) & \dots & \mathrm{Tr}_M(a_2a_n) \\ \dots & \dots & \dots & \dots \\ \mathrm{Tr}_M(a_na_1) & \mathrm{Tr}_M(a_na_2) & \dots & \mathrm{Tr}_M(a_na_n) \end{vmatrix},$$

where $\{a_1, a_2, \dots, a_n\}$ is a basis of A, is called the *discriminant of the module M*. Clearly, Δ_M is defined up to a square of a non-zero element of K. If the form is non-degenerate, i. e. if $\Delta_M \neq 0$, we call the module M *non-degenerate*.

The preceding definitions allow us to formulate the following criterion of separability.

Theorem 6.3.4. *A K-algebra A is separable if and only if there exists a non-degenerate A_L-module M for some extension L of the field K. Moreover, the field L can be chosen separable and the module M proper.*

Proof. If $a \in \operatorname{rad} A_L$, then, by Corollary 6.3.2, $B_M(a,b) = \mathrm{Tr}_M(ab) = 0$ for every $b \in A_L$ and the form B_M is degenerate. Hence, if M is a non-degenerate A_L-module, then the algebra A_L is semisimple. Now, Proposition 6.3.3 implies that the discriminant of the A_F-module M is equal to Δ_M for any extension F of the field L. This means that M_F is a non-degenerate A_F-module and the algebra A_F is semisimple. Hence A_L and, by Corollary 6.1.7, also the algebra A is separable.

Conversely, let the algebra A be separable and L be its splitting field. Then $A_L \simeq A_1 \times A_2 \times \dots \times A_s$, where $A_k = M_{n_k}(L)$. Let U_k be a simple

A_k-module. Then, for every matrix $a_k \in A_k$, the polynomial $\chi_{U_k,a_k}(x)$ is a characteristic polynomial of the matrix a_k. Denote by M the direct sum $U_1 \oplus U_2 \oplus \ldots \oplus U_s$. By Proposition 6.3.3, the polynomial $\chi_{M,b}(x)$, where $b = (a_1, a_2, \ldots, a_s)$ belongs to $A_1 \times A_2 \times \ldots \times A_s$, is a product of the characteristic polynomials of the matrices a_k. In particular, $\mathrm{Tr}_M(b) = \sum_{k=1}^{s} \mathrm{tr}\, a_k$ and $\mathrm{N}_M(b) = \prod_{k=1}^{s} \det a_k$. Considering the basis of A_L consisting of the matrix units e_{ij}^k (here, $k = 1, 2, \ldots, s$ indicates the component index and $i, j = 1, 2, \ldots, n$), we have, trivially, $\mathrm{Tr}_M(e_{ii}^k) = 1$ and $\mathrm{Tr}_M(e_{ij}^k) = 0$ for $i \neq j$. From here,

$$\mathrm{Tr}_M(e_{ij}^k e_{rt}^\ell) = \begin{cases} 1 & \text{if } k = \ell,\ i = t \text{ and } j = r, \\ 0 & \text{otherwise.} \end{cases}$$

Thus, there is precisely one 1 in every row and every column of the determinant Δ_M and all other entries are 0. We conclude that $\Delta_M = \pm 1 \neq 0$ and the module M is non-degenerate.

The fact that L can be chosen separable follows from Corollary 6.1.3. We are going to show that M is proper, i.e. that the coefficients of $\chi_{M,a}(x)$ belong to the field K for every $a \in A$. First, observe that $\chi_{M,a}(x)$ does not depend on the choice of the splitting field L: This is a consequence of Proposition 6.3.3 for every field containing L; if F is any other splitting field, then one can always construct a field containing both F and L. Therefore, in view of Proposition 5.4.5, L can be assumed normal. Write $G = G(L/K)$.

The group G acts on the algebra A_L by the formula $\sigma(a \otimes \lambda) = a \otimes \sigma(\lambda)$, where $a \in A$, $\lambda \in L$. We are going to show that for an element $b \in A_L$,

$$\chi_{M,\sigma(b)}(x) = \sigma\left(\chi_{M,b}(x)\right), \tag{6.3.1}$$

where $\sigma(f(x))$ denotes the polynomial whose coefficients are σ-images of the coefficients of $f(x)$. Once this formula is established, we obtain, for arbitrary $a \in A$, that $\sigma(\chi_{M,a}(x)) = \chi_{M,\sigma(a)}(x) = \chi_{M,a}(x)$ and therefore, by Theorem 5.4.4, all coefficients of $\chi_{M,a}(x)$ lie in the field K; thus, M is a proper module.

In order to establish (6.3.1), note first of all, that if $b = (a_1, a_2, \ldots, a_s) \in A_L$, then $\chi_{M,b}(x) = \chi_{M,b'}(x)$, where $b' = (a_{t_1}, a_{t_2}, \ldots, a_{t_s})$ for any permutation $(t_1, t_2, \ldots, t_s)$ of $(1, 2, \ldots, s)$. Since $A_L = \sigma(A_1) \times \sigma(A_2) \times \ldots \times \sigma(A_s)$, it follows by Theorem 2.5.2 that $\sigma(A_k) = A_{t_k}$ for a permutation $(t_1, t_2, \ldots, t_s)$. In particular, $n_k = n_{t_k}$. Denote by σ_k the restriction of σ to A_k and consider the isomorphism $\bar\sigma_k : A_k \to A_{t_k}$ mapping the matrix (λ_{ij}) to the matrix $(\sigma(\lambda_{ij}))$. The composition $\bar\sigma_k^{-1}\sigma_k$ is an automorphism of the algebra A_k which is identical on its center L. By the Skolem-Noether theorem (or, rather, Corollary 4.4.3) this is an inner automorphism and thus $\bar\sigma_k^{-1}\sigma_k(a) = u_k a u_k^{-1}$ for some $u_k \in A_k$. Therefore $\sigma_k(a) = v_k \bar\sigma_k(a) v_k^{-1}$, where $v_k = \bar\sigma_k(u_k)$ and σ has, up to a permutation of components, the form

$$(a_1, a_2, \ldots, a_s) \mapsto (v_1\bar\sigma_1(a_1)v_1^{-1}, v_2\bar\sigma_2(a_2)v_2^{-1}, \ldots, v_s\bar\sigma_s(a_s)v_s^{-1}).$$

Now, the characteristic polynomial of the matrix $\bar{\sigma}_k(a_k)$ and therefore also of the matrix $v_k\bar{\sigma}_k(a_k)v_k^{-1}$ is obviously obtained by applying σ to the coefficients of the characteristic polynomial of the matrix a_k, and the required formula (6.3.1) follows. The proof of the theorem is now complete. □

Let us remark that making use of an extension of the ground field in Theorem 6.3.5 is essential. In Exercise 9, an example of a separable algebra A is given such that every A-module is degenerate. If the field K is of characteristic 0 or if the algebra A is commutative, then a non-degenerate A-module always exists (see Exercise 6 and Example 1 below).

The polynomial $\chi_{M,a}(x)$, where M is the module constructed above, is called the *principal polynomial* of the element $a \in A$ and is denoted by $P_a(x)$. The trace $\mathrm{Tr}_M(a)$ and the norm $\mathrm{N}_M(a)$ are called, respectively, the *principal trace* and the *principal norm* of the element a and are denoted by $\mathrm{Tr}(a)$ and $\mathrm{N}(a)$. If there is a need to specify the algebra, one writes $P_{A/K,a}(x)$, $\mathrm{Tr}_{A/K}(a)$ and $\mathrm{N}_{A/K}(a)$. The bilinear from $B(a,b) = \mathrm{Tr}(ab)$ is called the *principal trace form* and its discriminant $\Delta(A/K)$ the *discriminant* of the separable algebra A (recall that it is determined up to the square of a non-zero element of K). In the course of the proof of Theorem 6.2.4, we have established the following fact.

Theorem 6.3.5. *The coefficients of a principal polynomial, and in particular the principal trace and the principal norm, belong to K. The principal trace form of a separable algebra is always non-degenerate, i. e. $\Delta(A/K) \neq 0$.*

In addition, since M is a proper module and every matrix is a root of its characteristic polynomial, we get the following proposition.

Proposition 6.3.6. *Every element of a separable algebra is a root of its principal polynomial.*

Furthermore, let us remark that the structure of M and the formula $(A \otimes F) \otimes_F L \simeq A \otimes L$ for $L \supset F$, immediately imply the following statement.

Proposition 6.3.7. *For every element $a \in A$, and every extension F of K,*

$$P_{A_F/F,a}(x) = P_{A/K,a}(x), \quad \mathrm{Tr}_{A_F/F}(a) = \mathrm{Tr}_{A/K}(a), \quad \mathrm{N}_{A_F/F}(a) = \mathrm{N}_{A/K}(a).$$

Now, we present two examples of computation of the principal polynomial.

Examples. 1. Let F be a separable extension of a field K. Then, for any splitting field L, $F \otimes L \simeq L^n$, where $n = [F:K]$, and the principal polynomial coincides with the characteristic polynomial of the regular module. Clearly, this holds also for any commutative separable algebra A.

2. Let A be a central simple algebra of dimension d^2, and L its maximal subfield. Then $A \otimes L \simeq M_d(L)$ and $P_a(x)$ is the characteristic polynomial of

the matrix corresponding to the element $a \otimes 1$. If $\chi_a(x)$ is the characteristic polynomial of the regular module, then $\chi_a(x) = (P_a(x))^d$ because the regular A_L-module is a direct sum of d simple modules. In particular, $\mathrm{N}_A(a) = (\mathrm{N}(a))^d$ and $\mathrm{Tr}(a) = d\mathrm{Tr}(a)$.

Exercises to Chapter 6

1. Given two modules M, N over an algebra A and an extension L of the ground field, prove that $\mathrm{Hom}_{A_L}(M_L, N_L) \simeq \mathrm{Hom}_A(M,N) \otimes L$. (Hint: One may use the theorem on the structure of solutions of a homogeneous system of linear equations.)

2. Call an A-module M *separable* if the A_L-module M_L is semisimple for every L. Prove that M is separable if and only if M is semisimple and the algebra $E_A(M)$ is separable.

3. Find necessary and sufficient conditions in order that
   a) the algebra A_L be simple for any L;
   b) the A_L-module M_L be simple for any L (such a module is called *absolutely simple*, and the corresponding representation *absolutely irreducible*).

4. Let F be a field of characteristic 2, $K = F(t)$ the field of rational functions over F, $A = K[x]/(x^4 - t^2)$. Find $R = \operatorname{rad} A$ and A/R. Verify that A has no subalgebra isomorphic to A/R. Construct a similar example for a field of arbitrary characteristic $p > 0$.

5. Let F and K be defined as in the previous exercise, $L = K[x]/(x^2 - t)$ and A be the L-algebra with a basis $\{1, r\}$, $r^2 = 0$. Considering A as a K-algebra, establish that $A/\operatorname{rad} A \simeq L$ and find two distinct subalgebras of A isomorphic to L (since A is commutative, these subalgebras are not conjugate in A). Construct a similar example for a field of arbitrary characteristic $p > 0$.

6. Prove that an algebra A over a field of characteristic 0 is semisimple if and only if its regular A-module is non-degenerate.

7. Using the result of the preceding exercise, deduce that there is a polynomial $F(x_{ij}^k)$ with integral coefficients in n^3 variables x_{ij}^k, $i,j,k = 1,2,\dots,n$, such that an algebra A over a field of characteristic 0 with structure constants γ_{ij}^k is semisimple if and only if $F(\gamma_{ij}^k) \neq 0$.

8. Let K be a field of characteristic p and $A = M_p(K)$. Verify that the regular A-module is degenerate. Carry over this result to an arbitrary central simple K-algebra of dimension p^2.

9. If D is a central division algebra of dimension p^2 over a field of characteristic p, prove that every D-module is degenerate (an example of such a division algebra is in Exercise 27 to Chap. 5). Thus, in Theorem 6.3.4 it is, indeed, necessary to consider A_L-modules and not only A-modules.

10. Let A be an algebra over a field K with a basis $\{a_1, a_2, \dots, a_n\}$ and structure constants γ_{ij}^k. Consider the algebra $\tilde{A}$ over the field $F = K(t_1, t_2, \dots, t_n)$ of

rational functions in n variables with the same basis and structure constants.[11] Let $\tilde{a} = \sum_{i=1}^{n} t_i a_i$ and $P(x, t_1, t_2, \ldots, t_n) = m_{\tilde{a}}(x)$ be the minimal polynomial of the element $\tilde{a}$ (obviously, this is a polynomial in $n+1$ variables $x, t_1, t_2, \ldots, t_n$). If $a = \sum_{i=1}^{n} \alpha_i a_i$ is an arbitrary element of the algebra A, then the polynomial $P_{A,a}(x) = P(x, \alpha_1, \alpha_2, \ldots, \alpha_n)$ is called the *principal polynomial* of the element a.

a) Prove that $P_{A,a}$ does not depend on the choice of a basis of the algebra A.
b) Verify that $P_{A_L, a\otimes 1}(x) = P_{A,a}(x)$ for every extension L of the field K.
c) Establish that for a separable algebra A, the present definition of a principal polynomial coincides with the one given in Sect. 6.3.

11. Keep the notation and definitions of the previous exercise. If $P_{A,a} = x^m + \beta_1 x^{m-1} + \ldots + \beta_m$, $\beta_i \in K$, put $\mathrm{Tr}_{A/K}(a) = -\beta_1$ and $\mathrm{N}_{A/K}(a) = (-1)^m \beta_m$, and call them the *principal trace* and the *principal norm*, respectively.
    a) Verify that the principal trace is a linear form on the space A and that $\mathrm{Tr}_{A/K}(ab) = \mathrm{Tr}_{A/K}(ba)$ and $\mathrm{N}_{A/K}(ab) = \mathrm{N}_{A/K}(a)\mathrm{N}_{A/K}(b)$.
    b) Prove that an algebra A is separable if and only if the bilinear form $\mathrm{Tr}_{A/K}(ab)$ on the space A is non-degenerate.

12. Let L be an extension of a field K. If a is an element of an L-algebra A, prove that
$$P_{A/K,a}(x) = \mathrm{N}_{L(x)/K(x)}\left(P_{A/L,a}(x)\right), \quad \mathrm{Tr}_{A/K}(a) = \mathrm{Tr}_{L/K}\left(\mathrm{Tr}_{A/L}(a)\right),$$
$$\mathrm{N}_{A/K}(a) = \mathrm{N}_{L/K}\left(\mathrm{N}_{A/L}(a)\right).$$

13. Prove that if an ideal I of an algebra A has a basis consisting of nilpotent elements, then $I \subset \operatorname{rad} A$. (Hint: Use the fact that the trace of a nilpotent matrix is 0.)

14. Prove that if $A/\operatorname{rad} A$ is separable, then $\operatorname{rad}(A \otimes B) = \operatorname{rad} A \otimes B + A \otimes \operatorname{rad} B$.

[11] We can define the tensor product of infinite dimensional algebras and see easily that $\tilde{A} = A \otimes F$.

7. Representations of Finite Groups

In this chapter we shall apply the general theory of semisimple algebras and their representations to obtain basic results of the classical theory of representations of finite groups.

7.1 Maschke's Theorem

A *representation of a group* G over a field K is a homomorphism of this group into the group $GL(V)$ of all invertible linear transformations of a vector space V over the field K. In other words, a representation T assigns to every element $g \in G$ an invertible linear operator $T(g) \in GL(V)$ in such a way that $T(gh) = T(g)T(h)$ for all $g, h \in G$. As in the case of representations of algebras, the concepts of similarity, reducibility, indecomposability, etc. are defined for group representations. In fact, the study of representations of a group G is equivalent to the study of representations of its group algebra (see Sect. 1.1, Example 6).

Recall that a basis of the group algebra KG consists of the elements of the group G with multiplication given by the group product. If $T : KG \to E(V)$ is a representation of the group algebra and $g \in G$, then $T(g)T(g^{-1}) = T(gg^{-1}) = T(1) = 1$. Therefore T is an invertible transformation and thus, restricting T to G, we get a representation of the group G. Conversely, let $T : G \to GL(V)$ be a representation of the group G. We extend T to the algebra KG "by linearity" defining $T\Big(\sum\limits_{g\in G} \alpha_g g\Big) = \sum\limits_{g\in G} \alpha_g T(g)$. Evidently, we obtain a representation $T : KG \to E(V)$ whose restriction to G coincides with the original representation. Thus, group representations and group algebra representations are essentially the same.

In this chapter, all groups under consideration will be finite. The following remarkable result, known as Maschke's theorem, plays a decisive role in the theory of representations of finite groups.

Theorem 7.1.1. *If K is a field whose characteristic does not divide the order of the group G, then the group algebra KG is separable.*

Proof. In view of Theorem 6.3.4, it is sufficient to show that there is a non-degenerate KG-module. In fact, in our situation, the regular KG-module is non-degenerate. Indeed, consider the basis $\{g_1, g_2, \ldots, g_n\}$ of the algebra KG consisting of all elements of the group G (the order of G equals n). If $g \neq 1$, then $g_i g \neq g_i$ for all $i = 1, 2, \ldots, n$ and therefore $\mathrm{Tr}(g) = 0$ (here, Tr denotes the trace of the regular representation). On the other hand, $\mathrm{Tr}(1) = [KG : K] = n$. Hence, $\mathrm{Tr}(g_i g_j) = 0$ for $g_j \neq g_i^{-1}$ and $\mathrm{Tr}(g_i g_j) = n$ for $g_i = g_j^{-1}$. Consequently, each row and each column of the discriminant Δ of the regular representation has exactly one non-zero element (note that $n = n1 \neq 0$ since K is of characteristic 0 or prime to n). This implies that $\Delta \neq 0$ (in fact, $\Delta = \pm n^n$) and the algebra KG is separable. □

Corollary 7.1.2. *If K is a field whose characteristic does not divide the order of the group G, then every representation of the group G over the field K is completely reducible.*

It turns out that a converse to Maschke's theorem holds, as well.

Theorem 7.1.3. *If K is a field whose characteristic divides the order of the group G, then the algebra KG is not semisimple.*

Proof. Consider the element $s = \sum_{x \in G} x$ of the algebra KG. Obviously, $gs = sg = s$ for every $g \in G$. Therefore s belongs to the center of the algebra KG. On the other hand, $s^2 = \sum_{x \in G} xs = ns = 0$ (since the order of G is divisible by the characteristic of K). According to Corollary 2.2.8, the algebra KG is not semisimple. □

As a consequence, the theory of group representations splits effectively into two fundamentally different theories: classical (when the field characteristic does not divide the group order) and modular (when the field characteristic divides the group order). In this chapter (with the exception of a few exercises) we shall deal only with the classical representation theory. Therefore, we have a standing assumption that K is a field whose characteristic does not divide the order of the group G.

7.2 Number and Dimensions of Irreducible Representations

Maschke's theorem and the theory of semisimple algebras and their representations yield relatively easily the following important results on the number and dimensions of irreducible representations.

Theorem 7.2.1. *If $d_1, d_2, \ldots, d_s$ are dimensions of all (pairwise non-isomorphic) representations of the group G over an algebraically closed field K, then $d_1^2 + d_2^2 + \ldots + d_s^2 = n$, where $n = (G : 1)$.*

Proof. By the Wedderburn-Artin theorem (or rather, by Corollary 2.4.4) and Theorem 2.6.2, $KG \simeq \prod_{i=1}^{s} M_{d_i}(K)$, where $d_1, d_2, \ldots, d_s$ are the dimensions of all irreducible representations of the algebra KG, and thus $n = [KG : K] = \sum_{i=1}^{s} d_i^2$. □

As before, let the field K be algebraically closed. Then the center of the algebra KG is, by Corollary 2.4.2, isomorphic to K^s, where s is the number of simple components of KG, or equivalently, the number of non-isomorphic irreducible representations. Hence, the number of irreducible representations and the dimension of the center of the algebra KG are equal. But an element $a = \sum_{x \in G} \alpha_x x$ belongs to the center of KG if and only if $ga = ag$, i. e. $gag^{-1} = a$, for every $g \in G$. Since $gag^{-1} = \sum_{x \in G} \alpha_x (gxg^{-1})$, this means that the coefficients of x and gxg^{-1} in the element a are equal.

Recall that the elements x and gxg^{-1} are called conjugate in the group G. The group G is partitioned into pairwise disjoint conjugacy classes $C_1, C_2, \ldots, C_s$. It follows from the above argument that the elements $c_i = \sum_{x \in C_i} x$, $i = 1, 2, \ldots, s$, form a basis of the center of the group algebra KG. We can therefore formulate the following theorem.

Theorem 7.2.2. *The number of irreducible representations of a finite group G over an algebraically closed field K is equal to the number of conjugacy classes of the group G.*

Corollary 7.2.3. *A group G is abelian if and only if all irreducible representations of G over an algebraically closed field are one-dimensional.*

Indeed, it is sufficient to remark that a group is abelian if and only if every conjugacy class consists of a single element and thus that the number of irreducible representations equals, by Theorem 7.2.2, the group order. Applying Theorem 7.2.1, we can see immediately that this is possible only when all irreducible representations are one-dimensional.

Corollary 7.2.4. *If G and H are abelian groups of the same order and K is an algebraically closed field, then the group algebras KG and KH are isomorphic.*

7.3 Characters

Let T be a representation of a group G over a field K and M the corresponding KG-module. Then the trace $\mathrm{Tr}_M(a)$ with respect to the module M is defined for every element $a \in KG$ (Sect. 6.3); it is the trace of the matrix $T(a)$ (in any basis). In particular, for every element $x \in G$, we get the field element $\chi(x) = \mathrm{Tr}_M(x)$. The function $\chi : G \to K$ is called the *character* of the representation T. If T is irreducible, then χ is called an *irreducible character*. The character of the regular representation is called the *regular character* and is denoted by χ_{reg}.

Proposition 7.3.1.

$$\chi_{\text{reg}}(x) = \begin{cases} n & \text{for } x = 1, \\ 0 & \text{for } x \neq 1. \end{cases}$$

The proof is obvious.

Proposition 7.3.2. *For every character, $\chi(gxg^{-1}) = \chi(x)$. In other words, a character is constant on each conjugacy class.*

Proof. For every representation T, $T(gxg^{-1}) = T(g)T(x)T(g)^{-1}$, and the similar matrices $T(x)$ and $T(gxg^{-1})$ have the same trace. □

Observe also that, as an immediate consequence of Corollary 2.6.3, we get the following theorem.

Theorem 7.3.3. *Let K be a field of characteristic 0. Then every representation is determined uniquely by its character, i. e. equality of characters implies similarity of representations.*

Now, let the field K be algebraically closed and $\chi_1, \chi_2, \dots, \chi_s$ be all the irreducible characters of the group G over the field K. Denote by χ_{ij} the element $\chi_i(g_j)$, where $g_j \in C_j$ ($C_1, C_2, \dots, C_s$ are the conjugacy classes of the group G). The square matrix $X = (\chi_{ij})$ is called the *character table* of the group G over the field K. Let us remark that $KG \simeq \bigoplus_{i=1}^{s} d_i M_i$, where M_i is the module of the ith irreducible representation and $d_i = [M_i : K]$; hence, $\chi_{\text{reg}} = \sum_{i=1}^{s} d_i \chi_i$.

As we have already pointed out, the elements $c_i = \sum_{x \in C_i} x$, $i = 1, 2, \dots s$, form a basis of the center C of the group algebra KG. On the other hand, $C \simeq K^s$ and therefore, if $1 = e_1 + e_2 + \dots + e_s$ is a decomposition of the identity of the algebra C, then $\{e_1, e_2, \dots, e_s\}$ is also a basis of C. Consequently, there are elements α_{ij} and β_{ij} in the field K such that $c_i = \sum_{j=1}^{s} \alpha_{ij} e_j$ and

$e_i = \sum_{j=1}^{s} \beta_{ij}c_j$; and thus the matrices $A = (\alpha_{ij})$ and $B = (\beta_{ij})$ are reciprocal. It turns out that the coefficients α_{ij} and β_{ij} are closely related to the character table.

Proposition 7.3.4. *Denote by d_i the dimension of the irreducible representation with character χ_i and h_j the number of elements in the class C_j. Then*

$$\alpha_{ij} = \frac{h_i}{d_j}\chi_{ji}\,, \quad \beta_{ij} = \frac{d_i}{n}\chi_i(g_j^{-1}), \quad \textit{where} \quad g_j \in C_j\,.$$

Proof. Observe that the element e_j acts on the jth irreducible representation as identity, while the elements e_k $(k \neq j)$ act on it trivially. Therefore $\chi_j(e_k) = 0$ for $k \neq j$ and $\chi_j(e_j) = d_j$. From here,

$$\chi_j(c_i) = \chi_j\Big(\sum_{k=1}^{s} \alpha_{ik}e_k\Big) = \sum_{k=1}^{s} \alpha_{ik}\chi_j(e_k) = d_j\alpha_{ij}.$$

On the other hand, $\chi_j(c_i) = h_i\chi_{ji}$ and the formula for α_{ij} follows.

In order to compute β_{ij}, we use the fact that $\chi_{\mathrm{reg}} = \sum_{i=1}^{s} d_i\chi_i$. Observe that $\chi_{\mathrm{reg}}(c_kg) = 0$ if $g^{-1} \notin C_k$ and $\chi_{\mathrm{reg}}(c_kg) = n$ if $g^{-1} \in C_k$ (this follows from Corollary 7.3.1). Therefore, if $g_j \in C_j$, then $\chi_{\mathrm{reg}}(e_ig_j^{-1}) = \chi_{\mathrm{reg}}\Big(\sum_{k=1}^{s} \beta_{ik}c_kg_j^{-1}\Big) = n\beta_{ij}$. On the other hand, $\chi_{\mathrm{reg}}(e_ig_j^{-1}) = \sum_{k=1}^{s} d_k\chi_k(e_ig_j^{-1})$ $= d_i\chi_i(g_j^{-1})$ because $\chi_k(e_ig_j^{-1}) = 0$ for $k \neq i$ and $\chi_i(e_ig_j^{-1}) = \chi_i(g_j^{-1})$. The formula for β_{ij} follows. □

Taking into account that the matrices A and B are reciprocal, we obtain immediately the following "orthogonality relations" for characters.

Theorem 7.3.5.

$$\frac{1}{n}\sum_{k=1}^{s} h_k\chi_i(g_k)\chi_j(g_k^{-1}) = \begin{cases} 0 & \textit{for } i \neq j, \\ 1 & \textit{for } i = j; \end{cases}$$

$$\frac{1}{n}\sum_{k=1}^{s} \chi_k(g_i)\chi_k(g_j^{-1}) = \begin{cases} 0 & \textit{for } i \neq j, \\ 1/h_i & \textit{for } i = j. \end{cases}$$

Corollary 7.3.6. *A representation T of a group G over an algebraically closed field of characteristic 0 is irreducible if and only if its character χ satisfies*

$$\frac{1}{n}\sum_{k=1}^{s} h_k\chi(g_k)\chi(g_k^{-1}) = 1.$$

Proof. Decompose the representation T into a direct sum of irreducible representations. Correspondingly, the character χ can be expressed as $\chi = \sum_{i=1}^{s} m_i x_i$, where $\chi_1, \chi_2, \ldots, \chi_s$ are irreducible characters. But then

$$\frac{1}{n}\sum_{k=1}^{s} h_k \chi(g_k)\chi(g_k^{-1}) = \frac{1}{n}\sum_{i,j} m_i m_j \sum_{k=1}^{s} h_k \chi_i(g_k)\chi_j(g_k^{-1}) = \sum_{i=1}^{s} m_i^2 ,$$

and this sum is equal to 1 if and only if $\chi = \chi_i$ for some i, i.e., in view of Theorem 7.3.3, if T is an irreducible representation. □

If $K = \mathbb{C}$ is the field of complex numbers, then the orthogonality relations can be given a slightly different form. To that end, we introduce the following lemma.

Lemma 7.3.7. *If χ is the character of a d-dimensional representation of a group G over the field of complex numbers, then, for every $g \in G$, $\chi(g)$ is a sum of d n-th roots of unity and $\chi(g^{-1}) = \overline{\chi(g)}$, where as usual, $\overline{z}$ is the complex conjugate of the number z.*

Proof. Since $g^n = e$, we get $\left(T(g)\right)^n = E$ for every element $g \in G$. Since the polynomial $x^n - 1$ has no multiple roots, it follows that the matrix $T(g)$ is similar to the diagonal matrix

$$T(g) \sim \begin{pmatrix} \varepsilon_1 & & & 0 \\ & \varepsilon_2 & & \\ & & \ddots & \\ 0 & & & \varepsilon_d \end{pmatrix}, \quad \text{where } \varepsilon_i^n = 1.$$

From here, $\chi(g) = \varepsilon_1 + \varepsilon_2 + \ldots + \varepsilon_d$ and

$$T(g^{-1}) \sim \begin{pmatrix} \varepsilon_1^{-1} & & & 0 \\ & \varepsilon_2^{-1} & & \\ & & \ddots & \\ 0 & & & \varepsilon_d^{-1} \end{pmatrix}.$$

This results in

$$\chi(g^{-1}) = \varepsilon_1^{-1} + \varepsilon_2^{-1} + \ldots + \varepsilon_d^{-1} = \overline{\varepsilon}_1 + \overline{\varepsilon}_2 + \ldots + \overline{\varepsilon}_d = \overline{\chi(g)}.$$

□

In particular, $\chi_i(g_j^{-1}) = \overline{\chi}_{ij}$ and the orthogonality relations of Theorem 7.3.5 take the form

$$\frac{1}{n}\sum_{k=1}^{s} h_k \chi_{ik}\overline{\chi}_{jk} = \begin{cases} 0 & \text{for } i \neq j, \\ 1 & \text{for } i = j; \end{cases}$$

$$\frac{1}{n}\sum_{k=1}^{s} \chi_{ki}\overline{\chi}_{kj} = \begin{cases} 0 & \text{for } i \neq j, \\ 1/h_i & \text{for } i = j. \end{cases}$$

7.4 Algebraic Integers

In this section we shall need some properties of algebraic integers. Recall that an *algebraic integer* is, by definition, a (complex) root of an equation $x^m + a_1x^{m-1} + \ldots + a_m = 0$ with integral coefficients a_i.

Proposition 7.4.1. *A rational number which is an algebraic integer is an integer.*

Proof. Let z be a root of an equation $x^m + a_1x^{m-1} + \ldots + a_m = 0$ with integers a_i and $z = p/q$ with relatively prime integers p and $q > 1$. Passing to a common denominator, we get $p^m = -a_1qp^{m-1} - a_2q^2p^{m-2} - \ldots - a_mq^m$. This is impossible because p and q are relatively prime. □

The following lemma provides a convenient criterion for a number z to be an algebraic integer.

Lemma 7.4.2. *In order that z be an algebraic integer, it is necessary and sufficient that there exist complex numbers $y_1, y_2, \ldots, y_t$ such that $zy_i = \sum_{j=1}^{t} a_{ij}y_j$, where all a_{ij} are integers and not all y_i are zero.*

Proof. If z is a root of an integral equation $x^m + a_1x^{m-1} + \ldots + a_m = 0$, then we may take, trivially, $y_1 = 1$, $y_2 = z$, ..., $y_m = z^{m-1}$.

Conversely, let $y_1, y_2, \ldots, y_t$ have the required property. Denote by A the matrix (a_{ij}) and by Y the column vector whose coordinates are $y_1, y_2, \ldots, y_t$. Then $(zE - A)Y = 0$ and thus $\det(zE - A) = 0$. However, the determinant $\det(zE - A) = z^t + a_1z^{t-1} + \ldots + a_t$, where a_i are integral linear combinations of products of elements of the matrix A and thus integers. We conclude that z is an algebraic integer. □

Corollary 7.4.3. *The sum and product of algebraic integers are algebraic integers. In other words, the algebraic integers form a ring.*

Proof. Let $y_1, y_2, \ldots, y_t$ be complex numbers such that $zy_i = \sum_{j=1}^{t} a_{ij}y_j$ (with integers a_{ij}) and $y'_1, y'_2, \ldots, y'_r$ such that $z'y'_i = \sum_{j=1}^{r} a'_{ij}y'_j$ (with integers a'_{ij}).

Then one can see easily that the numbers $\{y_i y'_j \mid i = 1, 2, \ldots, t;\ j = 1, 2, \ldots, r\}$ satisfy similar conditions for the numbers $z + z'$ and zz'. □

Since the roots of unity are obviously algebraic integers, we obtain the following corollary of Lemma 7.3.7.

Corollary 7.4.4. *If χ is a character of a group G over the field of complex numbers, then $\chi(x)$ is an algebraic integer for every $x \in G$.*

We shall now employ the notation of the previous section. In particular, let $X = (\chi_{ij})$ be the character table of a group G over the field of complex numbers.

Theorem 7.4.5. *All numbers $\alpha_{ij} = \dfrac{h_i}{d_j}\chi_{ji}$ are algebraic integers.*

Proof. Note that, for all i and j, $c_i c_j$ is an element of the center of the algebra $\mathbb{C}G$. On the other hand, $c_i c_j$ is an integral linear combination of the elements of the group G. It follows that $c_i c_j = \sum\limits_k \gamma_{ijk} c_k$, where γ_{ijk} are integers. Besides,

$$c_i c_j = \Big(\sum_p \alpha_{ip} e_p\Big)\Big(\sum_q \alpha_{jq} e_q\Big) = \sum_p \alpha_{ip}\alpha_{jp} e_p$$

and $c_k = \sum\limits_p \alpha_{kp} e_p$; thus $c_i c_j = \sum\limits_{k,p} \gamma_{ijk}\alpha_{kp} e_p$ and $\alpha_{ip}\alpha_{jp} = \sum\limits_k \gamma_{ijk}\alpha_{kp}$. Writing $z = \alpha_{ip}$, $y_j = \alpha_{jp}$ (for a fixed p), we can apply Lemma 7.4.2 and conclude that α_{ip} is an algebraic integer. □

Corollary 7.4.6. *The dimensions d_i of irreducible complex representations divide the order of the group.*

Proof. Rewrite the list of the orthogonal relations of Theorem 7.3.5 to the form

$$\sum_{k=1}^{s} \frac{h_k \chi_{ik}}{d_i} \chi_i(g_k^{-1}) = \frac{n}{d_i}.$$

Since $\dfrac{h_k \chi_{ik}}{d_i} = \alpha_{ki}$ and $\chi_i(g_k^{-1})$ are algebraic integers, also the number $\dfrac{n}{d_i}$ is an algebraic integer. As a rational number, it must be an integer, as required. □

7.5 Tensor Products of Representations

In addition to usual module theoretical constructions, one can define yet another operation for group representations, viz. the tensor (or Kronecker) product.

Let M and N be two modules over the group algebra KG. Considered as vector spaces, their tensor product can be endowed with KG-module structure by defining $(m \otimes n)g = mg \otimes ng$ for every element $g \in G$. The module constructed in this way is called the *tensor product of the KG-modules M and N* and the respective representation of the group G the *tensor product of the representations* corresponding to the modules M and N.

We are going to compute the character of a tensor product of representations. Let T be a representation corresponding to a module M which, in a basis $\{u_1, u_2, \dots, u_m\}$, has the form

$$T(g) = \begin{pmatrix} \varphi_{11}(g) & \varphi_{12}(g) & \dots & \varphi_{1m}(g) \\ \varphi_{21}(g) & \varphi_{22}(g) & \dots & \varphi_{2m}(g) \\ \dots & \dots & \dots & \dots \\ \varphi_{m1}(g) & \varphi_{m2}(g) & \dots & \varphi_{mm}(g) \end{pmatrix},$$

and S a representation corresponding to a module N which, in a basis $\{v_1, v_2, \dots, v_n\}$, has the form

$$S(g) = \begin{pmatrix} \psi_{11}(g) & \psi_{12}(g) & \dots & \psi_{1n}(g) \\ \psi_{21}(g) & \psi_{22}(g) & \dots & \psi_{2n}(g) \\ \dots & \dots & \dots & \dots \\ \psi_{n1}(g) & \psi_{n2}(g) & \dots & \psi_{nn}(g) \end{pmatrix}.$$

The tensor products $u_i \otimes v_j$, $i = 1, 2, \dots, m$; $j = 1, 2, \dots, n$, form a basis of $M \otimes N$, and

$$\begin{aligned}(u_i \otimes v_j)g = u_i g \otimes v_j g = \Big(\sum_k \varphi_{ik}(g)u_k\Big) \otimes \Big(\sum_\ell \psi_{j\ell}(g)v_\ell\Big) = \\ = \sum_{k,\ell} \varphi_{ik}(g)\psi_{j\ell}(g)(u_k \otimes v_\ell).\end{aligned}$$

Thus, the elements of the matrix $(T \otimes S)(g)$ corresponding to g in this representation[12] are all possible products of the elements of $T(g)$ and $S(g)$. In particular,

$$\operatorname{Tr}(T \otimes S)(g) = \sum_{i=1}^{m}\sum_{j=1}^{n} \varphi_{ii}(g)\psi_{jj}(g) = \big(\operatorname{Tr} T(g)\big)\big(\operatorname{Tr} S(g)\big).$$

We have just proved the following proposition.

Proposition 7.5.1. *The character of a tensor product of two representations is equal to the product of the characters of these representations.*

[12] This matrix is called the Kronecker or tensor product of the matrices $T(g)$ and $S(g)$ and is denoted by $T(g) \otimes S(g)$.

Corollary 7.5.2. *Let $\chi_1, \chi_2, \dots, \chi_s$ be the irreducible characters of a group G. Then there exist natural numbers n_{ijk} such that $\chi_i\chi_j = \sum_{k=1}^{s} n_{ijk}\chi_k$ for any i, j.*

Proof. Let $M_1, M_2, \dots, M_s$ be the simple $K(G)$-modules. Then $M_i \otimes M_j \simeq \bigoplus_k n_{ijk}M_k$ for some natural numbers n_{ijk} and from here, everything follows. □

Now, let $G = G_1 \times G_2$. Every representation T of one of the factors (say, of G_1) can be considered as a representation of the entire group G if we set $T(g_1, g_2) = T(g_1)$. In particular, if T is a representation of G_1 and S a representation of G_2, we may construct their tensor product $T \otimes S$ which is a representation of the group G and the following theorem holds.

Theorem 7.5.3. *Let K be an algebraically closed field. If T is an irreducible representation of G_1 and S an irreducible representation of G_2, then the representation $T \otimes S$ is an irreducible representation of $G = G_1 \times G_2$ and every irreducible representation of the group G is obtained this way.*

Proof. Let M be the KG_1-module and N the KG_2-module corresponding to the representations T and S, respectively. Since K is algebraically closed, $E_{KG_1}(M) = E_{KG_2}(N) = K$ and, by Theorem 2.6.7, the linear map $u \mapsto ua$ ($u \in M$) attached to every $a \in KG_1$ defines an epimorphism of the algebra KG_1 onto $E(M)$.

Choose a basis $\{u_1, u_2, \dots, u_m\}$ of the module M and consider a non-zero element $x = \sum_{i=1}^{m} u_i \otimes v_i$, with $v_i \in N$, from $M \otimes N$. Without loss of generality, assume that $v_1 \neq 0$. Let $a = \sum_{j=1}^{r} \alpha_j g_j$, $\alpha_j \in K$, $g_j \in G_1$ be an element of KG_1 such that the corresponding endomorphism of M maps u_1 into a prescribed element u and all the other $u_2, u_3, \dots, u_m$ into zero. Then

$$x\sum_{j=1}^{r} \alpha_j(g_j, 1) = \sum_{i=1}^{m}\sum_{j=1}^{r} \alpha_j u_i g_j \otimes v_i = \sum_{i=1}^{m} u_i a \otimes v_i = u \otimes v_1 .$$

Similarly, given $v \in N$, there is an element $b \in KG$ such that $(u \otimes v_1)b = u \otimes v$. Consequently, the submodule generated by the element x is the entire $M \otimes N$, i.e. $M \otimes N$ is a simple module.

Two elements (g_1, g_2) and (g_1', g_2') are conjugate in the group $G = G_1 \times G_2$ if and only if g_1 and g_1' are conjugate in G_1 and g_2, g_2' are conjugate in G_2. Therefore, if $C_1, C_2, \dots, C_s$ are the conjugacy classes of G_1 and $D_1, D_2, \dots, D_t$ are the conjugacy classes of G_2, then $C_i \times D_j$, $i = 1, 2, \dots, s$; $j = 1, 2, \dots, t$, are the conjugacy classes of $G_1 \times G_2$. In particular, the number of these classes is st, and therefore, if we show that, for simple modules, the isomorphism $M \otimes N \simeq M' \otimes N'$ implies $M \simeq M'$ and $N \simeq N'$, we can conclude, in view of Theorem 7.2.2, that every simple KG-module has the form $M \otimes N$.

Denote by χ, χ', ξ and ξ' the characters corresponding to the modules M, M', N and N', respectively. Without loss of generality, let $M \not\simeq M'$. Choose a representative g_i in the class C_i, f_j in the class D_j, and let n_i be the number of elements in G_i, h_i in the class C_i and k_j in the class D_j. Then the number of elements of G is $n_1 n_2$ and the number of elements in the class $C_i \times D_j$ is $h_i k_j$; moreover, (g_i, f_j) is a representative of the class $C_i \times D_j$. The character corresponding to $M \otimes N$ is $\chi\xi$ and the character corresponding to $M' \otimes N'$ is $\chi'\xi'$. Then

$$\frac{1}{n_1 n_2}\sum_{i,j} h_i k_j \chi\xi(g_i, f_j)\chi'\xi'(g_i^{-1}, f_j^{-1}) =$$

$$= \frac{1}{n_1 n_2}\sum_{i,j} h_i k_j \chi(g_i)\xi(f_j)\chi'(g_i^{-1})\xi'(f_j^{-1}) =$$

$$= \frac{1}{n_1}\sum_{i} h_i \chi(g_i)\chi'(g_i^{-1}) \cdot \frac{1}{n_2}\sum_{j} k_j \xi(f_j)\xi'(f_j^{-1}) = 0\,,$$

and thus, in view of Theorem 7.3.5 and the fact that $\chi\xi$ and $\chi'\xi'$ are irreducible characters, $\chi\xi \neq \chi'\xi'$ and hence $M \otimes N \not\simeq M' \otimes N'$. The proof is completed. □

Thus, if we know the representations of the groups G_1 and G_2, we can construct all representations of the direct sum $G_1 \times G_2$.

We shall apply the construction of the tensor product of representations to prove the following result which strengthens Corollary 7.4.6.

Theorem 7.5.4. *Let C be the center of a group G. The dimension of every irreducible representation of G over the field of complex numbers divides the index $(G : C)$.*

Proof. Let d be the dimension of an irreducible representation T and M the corresponding module.

If $g \in C$, then $T(g)$ commutes with all matrices of the representation T and, by Schur's lemma, it is scalar: $T(g) = \lambda(g)E$. Consider the representation T_m of the group $G \times G \times \ldots \times G$ given by $M \otimes M \otimes \ldots \otimes M$ (m times). If elements g_i belong to C, then $T_m(g_1, g_2, \ldots, g_m) = \lambda(g_1)\lambda(g_2)\ldots\lambda(g_m)E$. Thus, in particular, if $g_1 g_2 \ldots g_m = 1$, then $T_m(g_1, g_2, \ldots, g_m) = E$. Now, the elements $(g_1, g_2, \ldots, g_m)$ with $g_i \in C$ and $g_1 g_2 \ldots g_m = 1$ form a normal subgroup H of $G \times G \times \ldots \times G$. Consequently, T_m can be interpreted as a representation of the quotient group $(G \times G \times \ldots \times G)/H$ whose order is n^m/c^{m-1} (here $n = (G : 1)$ and $c = (C : 1)$). Now, by Corollary 7.4.6, the dimension d^m of the representation T_m divides n^m/c^{m-1}, i.e. $n^m/c^{m-1}d^m$ is an integer for every m. Denoting by q the rational number n/cd, this means that cq^m is an integer for every m. This is possible only if q is an integer and thus d divides n/c, as required. □

7.6 Burnside's Theorem

In this paragraph we are going to present an application of the theory of representations to establish the existence of normal subgroups and consequently to prove the non-simplicity and solvability of certain classes of groups. All along this section, representations are considered over the field of complex numbers.

Let T be an irreducible representation of dimension d of a group G and χ be its character. According to Lemma 7.3.7, the number $\chi(g)$ is, for every $g \in G$, a sum of d nth roots of unity, where $n = (G : 1)$. Besides, if the matrix $T(g)$ is not scalar, these roots are distinct and then

$$|\chi(g)| = |\varepsilon_1 + \varepsilon_2 + \ldots + \varepsilon_d| < |\varepsilon_1| + |\varepsilon_2| + \ldots + |\varepsilon_d| = d\,.$$

Denote by $\mathbb{Q}$ the field of rational numbers, by ε a primitive nth root of unity and $L = \mathbb{Q}[\varepsilon]$. Then L is a splitting field of the polynomial $x^n - 1$, and thus, by Theorem 5.4.4, a normal extension of the field $\mathbb{Q}$. Denote by Γ its Galois group. Note that for every element $\sigma \in \Gamma$ and every root ε_i of 1, $\sigma(\varepsilon_i)$ is also a root of 1. In particular, $\sigma(\chi(g))$ is also a sum of d roots of unity and therefore $|\sigma(\chi(g))| \leq d$. These considerations yield the following result.

Theorem 7.6.1. *Let C be a conjugacy class of G whose number of elements h is relatively prime to the dimension d of an irreducible representation T. Then either all matrices $T(g)$ $(g \in C)$ are scalar or the character χ of T satisfies $\chi(g) = 0$ for all $g \in C$.*

Proof. By Theorem 7.4.5, $\frac{h}{d}\chi(g)$, where $g \in C$, is an algebraic integer. At the same time, $\chi(g)$ is also an algebraic integer. Since d and h are relatively prime, there exist integers x and y such that $xd + yh = 1$. Then

$$z = y\frac{h}{d}\chi(g) + x\chi(g) = \frac{yh + xd}{d}\chi(g) = \frac{\chi(g)}{d}$$

is an algebraic integer. If $T(g)$ is not a scalar matrix, we have shown that $|z| < 1$. On the other hand, for every $\sigma \in \Gamma$, the number $\sigma(z)$ is an algebraic integer (satisfying the same equations as z does) and $|\sigma(z)| < 1$. Consequently, also the number $q = \prod_{\sigma \in \Gamma} \sigma(z)$ is an algebraic integer and $|q| < 1$. However, evidently, $\sigma(q) = q$ for all $\sigma \in \Gamma$ and thus $q \in \mathbb{Q}$ (by Theorem 5.4.4). In view of Proposition 7.4.1, q is an integer, and thus necessarily $q = 0$. Therefore also $z = 0$, as required. □

Let us point out that the scalar matrices form a normal subgroup of the group of non-singular matrices. Therefore, those elements $g \in G$ for which the matrices $T(g)$ are scalar, form a normal subgroup $N(T)$ of G. If T is irreducible and not one-dimensional, then $N(T) \neq G$. These arguments suggest an application of Theorem 7.6.1 to establish the existence of normal subgroups. We are going to prove two theorems of Burnside in this direction.

Theorem 7.6.2. *If there is a conjugacy class $C \neq \{1\}$ of G whose number of elements is $h = p^t$, where p is prime, then G is not simple, i. e. G contains a non-trivial normal subgroup.*

Proof. Let $T_1, T_2, \ldots, T_s$ be all irreducible representations of G, $d_1, d_2, \ldots, d_s$ their dimensions and $\chi_1, \chi_2, \ldots, \chi_s$ their characters. We shall assume that $T_1(g) = 1$ for all g. Then $\chi_1(g) = 1$ for all g. If there is yet another one-dimensional T_i, then its kernel is a non-trivial normal subgroup of G. Therefore, we may assume that $d_i > 1$ for all $i \neq 1$.

Let $g \in C$. If $T_i(g)$ is a scalar matrix, then again there is a non-trivial normal subgroup of G. Otherwise, if d_i is not a divisor of p, then $\chi_i(g) = 0$, by Theorem 7.6.1. If all d_i divide p, we shall use the formula $\chi_{\text{reg}} = \sum_{i=1}^{s} = d_i \chi_i$ and apply Proposition 7.3.1. We get

$$\chi_{\text{reg}}(g) = 0 = 1 + \sum_{i=2}^{s} d_i \chi_i(g) = 1 + pz\,,$$

where z is an algebraic integer. Since $z = -\dfrac{1}{p}$, we get, in view of Proposition 7.4.1, a contradiction. The proof of the theorem is completed. □

Theorem 7.6.3. *If $(G : 1) = p^a q^b$, where p and q are primes, then the group G is solvable.*[13]

Proof. The proof will be given by induction on the order of the group G. We shall make use of the following well-known results from the theory of finite groups:

a) If the order of G is a power of a prime, then G has a non-trivial center.

b) If the order of G is divisible by p^a, where p is a prime, then there is a subgroup of order p^a (Sylow's theorem).

Choose a subgroup H of order p^a in G and take $g \neq 1$ from the center of H. Denote by $\overline{H} = \{x \in G \mid xg = gx\}$ the normalizer of g in G. Evidently, $\overline{H} \supset H$ and therefore $(G : \overline{H})$ divides $(G : H) = q^b$. Now, the number of conjugates of g equals $(G : \overline{H})$ and thus is a power of q. Hence, by Theorem 7.6.2, there is a non-trivial normal subgroup N in G. By induction hypothesis, both N and G/N are solvable and therefore G is also solvable. This completes the proof of the theorem. □

[13] Recall that a group G is called solvable if there is a series of subgroups $G = G_0 \supset G_1 \supset \ldots \supset G_m = \{1\}$ such that G_{i+1} is a normal subgroup of G_i and the quotient group G_i/G_{i+1} is abelian for all $i = 0, 1, \ldots, m-1$.

Exercises to Chapter 7

Except in Exercises 18–22, K is always assumed to be a field whose characteristic does not divide the order of the group G.

1. Let $G = \{g_1, g_2, \dots, g_n\}$ be a finite group, M and N two KG-modules and $f : M \to N$ a linear transformation. Prove that the map $\bar{f} : M \to N$ given by the formula

$$\bar{f}(m) = \frac{1}{n} \sum_{i=1}^{n} f(mg_i^{-1})g_i$$

is a homomorphism of KG-modules.

2. Derive from Exercise 1 the fact that in this situation every submodule $N \subset M$ is a direct summand. (Hint: Apply the construction to a projector of the space M onto the subspace N and use Theorem 1.6.2.) This result provides a new proof of Maschke's theorem, independent of results in Chapter 6.

3. Establish the isomorphism $KG \simeq KG_1 \otimes KG_2$ if $G = G_1 \times G_2$.

In Exercises 4–6, the field K contains a primitive nth root of unity, where $n = (G:1)$ (i.e. K is a splitting field for the polynomial $x^n - 1$). The group G is always assumed to be abelian.

4. Prove that the group algebra KG is a split algebra and that the group G has n distinct irreducible representations which are all one-dimensional (i.e. all are homomorphisms $G \to K^*$, where K^* is the multiplicative group of the field K).

5. Denote by $\hat{G}$ the set of all irreducible representations of the group G over the field K (these are the characters of G over K). For arbitrary characters f_1 and f_2 put $(f_1 f_2)(g) = f_1(g) f_2(g)$, where $g \in G$.
   a) Verify that $f_1 f_2$ is also a character of G over K and that $\hat{G}$ is an abelian group of order n with respect to this operation.
   b) Prove that, for a fixed element $g \in G$, the map $\tilde{g} : \hat{G} \to K$ given by the formula $\tilde{g}(f) = f(g)$ is a character of the group $\hat{G}$.
   c) Prove that the map $\delta : G \to \hat{\hat{G}}$ given by $\delta(g) = \tilde{g}$ is a group homomorphism.
   d) Establish that $\operatorname{Ker} \delta = \{1\}$, i.e. that δ is a monomorphism and thus, since $(G:1) = (\hat{\hat{G}}:1)$, that δ is an isomorphism.

6. Using the fact that every abelian group G can be written as a direct product of cyclic groups, compute explicitly all its characters and show that $\hat{G} \simeq G$ (in contrast to the isomorphism δ of the previous exercise, this isomorphism depends substantially on an explicit decomposition of the group G into a product of cyclic groups).

7. The subset $\{e, i, j, k, -e, -i, -j, -k\}$ of the quaternion algebra (see Sect. 1.1, Example 4) is called the *quaternion group*. Verify that these eight elements indeed form a multiplicative group. Find all non-trivial representations of this group over the fields of real and complex numbers. (Hint: In the latter case, one can use the results of Exercise 3 to Chap. 1.)

8. The *dihedral group* D_n is a group generated by a and b subject to the defining relations $a^n = b^2 = 1, \ \ ba = a^{-1}b$.
   a) Prove that D_n is a group of order $2n$.
   b) Verify that the correspondence

$$a \mapsto \begin{pmatrix} \cos\frac{2\pi k}{n} & -\sin\frac{2\pi k}{n} \\ \sin\frac{2\pi k}{n} & \cos\frac{2\pi k}{n} \end{pmatrix}, \qquad b \mapsto \begin{pmatrix} 0 & 1 \\ 1 & 0 \end{pmatrix}$$

   (with an integer k) is a representation T_k of the dihedral group D_n and that for different k satisfying the inequality $0 < k < n/2$, these representations are irreducible and not similar.
   c) Find the one-dimensional representations of the group D_n and prove that these representations together with the representations T_k, $0 < k < n/2$, from part b) constitute all irreducible representations of D_n over the fields of complex and real numbers. (Hint: Use Theorem 7.2.1.)

The following exercises (9–13) deal with the representations of the *symmetric group* S_n, i. e. the group of all permutations of the set $\{1, 2, \ldots, n\}$. Recall some facts concerning the structure of this group. The permutation $(i_1, i_2, \ldots, i_k)$ which maps i_1 to i_2, i_2 to $i_3, \ldots, i_k$ to i_1 and all the other numbers into themselves is called a *cycle* of length k. Here, all numbers $i_1, i_2, \ldots, i_k$ are distinct. In case $k = 1$, the respective cycle is evidently the identity permutation. Two cycles $(i_1, i_2, \ldots, i_k)$ and $(j_1, j_2, \ldots, j_\ell)$ are said to be *independent* if the sets $\{i_1, i_2, \ldots, i_k\}$ and $\{j_1, j_2, \ldots, j_\ell\}$ are disjoint. Every permutation σ can be decomposed into a product of non-intersecting cycles $\sigma = (i_{11}, \ldots, i_{1k_1})(i_{21}, \ldots, i_{2k_2}) \ldots (i_{t1}, \ldots, i_{tk_t})$, where $k_1 + k_2 + \ldots + k_t = n$; in fact, this decomposition is unique (up to an order of the factors since, obviously, independent cycles commute). The collection of lengths $(k_1, k_2, \ldots, k_t)$ is called the *cycle type* of the permutation σ.

9. Prove that two permutations are conjugate in S_n if and only if they have the same cycle type. In this way, a conjugacy class of S_n is uniquely determined by a partition of n into a sum of natural summands $n = k_1 + k_2 + \ldots + k_t$.

In what follows, we always assume that $k_1 \geq k_2 \geq \ldots \geq k_t$. Such a cycle type is conveniently described by the so-called *Young diagram*, i. e. an arrangement of n cells into t rows with k_i cells in the ith row.

Examples (for $n = 5$):

5=4+1   5=3+1+1   5=2+2+1

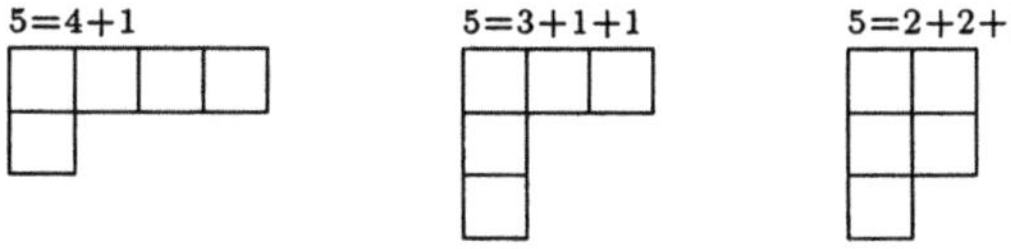

A *position* on a Young diagram is an arbitrary distribution of the numbers $\{1, 2, \ldots, n\}$ into the cells of that diagram. The Young diagram corresponding to the partition $(k_1, k_2, \ldots, k_s)$ is said to be *higher* than the diagram corresponding to $(\ell_1, \ell_2, \ldots, \ell_t)$ if $k_1 > \ell_1$, or $k_1 = \ell_1$, but $k_2 > \ell_2$, or $k_1 = \ell_1$, $k_2 = \ell_2$, but $k_3 > \ell_3$ etc. (lexicographical order).

10. Let D_1 and D_2 be positions on two Young diagrams, the first of which is higher than the second one.
    a) Prove that there are numbers $i \neq j$ such that both appear in the same row of the first diagram and in the same column of the second diagram.
    b) Prove that for any permutation σ, there are transpositions (i. e. cycles of length 2) $\tau_1 = (i_1, i_2)$ and $\tau_2 = (j_1, j_2)$ such that $\tau_1\sigma = \sigma\tau_2$ and the numbers i_1, i_2 are in the same row of the position D_1 and the numbers j_1, j_2 in the same column of the position D_2.

11. For a given position D on a Young diagram, denote by P_D the set of all permutations which map numbers of a given row only into numbers of that row, and by Q_D all the permutations which map numbers of a given column into the numbers of that column.
    a) Verify that P_D and Q_D are subgroups of S_n and that $P_D \cap Q_D = \{1\}$.
    b) Prove that if D_1 and D_2 are positions on the same Young diagram, then either there is a pair of numbers $i \neq j$ which are in the same row of D_1 and in the same column of D_2, or by applying a suitable permutation from P_{D_1} to D_1 and a suitable permutation from Q_{D_2} to D_2 one obtains the same position.
    c) Prove that, for an arbitrary position D and an arbitrary permutation σ, either there are transpositions $\tau_1 \in P_D$ and $\tau_2 \in Q_D$ such that $\tau_1\sigma = \sigma\tau_2$, or $\sigma = \xi\eta$, where $\xi \in P_D$ and $\eta \in Q_D$, and such a decomposition is unique.

12. Given a position D on a Young diagram, the element c_D of the group algebra $A = KS_n$ defined by the formula

$$c_D = \sum_{\substack{\xi \in P_D \\ \eta \in Q_D}} \operatorname{sgn}(\eta)\xi\eta ,$$

where $\operatorname{sgn}(\eta)$ is the signature of the permutation η (equal to 1 for η even and -1 for η odd), is called the *Young symmetrizer* corresponding to D.
Prove that if $\sigma \in P_D$, then $\sigma c_D = c_D$ and if $\sigma \in Q_D$, then $c_D\sigma = \operatorname{sgn}(\sigma)c_D$. Conversely, if $a \in A$ is an arbitrary element satisfying the above conditions, then $a = \alpha c_D$ for some $\alpha \in K$. (Hint: Use the result of Exercise 11.c).)

13. Write $M_D = c_D A$, where c_D is a Young symmetrizer and $A = KS_n$.
    a) Prove that $E_A(M_D) = K$. (Hint: Use the result of the preceding exercise.)
    b) Under the assumption of Exercise 10, prove that $\operatorname{Hom}_A(M_{D_1}, M_{D_2}) = 0$.
    c) Deduce the following statement: If D runs through all positions on Young diagrams, M_D runs through all simple A-modules; moreover, $M_{D_1} \simeq M_{D_2}$ if and only if D_1 and D_2 are positions on the same diagram.

14. Let $A = KG$, M and N be two arbitrary A-modules and η and ψ the characters of the corresponding representations. Using the notation of Theorem 7.3.5, prove that

$$\frac{1}{n}\sum_{k=1}^{s} h_k\chi(g_k)\psi(g_k^{-1}) = \dim\operatorname{Hom}_A(M, N).$$

15. Using Corollary 7.2.3 and 7.4.6, deduce that every group of order p^2, where p is a prime, is abelian.

16. Let M be a module over the group algebra KG, and M^* the space of all linear forms on M, i. e. $M^* = \operatorname{Hom}_K(M, K)$. Defining $(fg)(m) = f(mg^{-1})$ for arbitrary $f \in M^*$, $m \in M$, $g \in G$, verify that M^* turns into a KG-module. If T is the representation corresponding to M, then the representation T^* corresponding to M^* satisfies $T^*(g) = T(g^{-1})'$, where $'$ denotes the transpose of a

matrix. In particular, if χ is the character of the representation T and χ^* the caharcter of the representation T^*, then $\chi^*(g) = \chi(g^{-1})$, and if $K = \mathbb{C}$, then $\chi^*(g) = \overline{\chi(g)}$.

17. A representation T of a group G over the field of complex (or real) numbers is called *unitary* if all matrices $T(g)$ are unitary.
    a) Prove that every complex (real) representation of a finite group $G = \{g_1, g_2, \dots, g_n\}$ is similar to a unitary one. (Hint: On the corresponding module M, choose a scalar product (u,v) turning M into a unitary (Euclidean) space and put $\langle u,v\rangle = \sum_{i=1}^{n}(ug_i, vg_i)$. Then M is a unitary space with respect to the scalar product $\langle u,v\rangle$; furthermore, $\langle ug,vg\rangle = \langle u,v\rangle$ for all $g \in G$.)
    b) From here, deduce yet another proof of the fact that every representation of G over $\mathbb{R}$ or over $\mathbb{C}$ is completely reducible.
    c) Considering an infinite cyclic group, show that the conclusions of part a) do not hold for infinite groups.

In the final exercises, we shall assume that the characteristic p of the field K divides the order n of the group G.

18. Let H be a subgroup of G such that the index $(G : H)$ and p are relatively prime; let N be a submodule of a KG-module M which, as a KH-module has a complement. Prove that N has a complement as a KG-module. (Hint: Choose representatives of the cosets of H in G and proceed as in Exercise 1 and 2.)

19. Assume that G is a p-group, i.e. $n = p^k$. Write $I = \left\{\sum_{g\in G} a_g g \mid \sum_g a_g = 0\right\}$. Prove that $I = \operatorname{rad} KG$ and $KG/I \simeq K$. (Hint: Use the results of Exercise 13 to Chap. 6.)

20. a) Let M be an irreducible representation of a p-group G. Prove that $[M : K] = 1$ and $mg = m$ for all $m \in M$, $g \in G$.
    b) Prove that every representation of G over K is similar to a *unipotent triangular representation*, i.e. to a representation of the form

$$T(g) = \begin{pmatrix} 1 & & & * \\ & 1 & & \\ & & \ddots & \\ 0 & & & 1 \end{pmatrix}.$$

    c) Deduce from part b) that every finite p-group G is isomorphic to a group of unipotent upper triangular matrices over the field of integers modulo p.
    d) Prove that every finite p-group G is *nilpotent*, i.e. there is a chain of normal subgroups $G = G_0 \supset G_1 \supset \dots \supset G_k = \{1\}$ such that the factor group G_{i+1}/G_i is in the center of G/G_i for each $i = 1, 2, \dots, k$.

21. Describe the indecomposable representations of a cyclic p-group over a field of characteristic p; check that the number of these representations equals the order of the group.

22. Let G be a non-cyclic group of order p^2 (i.e. a direct product of two cyclic groups of order p). For arbitrary even d, construct an indecomposable representation of dimension d of the group G. (Hint: If a and b are generators of the cyclic summands of G, set

$$T(a) = \begin{pmatrix} E & E \\ 0 & E \end{pmatrix}, \quad T(b) = \begin{pmatrix} E & X \\ 0 & E \end{pmatrix},$$

where E is a unit and X an arbitrary matrix.) If K is infinite, verify that there is an infinite number of non-similar representations of dimension d of the group G. Translate this result to an arbitrary non-cyclic p-group.

23. Let G be a direct product of three cyclic groups of order p with generators a, b and c. Taking

$$T(a) = \begin{pmatrix} E & E \\ 0 & E \end{pmatrix}, \quad T(b) = \begin{pmatrix} E & X \\ 0 & E \end{pmatrix}, \quad T(c) = \begin{pmatrix} E & Y \\ 0 & E \end{pmatrix},$$

where X and Y are arbitrary square matrices, we get a representation $T = T_{X,Y}$ of the group G; moreover, $T_{X,Y}$ and $T_{X',Y'}$ are similar if and only if the pairs of matrices X, Y and X', Y' are similar, i. e. if there is a matrix C such that $X' = CXC^{-1}$ and $Y' = CYC^{-1}$. Let us remark that S.A. Krugljak has constructed for $p > 2$ a representation $S_{X,Y}$ for an arbitrary non-cyclic p-group G which depends on a pair of matrices X, Y and such that $S_{X,Y}$ and $S_{X',Y'}$ are similar if and only if the pairs of matrices X, Y and X', Y' are similar. The classification of pairs of matrices with respect to similarity is one of the most difficult problems of linear algebra which has not been solved so far.

8. The Morita Theorem

In Sect. 2.3 we have noted that modules over a division algebra D and modules over the simple algebra $M_n(D)$ are "equally structured". Results of Sect. 2.6 show that, in general, modules over isotypic semisimple algebras possess the same properties: such modules have isomorphic endomorphism rings, etc. In Sect. 3.5 these results have been extended to projective modules over arbitrary isotypic algebras (Lemma 3.5.5). It turns out that one can remove the requirement of projectivity: All modules over isotypic algebras are equally structured. However, in order to formulate this statement properly, it is necessary to introduce a number of concepts which presently play an important role in various areas of mathematics. Above all, it is the concept of a category and a functor, as well as the notion of an equivalence of categories, which appears to be a mathematical formulation of the expression "equally structured".

The Morita theorem which we are going to prove in this chapter just asserts that two algebras are isotypic if and only if their module categories are equivalent. Techniques applied to proving the theorem (tensor product, exact sequences) turn out to be useful also for many other problems. In particular, in Sect. 8.5 we shall construct the tensor algebra of a bimodule generalizing the concept of the path algebra of a diagram and playing a similar role in describing non-semisimple algebras (which are not necessarily split).

8.1 Categories and Functors

A *category* $\mathcal{C}$ consists of the following data:

1) a set $\operatorname{Ob}\mathcal{C}$ whose elements are called the *objects* of the category $\mathcal{C}$;

2) a set $\operatorname{Mor}\mathcal{C}$ whose elements are called the *morphisms* of the category $\mathcal{C}$;

3) there is an ordered pair of objects (X, Y) of the category $\mathcal{C}$ associated with every morphism $f \in \operatorname{Mor}\mathcal{C}$ (we write $f : X \to Y$ and say that f is a morphism from the object X to the object Y; X is the *initial object* and Y is the *terminal object* of the morphism f; the set of all morphisms from X to Y is denoted by $\operatorname{Hom}(X, Y)$ or, if one needs to specify the category, by $\operatorname{Hom}_{\mathcal{C}}(X, Y)$);

4) for every triplet of objects $X, Y, Z \in \operatorname{Ob}\mathcal{C}$ and every pair of morphisms $f : X \to Y$ and $g : Y \to Z$ there is a uniquely defined morphism $gf : X \to Z$ which is called the *composition* or *product* of the morphisms f and g;

5) multiplication of morphisms is associative, i.e. for every triplet of morphisms f, g, h we have $h(gf) = (hg)f$, provided that the products are defined;[14]

6) for every object $X \in \mathrm{Ob}\,\mathcal{C}$, there exists a morphism $1_X \in \mathrm{Hom}(X, X)$ such that $f1_X = f$ and $1_X g = g$ for all morphisms $f : X \to Y$ and $g : Z \to X$.

It is easy to see that a morphism 1_X with the above properties is unique. It is called the *identity morphism* of the object X.

Examples of Categories. 1. The *category* Sets *of sets.* Objects of this category are sets and morphisms $f : X \to Y$ are maps of the set X into the set Y. Composition of morphisms is the usual composition of maps. It is evident that all category axioms are satisfied.[15]

2. The *category* Gr *of groups.* Objects of this category are groups, morphisms $f : X \to Y$ are homomorphisms of the group X into the group Y and composition is the usual product of homomorphisms.

3. The *category of vector spaces* over a field K (denoted by Vect or, specifying the field, by Vect_K), the *category of K-algebras* Alg (or Alg_K), the *category* mod-A *of right modules* and the *category* A-mod *of left modules* over the algebra A, etc. are defined analogously. In all these examples, morphisms are some maps of the sets with the usual composition. However, the following examples show that there are categories of a different kind.

4. Every semigroup P (with identity) can be regarded as a set of morphisms of a category consisting of a single object. Here, composition of morphisms naturally coincides with their product in the semigroup P.

5. The *category* Mat *of matrices.* Objects of this category are natural numbers; the set of morphisms $\mathrm{Hom}(m, n)$ is defined to be the set of all $n \times m$ matrices with entries from a field K. Composition of the morphisms is the usual product of matrices. Here a verification of all axioms is also trivial.

6. Let M be a partially ordered set. Consider it as the set of objects of a category in which $\mathrm{Hom}(x, y)$ consists of a single element when $x \leq y$ and it is empty otherwise. Composition of the morphisms is defined in a natural manner.

7. The *path category.* Let $\mathcal{D}$ be a diagram (see Sect. 3.6). One can associate with $\mathcal{D}$ the following category $\mathcal{C}_{\mathcal{D}}$. Put $\mathrm{Ob}\,\mathcal{C}_{\mathcal{D}} = \mathcal{D}$ and for $i, j \in \mathcal{D}$, let $\mathrm{Hom}(i, j)$ be the set of all paths from i to j. Composition of the paths is defined, as in Chapter 3, by concatenation and 1_i is the "empty" path with both starting and terminal object at i (see Chap. 3). Again, we get a category which is called the path category of the diagram $\mathcal{D}$.

[14] It is easy to see that if one of the sides of this equality is defined, so is the other; this happens if and only if the terminal object of f coincides with the initial object of g and the terminal object of g with the initial object of h.

[15] Of course, in this definition Ob(Sets) and Mor(Sets) are not sets. However, for all practical purposes this is not essential: We can always restrict ourselves to the subsets of a fixed set (and their maps). This remark also refers to the other analogous examples.

8. The *dual category*. For any category $\mathcal{C}$, one can construct a new category $\mathcal{C}^o$ in the following way: $\mathrm{Ob}\,\mathcal{C}^o = \mathrm{Ob}\,\mathcal{C}$, $\mathrm{Mor}\,\mathcal{C}^o = \mathrm{Mor}\,\mathcal{C}$ and the initial (terminal) object of a morphism f in the category $\mathcal{C}^o$ is its terminal (initial) object in the category $\mathcal{C}$. The product gf in the category $\mathcal{C}^o$ is defined to be the product fg in the category $\mathcal{C}$. The category $\mathcal{C}^o$ is said to be dual (opposite) to the category $\mathcal{C}$. Evidently, $\mathcal{C}^{oo} = \mathcal{C}$.

In order to avoid any confusion, objects and morphisms of the category $\mathcal{C}^o$ are usually marked by a little circle: X^o, f^o, etc. Then the above definitions can be written in the form $\mathrm{Ob}\,\mathcal{C}^o = (\mathrm{Ob}\,\mathcal{C})^o$, $\mathrm{Mor}\,\mathcal{C}^o = (\mathrm{Mor}\,\mathcal{C})^o$, $\mathrm{Hom}_{\mathcal{C}^o}(X^o, Y^o) = \mathrm{Hom}_{\mathcal{C}}(Y, X)^o$ and $g^o f^o = (fg)^o$.

In every category one can define the concept of an *isomorphism*. Indeed, a morphism $f : X \to Y$ is said to be an isomorphism if and only if there is a morphism $f^{-1} : Y \to X$ such that $f^{-1}f = 1_X$ and $ff^{-1} = 1_Y$. Evidently, these conditions define the morphism f^{-1} uniquely. The morphism f^{-1} is called the *inverse* of f. Of course, f^{-1} is also an isomorphism and $(f^{-1})^{-1} = f$. Moreover, it is easy to see that a composition of isomorphisms f and g (if defined) is again an isomorphism and that $(gf)^{-1} = f^{-1}g^{-1}$.

As much as the concept of a homomorphism plays an important role in the study of groups, algebras and modules, a central concept of category theory is that of a functor.

A *functor* F from a category $\mathcal{C}$ to a category $\mathcal{D}$ is a pair of maps $F_{\mathrm{ob}} : \mathrm{Ob}\,\mathcal{C} \to \mathrm{Ob}\,\mathcal{D}$ and $F_{\mathrm{mor}} : \mathrm{Mor}\,\mathcal{C} \to \mathrm{Mor}\,\mathcal{D}$ satisfying the following conditions:

1) if $f : X \to Y$, then $F_{\mathrm{mor}}(f) : F_{\mathrm{ob}}(X) \to F_{\mathrm{ob}}(Y)$;

2) $F_{\mathrm{mor}}(1_X) = 1_{F_{\mathrm{ob}}(X)}$;

3) if gf is defined, then $F_{\mathrm{mor}}(gf) = F_{\mathrm{mor}}(g)F_{\mathrm{mor}}(f)$.

Usually, instead of $F_{\mathrm{mor}}(f)$ and $F_{\mathrm{ob}}(X)$ one simply writes $F(f)$ and $F(X)$.

Examples of functors. 1. Let $\mathcal{C}$ be a category. Fix an object $X \in \mathrm{Ob}\,\mathcal{C}$ and construct the functor $h_X : \mathcal{C} \to \mathrm{Sets}$ in the following way. If $Y \in \mathrm{Ob}\,\mathcal{C}$, define $h_X(Y) = \mathrm{Hom}(X, Y)$. If $f : Y \to Z$, then $h_X(f)$ is the map of the sets $\mathrm{Hom}(X, Y) \to \mathrm{Hom}(X, Z)$ assigning to every morphism $g : X \to Y$ the morphism $fg : X \to Z$. The conditions 1) and 2) are satisfied trivially and 3) follows from the associativity of multiplication of morphisms.[16]

If $\mathcal{C} = \text{mod-}A$ (or $A\text{-mod}$), where A is an algebra over K, then all sets $\mathrm{Hom}(X, Y)$ are vector spaces over K and one can see easily that, for any f, the map $h_X(f)$ is a homomorphism. Therefore h_X can be considered in this case as a functor to the category Vect of vector spaces over the field K.

2. *Forgetful functors*. Let $\mathcal{C} = \mathrm{Gr}$, $\mathcal{D} = \mathrm{Sets}$. Define the functor $F : \mathcal{C} \to \mathcal{D}$ by $F(X) = X$ and $F(f) = f$ for every $X \in \mathcal{C}$ and $f \in \mathrm{Mor}\,\mathcal{C}$. In other words, we forget the group structure on X and consider X simply as a set and homomorphisms as set maps. This functor is called the forgetful functor from the category of groups to the category of sets.

[16] The reader not familiar with category techniques is advised to verify the conditions.

In a similar way, we may construct a variety of examples of forgetful functors taking for $\mathcal{C}$ a category of sets with "more" structure and for $\mathcal{D}$ a category of sets with "less" structure.

Take, for example: a) $\mathcal{C} = \mathrm{Alg}_K$, $\mathcal{D} = \mathrm{Vect}_K$; b) $\mathcal{C} = \text{mod-}A$, $\mathcal{D} = \mathrm{Vect}$; c) $\mathcal{C} = \mathrm{Alg}_L$, $\mathcal{D} = \mathrm{Alg}_K$, where L is an extension of the field K, etc.

3. Let A be an algebra, $B = M_n(A)$. Construct a functor $G : \text{mod-}A \to \text{mod-}B$ in the following way. For every A-module M, put $G(M) = nM$. We endow $G(M)$ with a B-module structure in a natural way: Considering an element $x \in G(M)$ as an n-dimensional vector with coordinates from M, define xb for $b \in B$ using the ordinary matrix multiplication rule. If $f : M \to N$ is an A-module homomorphism, define $G(f) : G(M) \to G(N)$ coordinatewise: For $x = (x_1, x_2, \ldots, x_n)$ we put $G(f)x = (fx_1, fx_2, \ldots, fx_n)$. It is easy to verify that $G(f)$ is a homomorphism of B-modules and that this construction indeed defines a functor.

4. If L is an extension of a field K, then it is possible to construct a functor $F : \mathrm{Alg}_K \to \mathrm{Alg}_L$ defining $F(A)$ to be the L-algebra $A_L = A \otimes L$ and $F(f)$, where $f : A \to B$, to be the L-algebra homomorphism $f \otimes 1 : A_L \to B_L$.

5. Let $\mathcal{C}$ be a semigroup with identity regarded as a category with a single object (Example 4 of a category). Let us clarify the meaning of a functor from the category $\mathcal{C}$ into the category Vect_K. Since $\mathrm{Ob}\,\mathcal{C}$ consists of a single element, F_{ob} is determined by a single vector space V. Then, for every element a of the semigroup, $F(a) \in E(V)$; moreover, $F(1) = 1_V$ and $F(ab) = F(a)F(b)$. Hence, F_{mor} is a representation of the semigroup $\mathcal{C}$ on a vector space V.

6. If $\mathcal{C}^o$ is the dual (opposite) category of a category $\mathcal{C}$, the functors $F : \mathcal{C}^o \to \mathcal{D}$ are called *contravariant functors* from the category $\mathcal{C}$ to the category $\mathcal{D}$ (and in order to emphasize that it preserves the direction of arrows, the ordinary functors from $\mathcal{C}$ to $\mathcal{D}$ are called *covariant functors*). Since there is a one-to-one correspondence between $\mathrm{Ob}\,\mathcal{C}^o$ and $\mathrm{Ob}\,\mathcal{C}$, and also between $\mathrm{Mor}\,\mathcal{C}^o$ and $\mathrm{Mor}\,\mathcal{C}$, the maps F_{ob} and F_{mor} for a contravariant functor can be interpreted also as maps $\mathrm{Ob}\,\mathcal{C} \to \mathrm{Ob}\,\mathcal{D}$ and $\mathrm{Mor}\,\mathcal{C} \to \mathrm{Mor}\,\mathcal{D}$. However, then the axioms in the definition of a functor take on the following form:

1°) if $f : X \to Y$, then $F(f) : F(Y) \to F(X)$ (i.e. F "reverses the arrows");

2°) $F(1_X) = 1_{F(X)}$;

3°) $F(gf) = F(f)F(g)$ (i.e. F "reverses the order of the arrows").

An important example of a contravariant functor is obtained in analogy to Example 1. For a fixed object $X \in \mathrm{Ob}\,\mathcal{C}$, one can construct the functor $h^o_X : \mathcal{C}^o \to \mathrm{Sets}$ by setting $h^o_X(Y^o) = \mathrm{Hom}(Y, X)$ and defining $h^o_X(f^o)$ with $f : Y \to Z$ to be the map $\mathrm{Hom}(Z, X) \to \mathrm{Hom}(Y, X)$ assigning to a morphism $g : Z \to X$ the morphism $gf : Y \to X$. If $\mathcal{C} = \text{mod-}A$ (or A-mod), then h^o_X can be interpreted as a functor $\mathcal{C}^o \to \mathrm{Vect}$.

Categories of modules over algebras (and many other categories) have an additional structure: they are linear in the following sense.

A category $\mathcal{C}$ is called a *linear category over a field* K (or K*-linear* or simply *linear* if there is no danger of misunderstanding) if, for every pair

(X,Y) of its objects, the set of morphisms $\text{Hom}(X,Y)$ is endowed with the structure of a vector space over K and the composition of morphism is K-linear, i. e.

$$(f+g)h = fh + gh,$$
$$f(g+h) = fg + fh \quad \text{and}$$
$$(\lambda f)g = f(\lambda g) = \lambda(fg)$$

for any morphisms f, g, h such that the corresponding formulae make sense, and for any $\lambda \in K$.

Of course, a K-linear category with one object is just a K-algebra (cf. Example 4 above). If a category $\mathcal{C}$ is K-linear, then so is its dual $\mathcal{C}^o$ (with the same linear structure).

A functor $F : \mathcal{C} \to \mathcal{D}$ between two linear categories is said to be *linear* (K*-linear* if we need to specify the field K) if

$$F(f+g) = F(f) + F(g) \quad \text{and}$$
$$F(\lambda f) = \lambda F(f)$$

for any morphisms f, g such that $f+g$ is defined, and for any $\lambda \in K$. One can easily check that for every object X of a linear category $\mathcal{C}$ the functors h_X and h_X^o (considered as functors to Vect) are linear.

An important property of linear functors is the fact that they preserve direct sums. Namely, we have the following "categorical" characterization of direct sums of modules.

Proposition 8.1.1. *$M \simeq M_1 \oplus M_2 \oplus \ldots \oplus M_n$ if and only if there exist morphisms $i_k : M_k \to M$ and $p_k : M \to M_k$ for all $k = 1, 2, \ldots, n$ such that $p_k i_k = 1_{M_k}$, $p_k i_\ell = 0$ if $k \neq \ell$ and $i_1 p_1 + i_2 p_2 + \ldots + i_n p_n = 1_M$.*

Proof. If $M \simeq M_1 \oplus M_2 \oplus \ldots \oplus M_n$, we can take for i_k and p_k the natural embedding $M_k \to M$ and projection $M \to M_k$, respectively. On the other hand, given i_k and p_k, the homomorphisms

$$\begin{pmatrix} p_1 \\ p_2 \\ \cdot\cdot \\ p_n \end{pmatrix} : M \to \bigoplus_{k=1}^{n} M_k \quad \text{and} \quad (i_1, i_2, \ldots, i_n) : \bigoplus_{k=1}^{n} M_k \to M$$

are mutually inverse isomorphisms. □

Now we are able to define a direct sum of objects $M_1, M_2, \ldots, M_k$ of any linear category $\mathcal{C}$ as an object M such that morphisms $i_k : M_k \to M$ and $p_k : M \to M_k$ satisfying the relations of Proposition 8.1.1 exist. One can easily verify (and we recommend to do it) that such M is defined up to an isomorphism (in $\mathcal{C}$).

Corollary 8.1.2 *Let $F : \mathcal{C} \to \mathcal{D}$ be a linear functor between two linear categories and $M \simeq \bigoplus_{k=1}^{n} M_k$ in $\mathcal{C}$. Then $F(M) \simeq \bigoplus_{k=1}^{n} F(M_k)$ in $\mathcal{D}$.*

In what follows, all categories and functors will be assumed to be linear and we shall use Corollary 8.1.2 frequently without any reference.

8.2 Exact Sequences

In what follows, we shall often consider situations when one deals simultaneously with a number of modules and their homomorphisms related by various conditions. In order to describe such situations, "the language of diagrams and exact sequences" becomes very convenient. For instance, let M_i and N_i $(i = 1, 2, 3)$ be modules and $f_i : M_i \to N_i$ $(i = 1, 2, 3)$, $g : M_1 \to M_2$, $h : M_2 \to M_3$, $g' : N_1 \to N_2$ and $h' : N_2 \to N_3$ homomorphisms. In this case, we speak about a *diagram of modules*

$$\begin{array}{ccccc} M_1 & \xrightarrow{g} & M_2 & \xrightarrow{h} & M_3 \\ {\scriptstyle f_1}\downarrow & & {\scriptstyle f_2}\downarrow & & {\scriptstyle f_3}\downarrow \\ N_1 & \xrightarrow{g'} & N_2 & \xrightarrow{h'} & N_3 \end{array} \tag{8.2.1}$$

The diagram (8.2.1) is called *commutative* if $f_2 g = g' f_1$ and $f_3 h = h' f_2$. In other words, given two paths connecting a pair of modules in the diagram, the products of the homomorphisms taken along each of these paths are equal. The concept of a commutative diagram in a general case is defined similarly. It is clear that such a terminology allows to describe efficiently rather complex situations.

Consider a sequence (finite or infinite) of modules and homomorphisms

$$\cdots \longrightarrow M_{i-1} \xrightarrow{f_{i-1}} M_i \xrightarrow{f_i} M_{i+1} \longrightarrow \cdots \tag{8.2.2}$$

We say that such a sequence is *exact at* M_i if $\operatorname{Ker} f_i = \operatorname{Im} f_{i-1}$. If the sequence (8.2.2) is exact at all M_i's, then it is called *exact*.

We are going to give some examples illustrating this terminology.

Examples. 1. A sequence $0 \to N \xrightarrow{f} M$ (the first morphism is obviously zero) is exact if and only if $\operatorname{Ker} f = 0$, i.e. if f is a monomorphism. Similarly, a sequence $M \xrightarrow{g} N \to 0$ is exact if and only if g is an epimorphism.

2. We shall clarify the meaning of a sequence $0 \to N \xrightarrow{f} M \xrightarrow{g} L$ to be exact. As before, since the sequence is exact at N, f is a monomorphism. In other words, N can be considered (identifying it with $\operatorname{Im} f$) as a submodule of M. Since it is exact at M, $\operatorname{Im} f = \operatorname{Ker} g$, and thus N can be identified with the kernel of the homomorphism of g.

Similarly, the fact that a sequence $N \xrightarrow{f} M \xrightarrow{g} L \to 0$ is exact means that g is an epimorphism and L can be identified with the factor module $M/\operatorname{Im} f$ (this factor module is called the *cokernel* of the homomorphism f and is denoted by $\operatorname{Coker} f$).

3. Finally, a sequence

$$0 \longrightarrow N \xrightarrow{f} M \xrightarrow{g} L \longrightarrow 0$$

is exact if and only if f is a monomorphism, g is an epimorphism and N can be identified with a submodule of M and L with the factor module M/N.

4. Let us reformulate one of the definitions of a projective module (see Theorem 3.3.5) in the language of diagrams and exact sequences: A module P is projective if and only if every diagram

$$\begin{array}{ccccc} & & P & & \\ & & f\downarrow & & \\ M & \xrightarrow{g} & N & \longrightarrow & 0 \end{array}$$

whose row is exact can be completed to a commutative diagram

$$\begin{array}{ccccc} & & P & & \\ & \bar{f}\swarrow & f\downarrow & & \\ M & \xrightarrow{g} & N & \longrightarrow & 0 \end{array}$$

(recall that exactness means that g is an epimorphism and commutativity means that $f = g\bar{f}$).

An exact sequence

$$0 \longrightarrow N \xrightarrow{f} M \xrightarrow{g} L \longrightarrow 0$$

is said to be *split* if there are homomorphisms $\bar{f} : M \to N$ and $\bar{g} : L \to M$ such that $\bar{f}f = 1_N$ and $g\bar{g} = 1_L$.

In view of Proposition 1.6.2, it is sufficient to require the existence of $\bar{f}$ (or $\bar{g}$) only; in this case, M can be identified with the direct sum $N \oplus L$, f is the canonical inclusion $N \to N \oplus L$ (mapping $x \in N$ into $(x, 0)$) and g the canonical projection of $N \oplus L$ onto the second summand.

Finally, let us formulate a diagram lemma which will be often used in the sequel.

Lemma 8.2.1 (Five Lemma). *Let*

$$\begin{array}{ccccccccc} M_1 & \xrightarrow{f_1} & M_2 & \xrightarrow{f_2} & M_3 & \xrightarrow{f_3} & M_4 & \xrightarrow{f_4} & M_5 \\ \varphi_1\downarrow & & \varphi_2\downarrow & & \varphi_3\downarrow & & \varphi_4\downarrow & & \varphi_5\downarrow \\ N_1 & \xrightarrow{g_1} & N_2 & \xrightarrow{g_2} & N_3 & \xrightarrow{g_3} & N_4 & \xrightarrow{g_4} & N_5 \end{array} \tag{8.2.3}$$

be a commutative diagram with exact rows and isomorphisms φ_i, $i = 1,2,4,5$. Then φ_3 is also an isomorphism.

Proof. Let $x \in M_3$ belong to the kernel of φ_3, i.e. let $\varphi_3 x = 0$. Then $\varphi_4 f_3 x = g_3 \varphi_3 x = 0$ and thus, since φ_4 is an isomorphism, $f_3 x = 0$, i.e. $x \in \operatorname{Ker} f_3$.

Now, in view of the exactness at M_3, $\operatorname{Ker} f_3 = \operatorname{Im} f_2$. This means that there is an element $y \in M_2$ such that $x = f_2 y$. In addition, $g_2\varphi_2 y = \varphi_3 f_2 y = \varphi_3 x = 0$. Thus, $\varphi_2 y \in \operatorname{Ker} g_2 = \operatorname{Im} g_1$, i.e. $\varphi_2 y = g_1 z$ for some $z \in N_1$. However, φ_1 is also an isomorphism and therefore $z = \varphi_1 u$ with $u \in M_1$ and $\varphi_2 f_1 u = g_1 \varphi_1 u = g_1 z = \varphi_2 y$; from here $f_1 u = y$ and $x = f_2 y = f_2 f_1 u = 0$ (exactness at M_2). Consequently, $\operatorname{Ker} \varphi_3 = 0$ and so φ_3 is a monomorphism.

Now, choose an element $a \in N_3$. Since φ_4 is an isomorphism, there is $b \in M_4$ such that $\varphi_4 b = g_3 a$. Moreover, $\varphi_5 f_4 b = g_4 \varphi_4 b = g_4 g_3 a = 0$ and thus $f_4 b = 0$ and $b \in \operatorname{Ker} f_4 = \operatorname{Im} f_3$. Hence $b = f_3 c$, where $c \in M_3$. Put $\bar{a} = a - \varphi_3 c$. Since $g_3 \varphi_3 c = \varphi_4 f_3 c = \varphi_4 b = g_3 a$, $g_3 \bar{a} = 0$ and $\bar{a} \in \operatorname{Ker} g_3 = \operatorname{Im} g_2$. Thus $\bar{a} = g_2 d$ for some $d \in N_2$. Furthermore, $d = \varphi_2 \bar{c}$ for $\bar{c} \in M_2$. Then $\varphi_3 f_2 \bar{c} = g_2 \varphi_2 \bar{c} = g_2 d = \bar{a}$ and we get $a = \bar{a} + \varphi_3 c = \varphi_3(f_2 \bar{c} + c) \in \operatorname{Im} \varphi_3$. It follows that φ_3 is an epimorphism, and therefore an isomorphism. □

The following are the most common applications of the Five lemma.

1) Given a commutative diagram

$$\begin{array}{ccccccccc}
0 & \longrightarrow & M_1 & \longrightarrow & M_2 & \longrightarrow & M_3 & \longrightarrow & 0 \\
 & & {\scriptstyle\varphi_1}\big\downarrow & & {\scriptstyle\varphi_2}\big\downarrow & & {\scriptstyle\varphi_3}\big\downarrow & & \\
0 & \longrightarrow & N_1 & \longrightarrow & N_2 & \longrightarrow & N_3 & \longrightarrow & 0
\end{array}$$

with exact rows and isomorphisms φ_1 and φ_3, then φ_2 is also an isomorphism. This follows immediately from Lemma 8.2.1 if we complete the diagram by the (zero) homomorphisms of the zero modules to the form of diagram (8.2.3).

2) Given a commutative diagram

$$\begin{array}{ccccccc}
M_1 & \longrightarrow & M_2 & \longrightarrow & M_3 & \longrightarrow & 0 \\
{\scriptstyle\varphi_1}\big\downarrow & & {\scriptstyle\varphi_2}\big\downarrow & & {\scriptstyle\varphi_3}\big\downarrow & & \\
N_1 & \longrightarrow & N_2 & \longrightarrow & N_3 & \longrightarrow & 0
\end{array}$$

with exact rows and isomorphisms φ_1 and φ_2, then φ_3 is an isomorphism, as well. Clearly, one can complete this diagram to the diagram

$$\begin{array}{ccccccccc}
M_1 & \longrightarrow & M_2 & \longrightarrow & M_3 & \longrightarrow & 0 & \longrightarrow & 0 \\
{\scriptstyle\varphi_1}\big\downarrow & & {\scriptstyle\varphi_2}\big\downarrow & & {\scriptstyle\varphi_3}\big\downarrow & & \big\downarrow & & \big\downarrow \\
N_1 & \longrightarrow & N_2 & \longrightarrow & N_3 & \longrightarrow & 0 & \longrightarrow & 0
\end{array}$$

by the zero homomorphism.

3) Similarly, given a commutative diagram

$$\begin{array}{ccccccc}
0 & \longrightarrow & M_1 & \longrightarrow & M_2 & \longrightarrow & M_3 \\
 & & {\scriptstyle\varphi_1}\big\downarrow & & {\scriptstyle\varphi_2}\big\downarrow & & {\scriptstyle\varphi_3}\big\downarrow \\
0 & \longrightarrow & N_1 & \longrightarrow & N_2 & \longrightarrow & N_3
\end{array}$$

with exact rows and isomorphisms φ_2 and φ_3, then φ_1 is an isomorphism, as well.

8.3 Tensor Products

In this section we shall introduce a new important functor on a module category, namely the tensor product of modules.

Let A be an algebra, M a right and N a left A-module. In the vector space $M \otimes N$, consider the subspace T generated by all elements of the form $xa \otimes y - x \otimes ay$, where $x \in M$, $y \in N$ and $a \in A$. The factor space $(M \otimes N)/T$ is called the *tensor product* of the modules M and N over the algebra A and is denoted by $M \otimes_A N$. Denote by π the canonical projection $M \otimes N \to M \otimes_A N$. The composition of π with the bilinear map $\otimes : M \times N \to M \otimes N$ (assigning to (x, y) the element $x \otimes y$) results in a bilinear map $\otimes_A : M \times N \to M \otimes_A N$. The image of (x, y) is $x \otimes_A y = \pi(x \otimes y)$.

The map $\otimes_A$ possesses an additional property which will be called the "inner A-bilinearity": $xa \otimes_A y = x \otimes_A ay$. Moreover, since $\otimes$ is a universal bilinear map, $\otimes_A$ is a universal inner A-bilinear map in the sense of the following statement.

Theorem 8.3.1. *Let $F : M \times N \to V$ be an inner A-bilinear map into a vector space V. Then there is a unique linear map $f : M \otimes_A N \to V$ such that $F(x, y) = f(x \otimes_A y)$ for any $x \in M$ and $y \in N$.*

Proof. Since F is bilinear, there is a unique linear map $\varphi : M \otimes N \to V$ such that $F(x, y) = \varphi(x \otimes y)$ for any $x \in M$ and $y \in N$ (Theorem 4.2.1). However, $\varphi(xa \otimes y - x \otimes ay) = \varphi(xa \otimes y) - \varphi(x \otimes ay) = F(xa, y) - F(x, ay) = 0$ in view of inner A-bilinearity of F. Therefore $T \subset \operatorname{Ker} \varphi$ and φ induces a unique map $f : M \otimes_A N \to V$ such that $\varphi = f\pi$, and thus $f(x \otimes_A y) = f\pi(x \otimes y) = \varphi(x \otimes y) = F(x, y)$. □

Obviously, a universal inner A-bilinear map is unique up to a canonical isomorphism. The universality permits rather easily to establish basic properties of tensor products.

Proposition 8.3.2. *For every pair of A-module homomorphisms $f : M \to M'$ and $g : N \to N'$ there is a unique linear map $f \otimes_A g : M \otimes_A N \to M' \otimes_A N'$ such that $(f \otimes_A g)(x \otimes_A y) = fx \otimes_A gy$. If $f' : M' \to M''$ and $g' : N' \to N''$ is another pair of homomorphisms, then $(f' \otimes_A g')(f \otimes_A g) = f'f \otimes_A g'g$.*

Proof. Consider the map $F : M \times N \to M' \otimes_A N'$ such that $F(x, y) = fx \otimes_A gy$. It is easy to see that F is an inner A-bilinear map. Therefore, there is a unique map $f \otimes_A g : M \otimes_A N \to M' \otimes_A N'$ such that $(f \otimes_A g)(x \otimes_A y) = F(x, y) = fx \otimes_A gy$. The second statement is trivial. □

The preceding property allows us to consider tensor product as a functor on a module category. More precisely, let us fix a left A-module N and construct the functor $- \otimes_A N : \text{mod-}A \to \text{Vect}$ as follows. Assign to every right A-module M the vector space $M \otimes_A N$ and to every homomorphism

$f : M \to M'$ the linear transformation $f \otimes_A 1 : M \otimes_A N \to M' \otimes_A N$. Proposition 8.3.2 shows that the axioms for a functor are satisfied. Similarly, we may construct the functor $M \otimes_A -$: A-mod $\to$ Vect (for a fixed right A-module M).

The fact that tensor product is a functor enables us to turn sometimes $M \otimes_A N$ again into a module. Let, for instance, N be an A-B-bimodule (or, as one often says, consider a situation M_A, ${}_AN_B$). Then every element $b \in B$ induces an A-module homomorphism $N \to N$ assigning to every $y \in N$ the element yb, and thus a vector space transformation $M \otimes_A N \to M \otimes_A N$ assigning to every $x \otimes_A y$ the element $x \otimes_A yb$. Clearly, in this way $M \otimes_A N$ turns into a B-module. A similar situation ${}_BM_A$, ${}_AN$ (i. e. M is a B-A-bimodule and N a left A-module) defines on $M \otimes_A N$ a left B-module structure by $b(x \otimes_A y) = bx \otimes_A y$. Finally, in a situation ${}_BM_A$, ${}_AN_C$, the tensor product $M \otimes_A N$ becomes a B-C-bimodule. This allows to iterate the tensor product operation and define (in an appropriate situation) a product of three or more modules. As the following statement shows, the order in which the products are taken is immaterial.

Proposition 8.3.3. *In a situation* M_A, ${}_AN_B$, ${}_BL$ *there is a canonical isomorphism*

$$(M \otimes_A N) \otimes_B L \xrightarrow{\sim} M \otimes_A (N \otimes_B L)$$

assigning to $(x \otimes_A y) \otimes_B z$ *the element* $x \otimes_A (y \otimes_B z)$.

Proof. Fix an element $z \in L$ and define the map $F_z : M \times N \to M \otimes_A (N \otimes_B L)$ by $F_z(x, y) = x \otimes_A (y \otimes_B z)$. Clearly, this is an inner A-bilinear map and therefore there is a unique linear transformation $f_z : M \otimes_A N \to M \otimes_A (N \otimes_B L)$ assigning to $x \otimes_A y$ the element $x \otimes_A (y \otimes_B z)$. Varying z, we obtain an inner B-linear map $F : (M \otimes_A N) \times L \to M \otimes_A (N \otimes_B L)$ assigning to a pair $(x \otimes_A y, z)$ the element $x \otimes_A (y \otimes_B z)$. In turn, F defines a unique linear transformation $f : (M \otimes_A N) \otimes_B L \to M \otimes_A (N \otimes_B L)$ such that $f\big((x \otimes_A y) \otimes_B z\big) = x \otimes_A (y \otimes_B z)$. In a similar manner, we can construct a linear transformation $g : M \otimes_A (N \otimes_B L) \to (M \otimes_A N) \otimes_B L$ such that $g\big(x \otimes_A (y \otimes_B z)\big) = (x \otimes_A y) \otimes_B z$. Since all possible elements of the form $x \otimes_A (y \otimes_B z)$ (respectively, $(x \otimes_A y) \otimes_B z$) generate the space $M \otimes_A (N \otimes_B L)$ (respectively, $(M \otimes_A N) \otimes_B L$), f is an inverse of g, as required. □

We can see immediately that in a situation ${}_CM_A$, ${}_AN_B$, ${}_BL_D$ the above isomorphism is, in fact, a C-D-bimodule isomorphism.

We can also establish a relationship between the functors $\otimes$ and Hom.

For instance, observe that in a situation ${}_BM_A$, N_A the space $\mathrm{Hom}_A(M, N)$ can be turned into a B-module by $(fb)m = f(bm)$. Similarly, in a situation M_A, ${}_BN_A$, $\mathrm{Hom}_A(M, N)$ becomes a left B-module and in a situation ${}_BM_A$, ${}_CN_A$, it becomes a C-B-bimodule. Moreover, we have the following "adjoint formula".

Proposition 8.3.4 (Adjoint isomorphism). *In a situation M_A, ${}_AN_B$, L_B, there is a canonical isomorphism*

$$\mathrm{Hom}_B(M \otimes_A N, L) \xrightarrow{\sim} \mathrm{Hom}_A(M, \mathrm{Hom}_B(N, L)),$$

assigning to a homomorphism $\varphi : M \otimes_A N \to L$ the homomorphism $\bar{\varphi} : M \to \mathrm{Hom}_B(N, L)$ such that $\bar{\varphi}(x)(y) = \varphi(x \otimes_A y)$.

Proof. The fact that $\bar{\varphi}$ is an A-module homomorphism is trivial. We shall construct an inverse map. Let $\psi : M \to \mathrm{Hom}_B(N, L)$ be an A-module homomorphism. Then, as we can see immediately, the map $M \times N \to L$ sending (x, y) into $\psi(x)(y)$ is an inner A-bilinear map, and therefore defines a unique map $\bar{\psi} : M \otimes_A N \to L$ such that $\bar{\psi}(x \otimes_A y) = \psi(x)(y)$. Now, $\bar{\psi}$ is clearly a B-module homomorphism and the constructed maps $\mathrm{Hom}_B(M \otimes_A N, L) \rightleftarrows \mathrm{Hom}_A(M, \mathrm{Hom}_B(N, L))$ are mutually inverse isomorphisms. □

An important property of the functors $\otimes$ and Hom is their "exactness".

Proposition 8.3.5. 1) *A sequence of A-modules*

$$0 \longrightarrow M_1 \xrightarrow{f} M_2 \xrightarrow{g} M_3 \tag{8.3.1}$$

is exact if and only if, for any A-module N, the sequence

$$0 \longrightarrow \mathrm{Hom}_A(N, M_1) \xrightarrow{h_N(f)} \mathrm{Hom}_A(N, M_2) \xrightarrow{h_N(g)} \mathrm{Hom}_A(N, M_3) \tag{8.3.2}$$

is exact.

2) *A sequence of A-modules*

$$M_1 \xrightarrow{f} M_2 \xrightarrow{g} M_3 \longrightarrow 0 \tag{8.3.1$'$}$$

is exact if and only if, for any A-module N, the sequence

$$0 \longrightarrow \mathrm{Hom}_A(M_3, N) \xrightarrow{h_N^\circ(g)} \mathrm{Hom}_A(M_2, N) \xrightarrow{h_N^\circ(f)} \mathrm{Hom}_A(M_1, N) \tag{8.3.2$'$}$$

is exact.

Proof. Assume that the sequence (8.3.1) is exact. Then f is a monomorphism, and if $h_N(f)(\varphi) = f\varphi = 0$, where $\varphi : N \to M_1$, then also $\varphi = 0$. Thus, $h_N(f)$ is a monomorphism and the sequence (8.3.2) is exact at $\mathrm{Hom}_A(N, M_1)$. Since $\mathrm{Im}\, f = \mathrm{Ker}\, g$, $gf = 0$, and therefore $h_N(g)h_N(f) = h_N(gf) = 0$. Consequently, $\mathrm{Im}\, h_N(f) \subset \mathrm{Ker}\, h_N(g)$. Now, let $\varphi \in \mathrm{Ker}\, h_N(g)$, where $\varphi : N \to M_2$. In other words, $h_N(g)(\varphi) = g\varphi = 0$. Then $\mathrm{Im}\, \varphi \subset \mathrm{Ker}\, g = \mathrm{Im}\, f$. In view of the isomorphism $M_1 \simeq \mathrm{Im}\, f$, φ can be written as a composition $f\psi$, where $\psi : N \to M_1$. Hence, $\varphi = h_N(f)(\psi) \in \mathrm{Im}\, h_N(f)$ and the sequence (8.3.2) is exact at $\mathrm{Hom}_A(N, M_2)$.

Conversely, let the sequence (8.3.2) be exact for every N. Taking $N = \mathrm{Ker}\, f$, we see that the map $\mathrm{Hom}_A(\mathrm{Ker}\, f, M_1) \to \mathrm{Hom}_A(\mathrm{Ker}\, f, M_2)$ is a monomorphism. Thus, if φ is the embedding of $\operatorname{Ker} f$ into M_1, $f\varphi = 0$ and $\varphi = 0$. Therefore $\operatorname{Ker} f = 0$ and f is a monomorphism.

Now, let $N = M_1$. Then $f = f1_N = h_N(f)(1_N) \in \operatorname{Im} h_N(f) = \operatorname{Ker} h_N(g)$. Thus $gf = h_N(g)(f) = 0$ and $\operatorname{Im} f \subset \operatorname{Ker} g$. Finally, taking $N = \operatorname{Ker} g$ and denoting by φ the embedding of N in M_2, $h_N(g)(\varphi) = g\varphi = 0$. Thus $\varphi \in \operatorname{Ker} h_N(g) = \operatorname{Im} h_N(f)$. Therefore $\varphi = f\psi$, $\operatorname{Ker} g = \operatorname{Im}\varphi \subset \operatorname{Im} f$, and the sequence (8.3.1) is exact.

The other statement, 2), is proved similarly. □

Using the adjoint isomorphism, we can carry over exactness properties to tensor products.

Proposition 8.3.6. *If a sequence of right A-modules*

$$M_1 \to M_2 \to M_3 \to 0 \tag{8.3.3}$$

is exact, then, for any A-B-bimodule N, the sequence of B-modules

$$M_1 \otimes_A N \to M_2 \otimes_A N \to M_3 \otimes_A N \to 0 \tag{8.3.4}$$

is also exact.

Proof. In view of Proposition 8.3.5, we need to verify the exactness of the sequence

$$0 \to \operatorname{Hom}_B(M_3 \otimes_A N, L) \to \operatorname{Hom}_B(M_2 \otimes_A N, L) \to \operatorname{Hom}_B(M_1 \otimes_A N, L)$$

for any B-module L. By Proposition 8.3.4, the latter sequence can be rewritten as

$$0 \to \operatorname{Hom}_A\big(M_3, \operatorname{Hom}_B(N, L)\big) \to \operatorname{Hom}_A\big(M_2, \operatorname{Hom}_B(N, L)\big) \to \\ \operatorname{Hom}_A\big(M_1, \operatorname{Hom}_B(N, L)\big)$$

and thus its exactness follows immediately from Proposition 8.3.5. □

The above properties are often expressed by saying that the functor Hom is *left exact* and that $\otimes$ is *right exact*, or more precisely, that h_N and h_N^o are left exact and $- \otimes_A N$ is right exact. Of course, the functor $M \otimes_A -$ is also right exact (a proof is similar to the one of Proposition 8.3.6).

In conclusion, we record the following simple fact.

Proposition 8.3.7. *The map $M \to M \otimes_A A$ assigning to every $m \in M$ the element $m \otimes_A 1$ is an isomorphism of right A-modules. The map $N \to A \otimes_A N$ assigning to every $n \in N$ the element $1 \otimes_A n$ is an isomorphism of left A-modules.*

Proof. It is sufficient to observe that the map $M \times A \to M$, sending (m, a) into ma, is evidently an inner bilinear map and that the induced map $M \otimes_A A \to M$ is a homomorphism which is an inverse of the map $M \to M \otimes_A A$ under consideration. □

8.4 The Morita Theorem

In this section, we shall establish which algebras A and B have the property that their module categories mod-A and mod-B are "equally structured". To do that, we have to define first the meaning of being "equally structured", that is to say, to define which categories will be considered equal. To try to define an isomorphism of categories $\mathcal{C}$ and $\mathcal{C}'$ as a functor $F : \mathcal{C} \to \mathcal{C}'$ which possesses an inverse functor $G : \mathcal{C}' \to \mathcal{C}$, i. e. such that $GF = 1_{\mathcal{C}}$ and $FG = 1_{\mathcal{C}'}$, turns out to be not satisfactory. First of all, functors which appear naturally do not possess, as a rule, this property and secondly, some categories which are intuitively "equal" would not be isomorphic according to this definition. Take, for example, the category Mat of matrices and the category Vect of vector spaces. The situation is quite clear: The category of matrices describes the vector spaces "up to an isomorphism" whereas there are many "isomorphic copies" of each space in the category Vect. Thus, no bijective correspondence between these two categories is possible.

The following approach utilizing the concept of an isomorphism of functors rather than their equality, appears to be more natural. We are going to introduce rigorous definitions.

Let F and G be two functors from a category $\mathcal{C}$ to a category $\mathcal{C}'$. A *morphism* from the functor F to the functor G is a map φ which assigns to every object $X \in \operatorname{Ob}\mathcal{C}$ a morphism $\varphi(X) : F(X) \to G(X)$ (of the category $\mathcal{C}'$!) in such a way that, for every morphism $f : X \to Y$ of the category $\mathcal{C}$, the following diagram commutes:

$$\begin{array}{ccc} F(X) & \xrightarrow{\varphi(X)} & G(X) \\ {\scriptstyle F(f)}\downarrow & & \downarrow{\scriptstyle G(f)} \\ F(Y) & \xrightarrow{\varphi(Y)} & G(Y) \end{array} .$$

We shall write $\varphi : F \to G$.

If $H : \mathcal{C} \to \mathcal{C}'$ is another functor and $\psi : G \to H$ is a functor morphism, then the composition $\psi\varphi : F \to H$ is defined by setting $(\psi\varphi)(X) = \psi(X)\varphi(X)$. It is easy to see that the set of functors from a category $\mathcal{C}$ to a category $\mathcal{C}'$ together with their morphisms forms, with respect to this definition, a category: the *functor category* $\operatorname{Func}(\mathcal{C},\mathcal{C}')$. In addition, a morphism $\varphi : F \to G$ is an isomorphism in this category if and only if, for every object $X \in \operatorname{Ob}\mathcal{C}$, the morphism $\varphi(X)$ is an isomorphism. In this case, we say that φ is an isomorphism of the functors and write $\varphi : F \xrightarrow{\sim} G$, or $F \simeq G$.

It is not difficult to see that the isomorphisms constructed in Proposition 8.3.3 and 8.3.4 are, in fact, isomorphisms of the corresponding functors.

An *equivalence of the categories* $\mathcal{C}$ and $\mathcal{C}'$ is a pair of functors $F : \mathcal{C} \to \mathcal{C}'$ and $G : \mathcal{C}' \to \mathcal{C}$ such that $GF \simeq 1_{\mathcal{C}}$ and $FG \simeq 1_{\mathcal{C}'}$. If there is such an equivalence, the categories $\mathcal{C}$ and $\mathcal{C}'$ are called *equivalent*.

In the sequel, we shall find useful the following obvious properties of equivalent categories.

Proposition 8.4.1. *If a pair of functors* $F : \mathcal{C} \to \mathcal{C}'$ *and* $G : \mathcal{C}' \to \mathcal{C}$ *is an equivalence of categories, then*

1) *the correspondence* $\mathrm{Hom}_{\mathcal{C}}(X,Y) \to \mathrm{Hom}_{\mathcal{C}'}(F(X),F(Y))$ *mapping* f *to* $F(f)$ *is bijective;*
1′) *the correspondence* $\mathrm{Hom}_{\mathcal{C}'}(U,V) \to \mathrm{Hom}_{\mathcal{C}}(G(U),G(V))$ *mapping* g *to* $G(g)$ *is bijective;*
2) *a morphism* $f \in \mathrm{Mor}\,\mathcal{C}$ *is an isomorphism if and only if* $F(f)$ *is an isomorphism;*
2′) *a morphism* $g \in \mathrm{Mor}\,\mathcal{C}'$ *is an isomorphism if and only if* $G(g)$ *is an isomorphism;*
3) *every object* $X \in \mathrm{Ob}\,\mathcal{C}$ *is isomorphic to an object of the form* $G(U)$, *where* $U \in \mathrm{Ob}\,\mathcal{C}'$*;*
3′) *every object* $U \in \mathrm{Ob}\,\mathcal{C}'$ *is isomorphic to an object of the form* $F(X)$, *where* $X \in \mathrm{Ob}\,\mathcal{C}$.

Proof. We shall prove the assertions 1) and 1′), leaving the other statements as an easy exercise to the reader.

Let $f : X \to Y$ be a morphism of the category $\mathcal{C}$. Denote by φ an isomorphism of the functors $GF \simeq 1_{\mathcal{C}}$ and consider the commutative diagram

$$\begin{array}{ccc} GF(X) & \xrightarrow{\varphi(X)} & X \\ {\scriptstyle GF(f)}\downarrow & & \downarrow{\scriptstyle f} \\ GF(Y) & \xrightarrow{\varphi(Y)} & Y\ . \end{array}$$

Since $\varphi(X)$ is an isomorphism, $f = \varphi(Y)GF(f)\varphi^{-1}(X)$. It follows that $F(f) = F(f')$ implies $f = f'$, and thus the map from $\mathrm{Hom}_{\mathcal{C}}(X,Y)$ to $\mathrm{Hom}_{\mathcal{C}'}(F(X),F(Y))$ is injective. In a similar way, the map $\mathrm{Hom}_{\mathcal{C}'}(U,V) \to \mathrm{Hom}_{\mathcal{C}}(G(U),G(V))$ is injective.

Finally, consider an arbitrary monomorphism $g : F(X) \to F(Y)$. Let $f = \varphi(Y)G(g)\varphi^{-1}(X)$ and $g' = F(f)$. Then, as before, $f = \varphi(Y)G(g')\varphi^{-1}(X)$, and thus $G(g) = G(g')$. Consequently, $g = g' = F(f)$, and thus the map $\mathrm{Hom}_{\mathcal{C}}(X,Y) \to \mathrm{Hom}_{\mathcal{C}'}(F(X),F(Y))$ is bijective. □

Now, consider module categories. Let a pair of functors F, G be an equivalence of the categories mod-A and mod-B. Combining Proposition 8.4.1 with the exactness criterion (Proposition 8.3.5), we obtain the following proposition.

Proposition 8.4.2. 1) *A sequence of* A*-modules*

$$0 \longrightarrow M_1 \xrightarrow{f} M_2 \xrightarrow{g} M_3$$

is exact if and only if the sequence of B*-modules*

$$0 \longrightarrow F(M_1) \xrightarrow{F(f)} F(M_2) \xrightarrow{F(g)} F(M_3)$$

is exact.

2) *A sequence of A-modules*

$$M_1 \xrightarrow{f} M_2 \xrightarrow{g} M_3 \longrightarrow 0$$

is exact if and only if the sequence of B-modules

$$F(M_1) \xrightarrow{F(f)} F(M_2) \xrightarrow{F(g)} F(M_3) \longrightarrow 0$$

is exact.

3) *An A-module P is projective if and only if the B-module F(P) is projective.*

Proof. Here we shall prove 3), leaving the proofs of 1) and 2) to the reader. We shall use the "diagrammatical" definition of projectivity (see Sect. 8.2, Example 4). Assume that $F(P)$ is projective. Let

$$\begin{array}{ccccc} & & P & & \\ & & \downarrow{\scriptstyle f} & & \\ M & \xrightarrow{\pi} & N & \longrightarrow & 0 \end{array}$$

be a diagram of A-modules with exact row. Applying the functor F, we obtain the following diagram of B-modules

$$\begin{array}{ccccc} & & F(P) & & \\ & & \downarrow{\scriptstyle F(f)} & & \\ F(M) & \xrightarrow{F(\pi)} & F(N) & \longrightarrow & 0 \end{array}$$

with exact row (in view of statement 2)). It can be completed to a commutative diagram

$$\begin{array}{ccccc} & & F(P) & & \\ & {\scriptstyle g}\swarrow & \downarrow{\scriptstyle F(f)} & & \\ F(M) & \underset{F(\pi)}{\longrightarrow} & F(N) & \longrightarrow & 0, \end{array}$$

in which, by Proposition 8.4.1, $g = F(\bar{f})$ for some morphism $\bar{f} : P \to M$. But then $F(\pi\bar{f}) = F(\pi)F(\bar{f}) = F(\pi)g = F(f)$ and hence $\pi\bar{f} = f$. Consequently, the diagram

$$\begin{array}{ccccc} & & P & & \\ & {\scriptstyle \bar{f}}\swarrow & \downarrow{\scriptstyle f} & & \\ M & \xrightarrow{\pi} & N & \longrightarrow & 0 \end{array}$$

is commutative and P is projective.

Conversely, if P is projective, then the isomorphic module $GF(P)$ is projective, and by what we have just proved, $F(P)$ is projective. □

Corollary 8.4.3. *Let F, G be an equivalence of the categories* mod-A *and* mod-B *and $P = G(B)$. Then P is a projective A-module, $E_A(P) \simeq B$ and, for any A-module M, there is an epimorphism $nP \to M$ (for a suitable n).*

We say in this case that P is a *generating A-module* (a generator).

Proof. The fact that P is projective follows from the projectivity of the B-module B and Proposition 8.4.2. Moreover, by Proposition 8.4.1, $E_A(P) \simeq E_B(B) \simeq B$. Finally, every A-module M is isomorphic to a module of the form $G(N)$ for a suitable B-module N and there is an epimorphism $nB \to N$, which induces an epimorphism $G(nB) \simeq nP \to G(N) \simeq M$, as required. □

Now, let P be a projective A-module and $B = E_A(P)$. We shall say that B is a *minor* of the algebra A. We can define two functors F : mod-$A \to$ mod-B and G : mod-$B \to$ mod-A by $F(M) = \mathrm{Hom}_A(P, M)$, and $G(N) = N \otimes_B P$ (P is considered as a left B-module). Moreover, we can define also functor morphisms $\varphi : 1_{\text{mod-}B} \to FG$ and $\psi : GF \to 1_{\text{mod-}A}$ in the following way. For every B-module N, define $\varphi(N)$ to be the homomorphism $N \to \mathrm{Hom}_A(P, N \otimes_B P)$, mapping every element $x \in N$ into the A-homomorphism $\bar{x} : P \to N \otimes_B P$ such that $\bar{x}(p) = x \otimes_B p$. For every A-module M, define $\psi(M)$ to be the homomorphism $\mathrm{Hom}_A(P, M) \otimes_B P \to M$, mapping $f \otimes_B p$ (where $f \in \mathrm{Hom}_A(P, M)$ and $p \in P$) into $f(p) \in M$. It is easy to verify that φ and ψ are, indeed, functor morphisms.

Observe that in general, not every A-module is isomorphic to a module of the form $G(N)$. Indeed, there is always an epimorphism $f : nB \to N$. Let $N' = \mathrm{Ker}\, f$ and let g be an epimorphism $mB \to N'$. Then the sequence

$$mB \xrightarrow{g} nB \xrightarrow{f} N \longrightarrow 0$$

is exact. Consequently, the sequence

$$G(mB) \xrightarrow{G(g)} G(nB) \xrightarrow{G(f)} G(N) \longrightarrow 0$$

is exact. However, $G(nB) \simeq nP$ and $G(mB) \simeq mP$, and thus, if $M \simeq G(N)$, then there is an exact sequence of the form

$$mP \to nP \to M \to 0\,. \tag{8.4.1}$$

Therefore, it is natural to consider the category mod-P of all those A-modules for which there is an exact sequence of the form (8.4.1) together with all possible homomorphisms of such modules.

Theorem 8.4.4. *A pair of the functors $F = \mathrm{Hom}_A(P, -)$ and $G = - \otimes_B P$ is an equivalence of the categories* mod-P *and* mod-B.

Proof. We shall show that $\varphi : 1_{\text{mod-}B} \xrightarrow{\sim} FG$ and $\psi : GF \xrightarrow{\sim} 1_{\text{mod-}P}$. Indeed, $\varphi(B)$ is the natural isomorphism $B = \mathrm{Hom}_A(P, P) \xrightarrow{\sim} \mathrm{Hom}_A(P, B \otimes_B P) =$

$FG(B)$. Thus, also $\varphi(nB)$ is clearly an isomorphism. Furthermore, as we have seen above, for any B-module N, there is an exact sequence

$$mB \xrightarrow{g} nB \xrightarrow{f} N \longrightarrow 0.$$

Apply the functor FG. Since G is a right exact functor and F is exact (because of projectivity of P), we obtain the following commutative diagram with exact rows:

$$\begin{array}{ccccccc} mB & \xrightarrow{g} & nB & \xrightarrow{f} & N & \longrightarrow & 0 \\ \downarrow{\scriptstyle \varphi(mB)} & & \downarrow{\scriptstyle \varphi(nB)} & & \downarrow{\scriptstyle \varphi(N)} & & \\ FG(mB) & \xrightarrow{FG(g)} & FG(nB) & \xrightarrow{FG(f)} & FG(N) & \longrightarrow & 0. \end{array}$$

Now, since $\varphi(mB)$ and $\varphi(nB)$ are isomorphisms, it follows from the Five lemma (Lemma 8.2.1) that $\varphi(N)$ is also an isomorphism.

In a similar manner one can show that, for an module $M \in$ mod-P, the homomorphism $\psi(M)$ is an isomorphism. □

The following result follows from Theorem 8.4.4 and Corollary 8.4.3.

Theorem 8.4.5 (Morita). *The categories* mod-A *and* mod-B *are equivalent if and only if there is a projective generating A-module (progenerator) P such that $E_A(P) = B$. In this case, an equivalence of the categories is realized by a pair of functors $F = \mathrm{Hom}_A(P, -)$ and $G = - \otimes_B P$.*

The equivalence of categories has the following simple interpretation. Let $A \simeq k_1P_1 \oplus k_2P_2 \oplus \ldots \oplus k_tP_t$ be a decomposition of the regular A-module into a direct sum of principal A-modules subject to $P_i \not\simeq P_j$ for $i \neq j$. If P is a projective generating A-module, then there exists an epimorphism $nP \to A$, and thus $nP \simeq A \oplus M$. Consequently, by the Krull-Schmidt theorem, P has to contain all modules $P_1, P_2, \ldots, P_t$ as direct summands. Since every projective module is a direct sum of principal ones, we deduce that projective generators are those projective modules which contain every principal A-module as a direct summand. Taking into account Theorem 3.5.6, we can see that the module categories mod-A and mod-B are equivalent if and only if the algebras A and B are isotypic, that is to say, if and only if their basic algebras are isomorphic. In particular, the Morita theorem allows us to restrict the study of A-modules to the case when A is a basic algebra. Moreover, the results of Sect. 3.5 make it possible to oversee all algebras whose module categories are equivalent.

8.5. Tensor Algebras and Hereditary Algebras

In this section we are going to present a generalization of the construction of a "path algebra" (see Sect. 3.6). Such a generalization is a tensor algebra of a bimodule.

Let B be an algebra and V a bimodule over B. Then $V^{\otimes 2} = V \otimes_B V$ is again a B-bimodule. Iterating this procedure, we can construct B-bimodules $V^{\otimes k}$ for all $k \geq 2$ by putting $V^{\otimes k} = V^{\otimes k-1} \otimes_B V$. It is convenient to set $V^{\otimes 0} = B$ and $V^{\otimes 1} = V$. The associativity of tensor multiplication implies that $V^{\otimes k} \otimes_B V^{\otimes m} \simeq V^{\otimes (k+m)}$. In what follows, we shall identify $V^{\otimes k} \otimes_B V^{\otimes m}$ and $V^{\otimes (k+m)}$.

Now, consider the direct sum $T(V) = \bigoplus_{k=0}^{\infty} V^{\otimes k}$. Elements of $T(V)$ are finite sums $\sum_k t_k$, where $t_k \in V^{\otimes k}$. The isomorphism constructed above allows us to define, for any elements $t_k \in V^{\otimes k}$ and $t_m \in V^{\otimes m}$, the product by setting $t_k t_m = t_k \otimes_B t_m \in V^{\otimes (k+m)}$. This multiplication can be extended by linearity to the entire $T(V)$, and $T(V)$ becomes in this way an algebra (in general, infinite dimensional). It is called the *tensor algebra of the bimodule* V.

By construction, $T(V)$ contains the subalgebra $B = V^{\otimes 0}$ and the B-bimodule $V = V^{\otimes 1}$. Moreover, as the following theorem shows, $T(V)$ is a universal algebra with this property.

Theorem 8.5.1. *Let $\varphi : B \to A$ be an algebra homomorphism and $f : V \to A$ a homomorphism of B-bimodules.*[17] *Then there exists a unique algebra homomorphism $F : T(V) \to A$ such that the restriction of F on B and V coincides with φ and f, respectively.*

Proof. The homomorphism f induces B-bimodule homomorphisms $f^{\otimes k} : V^{\otimes k} \to A^{\otimes k}$. Moreover, multiplication in the algebra A induces a bimodule homomorphism $A^{\otimes k} \to A$ such that the image of $a_1 \otimes_B a_2 \otimes_B \ldots \otimes_B a_k$ is $a_1 a_2 \ldots a_k$. Thus, we obtain a family of homomorphisms $f_k : V^{\otimes k} \to A$ such that $f_k(v_1 \otimes_B v_2 \otimes_B \ldots \otimes_B v_k) = f(v_1)f(v_2)\ldots f(v_k)$. Evidently, we obtain in this way an algebra homomorphism $F : T(V) \to A$ (we put $f_0 = \varphi$ and $f_1 = f$). Moreover, F is unique; this follows immediately from the fact that $T(V)$ is generated (as an algebra) by the elements of B and V. □

As usual, the universal property which is formulated in Theorem 8.5.1, determines the algebra $T(V)$ up to an isomorphism.

There is a distinguished ideal in the algebra $T(V)$, namely, $J = J(V) = \bigoplus_{k=1}^{\infty} V^{\otimes k}$. We shall call it the *fundamental ideal* of the tensor algebra $T(V)$.

The ideals $I \subset T(V)$ such that $J^2 \supset I \supset J^k$ for some k, will be called *admissible* ideals.

The most important case in our study will be that of a semisimple (and also separable) algebra B. In this case, it turns out that tensor algebras of B-modules play the role of "universal covers", in analogy to what happens for path algebras in the split case (see Theorem 3.6.6).

Theorem 8.5.2. *Let A be a finite dimensional algebra, $R = \operatorname{rad} A$ and $B = A/R$. Assume that the algebra B is separable. Then the algebra A is isomorphic*

[17] A becomes a B-bimodule by setting $b_1 a b_2 = \varphi(b_1) a \varphi(b_2)$ for any $a \in A$, $b_1, b_2 \in B$.

to a quotient algebra of the algebra $T(V)$ by an admissible ideal, where $V = R/R^2$.

Proof. By the Wedderburn-Malcev theorem (Theorem 6.2.1), A has a subalgebra $\bar{A} \simeq B$. Hence, it is possible to define an algebra monomorphism $\varphi : B \to A$ and consider A as a B-module. In addition, $A = \bar{A} \oplus R$ (as a B-module). Since $B \otimes B^o$ is a semisimple algebra (Theorem 6.1.1), every B-bimodule is semisimple. In particular, R contains a B-subbimodule $\bar{R}$ such that $R = \bar{R} \oplus R^2$. Evidently, $\bar{R} \simeq V$ and we may use this isomorphism to define a B-bimodule monomorphism $f : V \to R$.

By Theorem 8.5.1, there is a homomorphism $F : T(V) \to A$ which restricts on B to φ and on V to f. Thus F induces an algebra isomorphism $T(V)/J^2 \simeq A/R^2$. This implies that $I = \operatorname{Ker} F \subset J^2$. On the other hand, $F(J) \subset R$, so $F(J^k) \subset R^k$ and, since R is nilpotent, $F(J^k) = 0$ for some k. As a consequence, $J^k \subset I$ and thus I is an admissible ideal.

Now, observe that $F(B) = \varphi(B) = \bar{A}$ and $F(V) = f(V) = \bar{R}$. Thus, every element $r \in R$ is of the form $r = F(x) + r'$, where $x \in J$ and $r' \in R^2$. It follows readily that, for every element $r \in R^i$, $r = F(x) + r'$ with $x \in J^i$ and $r' \in R^{i+1}$. In view of the nilpotency of the radical, we get the equality $F(J) = R$. Consequently, F is an epimorphism and $A \simeq T(V)/I$. The theorem follows. □

The pair (B, V), where $B = A/R$, $V = R/R^2$, will be called the *type of the algebra* A. If B is a separable algebra, we shall say that A is an *algebra of separable type.* Obviously, the type determines the diagram $\mathcal{D}(A)$ of the algebra A (see Sect. 3.6). We shall therefore call this diagram *the diagram of type* (B, V).

Clearly, if B is a semisimple algebra and V a finite dimensional B-bimodule, then every quotient algebra $T(V)/I$, where I is an admissible ideal, is finite dimensional and it is of type (B, V). In particular, the quotient algebra $T(V)/J^2$ is the least (dimensional) algebra of the given type.

As we have already mentioned, the algebra $T(V)$ is, in general, infinite dimensional. It is not difficult to give a condition under which it will be finite.

Proposition 8.5.3. *Let B be a semisimple algebra and V a finite dimensional B-bimodule. In order that the algebra $T(V)$ be finite dimensional, it is necessary and sufficient that the diagram of type (B, V) has no cycles.*

Proof. Evidently, $T(V)$ is finite dimensional if and only if $V^{\otimes m} = 0$ for some m. Decompose B into a direct product of simple algebras $B = B_1 \times B_2 \times \ldots \times B_n$. Let $1 = e_1 + e_2 + \ldots + e_n$ be the corresponding central decomposition of the identity and $V_{ij} = e_i V e_j$. In a diagram $\mathcal{D}$ of type (B, V), there is an arrow from the point i to the point j if and only if $V_{ij} \neq 0$. Observe that $V = \bigoplus_{i,j} V_{ij}$ (as B-bimodules) and that $V_{ij} \otimes_B V_{k\ell} \neq 0$ if and only if $V_{ij} \neq 0$, $V_{kl} \neq 0$ and $j = k$. Consequently, $e_i V^{\otimes 2} e_j \neq 0$ if and only if there is a path of length 2 from the point i to the point j. Similarly, $e_i V^{\otimes m} e_j \neq 0$ if and only if there is a path of length m from i to j, and this implies our assertion. □

We shall now apply the above technique to describe the hereditary algebras, i.e. the algebras whose right ideals are all projective (see Sect. 3.7).

Theorem 8.5.4. *A finite dimensional hereditary algebra A of separable type (B, V) is isomorphic to $T(V)$. Conversely, if there are no cycles in a diagram of type (B, V), then $T(V)$ is a finite dimensional hereditary algebra.*

Proof. First, verify that $T = T(V)$ is a hereditary algebra. Observe that, since $T(V)/J = B$ is a semisimple algebra and $J^m = 0$ for some m (by Proposition 8.5.3), $J = \operatorname{rad} T$ (by Proposition 3.1.13). Let $1 = e_1 + e_2 + \ldots + e_n$ be a minimal decomposition of the identity of the algebra B (clearly, it will be also a minimal decomposition of the identity of the algebra T).

If $P_i = e_i T = \bigoplus_{k=0}^{\infty} e_i V^{\otimes k}$, then $P_i J = e_i J = \bigoplus_{k=1}^{\infty} e_i V^{\otimes k} = e_i V \otimes_B T$. Since $e_i V \simeq \bigoplus_{j=1}^{n} s_j U_j$, where U_j are simple B-modules, $P_i J \simeq \bigoplus_{j=1}^{n} s_j (U_j \otimes_B T) \simeq \bigoplus_{j=1}^{n} s_j P_j$ is a projective module. Consequently, $J = \bigoplus_{i=1}^{n} e_i J$ is a projective module and T is a hereditary algebra (by Theorem 3.7.1).

In view of Theorem 8.5.2, it remains to prove that if $A = T/I$ is a hereditary algebra with an admissible ideal I, then $I = 0$ (note that, according to Corollary 3.7.3, there are no cycles in the diagram of a hereditary algebra). Write $R = J/I = \operatorname{rad} A$ and denote the principal modules by $\tilde{P}_i = P_i / P_i I$. By induction on k, we are going to prove that $R^k / R^{k+1} \simeq J^k / J^{k+1}$. For $k = 1$, this is immediate from the fact that $I \subset J^2$. Thus, assume that $R^{k-1}/R^k \simeq J^{k-1}/J^k$, $k \geq 2$ and consider the projective cover $\tilde{P} = P(R^{k-1}/R^k)$. Let $\tilde{P} \simeq \bigoplus_{i=1}^{n} s_i \tilde{P}_i$. By Theorem 3.3.7, $\tilde{P} = P(R^{k-1})$. Since R^{k-1} is a projective module, $\tilde{P} \simeq R^{k-1}$. Then $R^k \simeq \tilde{P} R$ and $R^k / R^{k+1} \simeq \tilde{P} R / \tilde{P} R^2$. However, since $R/R^2 \simeq J/J^2$ and $P = P(J^{k-1}/J^k) \simeq \bigoplus_{i=1}^{n} s_i P_i$, we get that $R^k / R^{k+1} \simeq PR/PR^2 \simeq J^k / J^{k+1}$, as required. Consequently, $I \subset J^k$ for any $k \geq 2$ and thus, since J is nilpotent, $I = 0$. The proof of the theorem is completed. □

Exercises to Chapter 8

1. Prove that each of the following categories is equivalent to its dual category:
   a) the category of finite dimensional vector spaces;
   b) the category of finite abelian groups. (Hint: Use Exercise 5 to Chap. 7.)

2. Let the commutative diagram

$$\begin{array}{ccccccc} M_1 & \longrightarrow & M_2 & \longrightarrow & M_3 & \longrightarrow & M_4 \\ {\scriptstyle\alpha_1}\big\downarrow & & {\scriptstyle\alpha_2}\big\downarrow & & {\scriptstyle\alpha_3}\big\downarrow & & {\scriptstyle\alpha_4}\big\downarrow \\ N_1 & \longrightarrow & N_2 & \longrightarrow & N_3 & \longrightarrow & N_4 \end{array}$$

   have exact rows. Prove the following statements:
   a) If α_1 is an epimorphism and both α_2 and α_4 are monomorphisms, then α_3 is a monomorphism.
   b) If α_4 is a monomorphism and both α_1 and α_3 are epimorphisms, then α_2 is an epimorphism.

3. (3×3 lemma) Prove that if in the commutative diagram

$$\begin{array}{ccccccccc} & & 0 & & 0 & & 0 & & \\ & & \downarrow & & \downarrow & & \downarrow & & \\ 0 & \to & M_1 & \to & N_1 & \to & L_1 & \to & 0 \\ & & \downarrow & & \downarrow & & \downarrow & & \\ 0 & \to & M_2 & \to & N_2 & \to & L_2 & \to & 0 \\ & & \downarrow & & \downarrow & & \downarrow & & \\ 0 & \to & M_3 & \to & N_3 & \to & L_3 & \to & 0 \\ & & \downarrow & & \downarrow & & \downarrow & & \\ & & 0 & & 0 & & 0 & & \end{array}$$

   all columns and any two of the three rows are exact, then the remaining row is exact.

4. A commutative diagram

$$\begin{array}{ccc} X & \xrightarrow{\bar{f}} & L \\ {\scriptstyle\bar{g}}\big\downarrow & & {\scriptstyle g}\big\downarrow \\ M & \xrightarrow{f} & N \end{array}$$

   is said to be a *Cartesian square* (pull-back) if for any commutative diagram

$$\begin{array}{ccc} Y & \xrightarrow{\xi} & L \\ {\scriptstyle\eta}\big\downarrow & & {\scriptstyle g}\big\downarrow \\ M & \xrightarrow{f} & N \end{array}$$

   there is a unique homomorphism $\varphi : Y \to X$ such that $\xi = \bar{f}\varphi$ and $\eta = \bar{g}\varphi$. Prove that a Cartesian square can be constructed for any given pair of homomorphisms $f : M \to N$ and $g : L \to N$. (Hint: Consider the submodule $\{(x, y) \mid f(x) = g(y)\}$ of $M \oplus L$.) Is such a square unique?

5. Let
$$0 \to L \to M \to N \to 0$$
be a given exact sequence, and $\varphi : N' \to N$ a homomorphism. Consider the diagram
$$\begin{array}{ccccccccc} 0 & \longrightarrow & L & \xrightarrow{f'} & M' & \xrightarrow{g'} & N' & \longrightarrow & 0 \\ & & {\scriptstyle i_L}\downarrow & & {\scriptstyle \varphi'}\downarrow & & {\scriptstyle \varphi}\downarrow & & \\ 0 & \longrightarrow & L & \xrightarrow{f} & M & \xrightarrow{g} & N & \longrightarrow & 0 \end{array}$$
where the right square is Cartesian and f' is given by relations $f = \varphi' f'$ and $g'f' = 0$ (in view of the uniqueness of the Cartesian square, f' is fully determined by these relations). Prove that the upper row of the diagram is exact (it is called a *lifting* of the given exact sequence along φ).

6. (Schanuel's lemma) Given two exact sequences
$$0 \longrightarrow N_1 \longrightarrow P_1 \xrightarrow{f_1} M \longrightarrow 0$$
and
$$0 \longrightarrow N_2 \longrightarrow P_2 \xrightarrow{f_2} M \longrightarrow 0$$
where P_1 and P_2 are projective modules, prove that $P_1 \oplus N_2 \simeq P_2 \oplus N_1$. (Hint: Consider a lifting of the first sequence along f_2 and the second one along f_1.)

7. Prove that the category of left projective A-modules is equivalent to the category dual to the category of right projective A-modules. (Hint: Apply the functor $h_A^o = \mathrm{Hom}_A(-, A)$.)

8. Prove that, for any A-module M, there is an exact sequence
$$P_1 \xrightarrow{f_1} P_0 \xrightarrow{f_0} M \longrightarrow 0$$
where P_1 and P_0 are projective modules, $\mathrm{Ker}\, f_0 \subset \mathrm{rad}\, P_0$, $\mathrm{Ker}\, f_1 \subset \mathrm{rad}\, P_1$ and such that, given another sequence
$$Q_1 \xrightarrow{g_1} Q_0 \xrightarrow{g_0} M \longrightarrow 0$$
with these properties, there is a commutative diagram
$$\begin{array}{ccccccc} P_1 & \xrightarrow{f_1} & P_0 & \xrightarrow{f_0} & M & \longrightarrow & 0 \\ {\scriptstyle \varphi_1}\downarrow & & {\scriptstyle \varphi_0}\downarrow & & {\scriptstyle 1_M}\downarrow & & \\ Q_1 & \xrightarrow{g_1} & Q_0 & \xrightarrow{g_0} & M & \longrightarrow & 0 \end{array}$$
where φ_0 and φ_1 are isomorphisms.

9. (M. Auslander) Given an A-module M, let
$$P_1 \xrightarrow{f_1} P_0 \xrightarrow{f_0} M \longrightarrow 0$$
be an exact sequence satisfying the properties listed in the preceding exercise. Put
$$\mathrm{Tr}\,(M) = \mathrm{Coker}\, h_A^o(f_1)$$
(see Example 6 of functors in Sect. 8.1 and Exercise 7).
   a) Prove that, up to an isomorphism, $\mathrm{Tr}\,(M)$ does not depend on the choice of the exact sequence with the above properties.
   b) Verify that the exact sequence
$$h_A^o(P_0) \longrightarrow h_A^o(P_1) \longrightarrow \mathrm{Tr}\,(M) \longrightarrow 0$$
possesses also the properties listed in Exercise 8.
   c) Prove that $\mathrm{Tr}\,(M)$ has no projective direct summands.
   d) Establish that for $M = P \oplus N$, where P is projective and N has no projective direct summands, $\mathrm{Tr}\,(\mathrm{Tr}\,(M)) \simeq N$.

10. Let U and U' be simple A-modules. Using the notation of Theorem 8.4.4, prove that if $F(U) \neq 0$, then $F(U)$ is a simple B-module. Moreover, $U \not\simeq U'$ implies that $F(U) \not\simeq F(U')$.

11. Let $1 = e_1 + e_2 + \ldots + e_n$ be a minimal decomposition of the identity of an algebra A, $n \geq 2$. Prove that the algebra A is semisimple (separable) if and only if for any pair of indices $i \neq j$, the algebra eAe, where $e = e_i + e_j$, is semisimple (separable).

12. Let A be a hereditary algebra, P a projective A-module and $B = E_A(P)$. Prove that B is hereditary.

13. Denote by $\mathcal{D}$ a diagram of the form

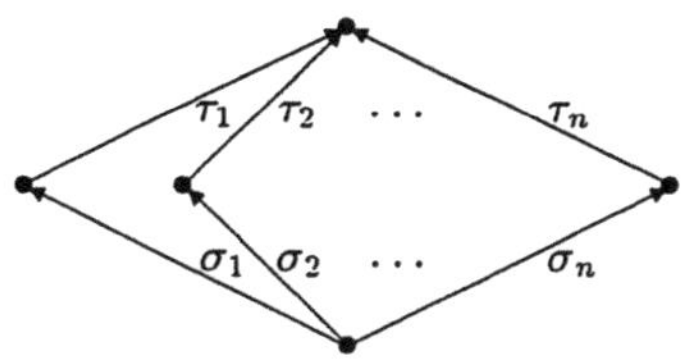

and let A be the quotient algebra of the path algebra $K(\mathcal{D})$ by the ideal generated by the element $\sum_{i=1}^{n} \sigma_i \tau_i$. Prove that the algebra A is not hereditary but that every algebra eAe, where e is an idempotent, $e \neq 1$, is hereditary. (Hint: If $\mathcal{D}'$ is a diagram consisting of some (but not all) vertices of the diagram $\mathcal{D}$ and of all arrows connecting them, and e is the sum of all paths of lengths 0 in $\mathcal{D}'$, then $eAe \simeq K(\mathcal{D}')$.)

The following set of exercises (14–19) deals with an application of tensor products, viz. the theory of induced representations of groups. Let H be a subgroup of a group G, and let N be a KH-module. Then the *induced* KG*-module* $\mathrm{Ind}_H^G(N)$ is defined to be $N \otimes_{KH} KG$. If T is the representation of the group H corresponding to the module N and χ its character, then the representation of G corresponding to the module $\mathrm{Ind}_H^G(N)$ is called the *representation induced from* T and its character $\mathrm{Ind}_H^G\chi$ the *character induced from* χ. Of course, every KG-module M can be viewed as a KH-module: As such, it will be denoted by $\mathrm{Res}_H^G(M)$ and called the *restriction* of M to a subgroup H.

14. (Frobenius reciprocity theorem) Let M be a KG-module, χ its character and N a KH-module and Ψ its character.
    a) Prove that $\mathrm{Hom}_{KG}(\mathrm{Ind}_H^G(N), M) \simeq \mathrm{Hom}_{KH}(N, \mathrm{Res}_H^G(M))$.
    b) If $n = (G:1)$, $m = (H:1)$, prove that

$$\frac{1}{n}\sum_{g\in G} \mathrm{Ind}_H^G\Psi(g)\chi(g^{-1}) = \frac{1}{m}\sum_{h\in H} \Psi(h)\mathrm{Res}_H^G\chi(h^{-1}).$$

    (Hint: Use Exercise 14 to Chap. 7.)
    c) If the characteristic of the field K does not divide the order of the group G and if M is a simple KG-module and N a simple KH-module, use a) to deduce that the multiplicity of M in $\mathrm{Ind}_H^G(N)$ is the same as the multiplicity of N in $\mathrm{Res}_H^G(M)$.

15. Let $F \supset H$ be two subgroups of G. Prove that $\mathrm{Ind}_H^G(N) \simeq \mathrm{Ind}_F^G(\mathrm{Ind}_H^F(N))$ for any KH-module N.

16. (Mackey's formula) Let F and H be two subgroups of G and $\sigma_1, \sigma_2, \ldots, \sigma_s$ a set of representatives of the double cosets of G by F and H (i.e. $G = \bigcup_{i=1}^{s} H\sigma_i F$, where the cosets are pairwise disjoint). Write $H_i = (\sigma_i^{-1} H \sigma_i) \cap F$. Every KH-module N can be considered as a KH_i-module by defining $x(\sigma_i^{-1} h \sigma_i) = xh$ for every h; denote this KH_i-module by N_i. Prove that

$$\mathrm{Res}_F^G(\mathrm{Ind}_H^G N) \simeq \bigoplus_{i=1}^{s} \mathrm{Ind}_{H_i}^F(N_i)$$

for every KH-module N. (Hint: As a KH-KF-bimodule, KG is a direct sum $\bigoplus_{i=1}^{s} V_i$, where V_i is the subspace with basis $\{h\sigma_i f \mid h \in H,\ f \in F\}$. Verify that $V_i \simeq \mathrm{Ind}_{H_i}^F(KH_i)$ as a KF-module and extend this isomorphism to all KH-modules along the lines of the proof of Theorem 8.4.4.)

17. a) Using the results of Exercise 14 and 16, prove that if K is a field of characteristic 0, then the KG-module $\mathrm{Ind}_H^G(N)$ is simple if and only if the KH-module N is simple and, for every i, the KH_i-modules N_i and $\mathrm{Res}_{H_i}^H(N)$ have no isomorphic direct summands (here H_i and N_i are defined as in Exercise 16 for $F = H$).
    b) If H is a normal subgroup, deduce that the representation $\mathrm{Ind}_H^G(T)$ is irreducible if and only if T is an irreducible representation which is not isomorphic to any representation T_σ , where $T_\sigma(h) = T(\sigma^{-1} h \sigma)$, $\sigma \notin H$.

18. a) Prove that every indecomposable representation of a subgroup H of a group G is isomorphic to a direct summand of a representation of the form $\mathrm{Res}_H^G(T)$, where T is an indecomposable representation of G.
    b) In the case that the characteristic p of the field K does not divide the index $(G : H)$, prove that every indecomposable representation of the group G is isomorphic to a direct summand of a representation of the form $\mathrm{Ind}_H^G(T)$, where T is an indecomposable representation of H. (Hint: For a KG-module M, construct an epimorphism $\tau : M \otimes_{KH} KG \to M$ of KG-modules which splits as an epimorphism of KH-modules and use Exercise 18 to Chap. 7.)

19. Using the results of the preceding exercise and those of Exercise 21 and 22 to Chap. 7, prove that a group has a finite number of indecomposable representations over a field of characteristic $p > 0$ if and only if its p-Sylow subgroup is cyclic (Higman's theorem).

9. Quasi-Frobenius Algebras

The duality which exists between the categories of the right and left modules plays an important role in the theory of finite dimensional algebras. In the present chapter we shall introduce this duality, investigate its properties and apply the obtained results to the study of two classes of algebras, viz. to quasi-Frobenius algebras introduced into the theory by T. Nakayama and to serial algebras, or principal ideal algebras, which were studied first by K. Asano.

9.1 Duality. Injective Modules

First, we shall establish the duality between the category of left modules and the category of right modules over a finite dimensional algebra A.

To every (finite dimensional) right A-module M we assign a left A-module M^* constructed as follows. As a vector space, M^* is the space $\mathrm{Hom}(M, K)$ of linear functionals on M (the *conjugate space*); operators from A act on M^* according to the formula $(af)(m) = f(ma)$ for all $a \in A$, $f \in M^*$ and $m \in M$. It is easy to verify that M^* becomes in this way a left A-module.

Similarly, if M is a left A-module, then the conjugate space M^* becomes a right A-module.

Every linear map $\varphi : M \to N$ induces a conjugate map $\varphi^* : N^* \to M^*$ defined by

$$(\varphi^* f)(m) = f(\varphi m).$$

We can check readily that if φ is a homomorphism, so is φ^*. Moreover, $(\varphi\psi)^* = \psi^*\varphi^*$ and $1^* = 1$.

Hence, assigning to every module M the module M^* and to every homomorphism φ the homomorphism φ^*, we get a contravariant functor (see Sect. 8.1). More precisely, we get two contravariant functors: one from the category mod-A of right A-modules to the category A-mod of left A-modules, and the other in the opposite direction. We shall call them the *duality functors*.

Recall that there exists a natural map $\delta_M : M \to M^{**}$ assigning to every vector $m \in M$ the linear functional $\delta_M(m) : M^* \to K$ such that $\delta_M(m)(f) = f(m)$. Evidently, δ_M defines a morphism from the identity functor to the composition of the duality functors defined above. It is a well-known fact from linear algebra that the map δ_M is an isomorphism for every finite dimensional space M. Thus, our observations can be formulated in the following way.

Theorem 9.1.1. *The duality functors define an equivalence of the categories* mod-A *and* $(A\text{-mod})^o$. *In particular, these functors are exact and* $M^* = 0$ *implies* $M = 0$.

Corollary 9.1.2. *A module M is simple if and only if the module M^{**} is simple. Moreover, if $M \simeq e\bar{A}$, where $\bar{A} = A/\operatorname{rad} A$ and e is a minimal idempotent of $\bar{A}$, then $M^* \simeq \bar{A}e$.*

Proof. Since $M \simeq M^{**}$, it is sufficient to verify that if M is not simple, neither is M^*. However, if N is a non-trivial submodule of M, consider the exact sequence

$$0 \to N \to M \to M/N \to 0,$$

and apply the duality functor. In the exact sequence

$$0 \to (M/N)^* \to M^* \to N^* \to 0,$$

both $(M/N)^*$ and N^* are non-zero modules, and thus M^* is not simple.

Now, let $M \simeq e\bar{A}$, where e is a minimal idempotent. Then $Me \neq 0$ and there is a functional $f \in M^*$ such that $f(Me) \neq 0$. Since $f(me) = (ef)m$, we get $eM^* \neq 0$. Consequently, $M^* \simeq \bar{A}e$. □

Corollary 9.1.3. *There is a bijective correspondence between the submodules of M and those of M^*, reversing the inclusion.*

Proof. In order to prove the corollary, we note that every submodule $N \subset M$ defines an epimorphism $\pi : M \to M/N$ and thus a monomorphism $\pi^* : (M/N)^* \to M^*$, that is, a submodule of M^*. This submodule has a simple interpretation: it coincides with the "orthogonal complement" $N^\perp = \{f \in M^* \mid f(N) = 0\}$. Moreover, $M^*/N^\perp \simeq N^*$. □

Let us mention the following obvious formulae:

$$(N_1 + N_2)^\perp = N_1{}^\perp \cap N_2{}^\perp, \quad (N_1 \cap N_2)^\perp = N_1{}^\perp + N_2{}^\perp. \tag{9.1.1}$$

A correspondence satisfying (9.1.1) is called an *anti-isomorphism of lattices.*

In what follows, we shall often deal with the submodule $(\operatorname{rad} M)^\perp \subset M^*$. Since $\operatorname{rad} M$ is the intersection of all maximal submodules of M, $(\operatorname{rad} M)^\perp$ is the sum of all minimal submodules of the module M^*. It is called the *socle* of the module M^* and is denoted by $\operatorname{soc} M^*$. Note that the socle of a module can be defined also by the formula

$$\operatorname{soc} M = \{m \in M \mid m(\operatorname{rad} A) = 0\}.$$

Furthermore, we can write $\operatorname{soc} M^* \simeq (M/\operatorname{rad} M)^*$.

We have already seen the importance of projective and, in particular, of principal modules in the study of the structure of algebras. It is natural to expect that their dual modules will be found also useful in the investigations. By "inverting the arrows" in the theorems on projective modules (Theorem 3.3.5 and Example 4 of Sect. 8.2), we get immediately the following result.

Theorem 9.1.4. *The following conditions for a module Q are equivalent:*

1) *Q^* is projective (in other words, $Q \simeq P^*$, where P is a projective module);*
2) *Q is a direct summand of the module nA^* for some n (A^* is the module dual to the regular module and will be called coregular);*
3) *$Q \simeq \bigoplus_i k_i P_i^*$, where P_i are principal modules;*
4) *every diagram of the form*

$$\begin{array}{ccccc} 0 & \longrightarrow & M & \xrightarrow{\varphi} & N, \\ & & {\scriptstyle\psi}\downarrow & & \\ & & Q & & \end{array}$$

in which the row is exact, can be completed to a commutative diagram

$$\begin{array}{ccccc} 0 & \longrightarrow & M & \xrightarrow{\varphi} & N; \\ & & {\scriptstyle\psi}\downarrow & \swarrow{\scriptstyle\bar\psi} & \\ & & Q & & \end{array}$$

in other words, the equation $x\varphi = \psi$ has a solution for every $\psi : M \to Q$ and for every monomorphism φ;

5) *every monomorphism $Q \to M$ splits, i. e. Q is a direct summand of any module which contains Q as a submodule.*

The modules satisfying the property 4) of Theorem 9.1.4 are called *injective*. In this way, Theorem 9.1.4 provides a characterization of the injective modules over finite dimensional algebras.

The above theorem implies, in particular, that indecomposable injective modules are just the modules dual to the principal ones. We shall call them *coprincipal* modules.

The following corollary is a consequence of Corollaries 9.1.3, 3.2.5 and 3.2.9.

Corollary 9.1.5. *If Q is a coprincipal A-module, then its socle is a simple module. Assigning to Q the simple module* $\operatorname{soc} Q$, *we establish a bijective correspondence between the coprincipal and the simple modules.*

In an analogy to the projective cover, we can introduce the concept of an *injective hull of M* as the least injective module containing a given module M as a submodule. More precisely, an injective module Q is the injective hull of a module M if there is a monomorphism $\varphi : M \to Q$ such that $\operatorname{Im}\varphi \supset \operatorname{soc} Q$, or equivalently, for any submodule X of Q, $\operatorname{Im}\varphi \cap X = 0$ implies $X = 0$. We shall write in this case $Q = Q(M)$.

The existence and properties of injective hulls follow immediately, in view of the duality, from the respective results on projective covers. We shall formulate the facts which will be needed in the sequel in the following theorem whose proof is left to the reader.

Theorem 9.1.6. 1) *If P is a projective cover of M^*, then P^* is an injective hull of M.*
2) $Q(M) = Q(\operatorname{soc} M)$.
3) $Q(M_1 \oplus M_2) \simeq Q(M_1) \oplus Q(M_2)$.
4) *If $\psi : M \to Q'$ is a monomorphism and Q' is an injective module, then $Q' = Q \oplus Q_1$, where $Q \simeq Q(M)$ and $\operatorname{Im}\psi \subset Q$.*

Corollary 9.1.7. *If $\operatorname{soc} M$ is a simple module and $\ell(M) \geq \ell(Q_i)$ for any coprincipal module Q_i, then M is a coprincipal module.*

Proof. The statement follows from the fact that, if $Q = Q(M)$, then $\operatorname{soc} Q \simeq \operatorname{soc} M$. Therefore Q is a coprincipal module. □

Observe that $\ell(Q_i) = \ell(Q_i^*)$, and Q_i^* is a principal module. Therefore the maximal length of right coprincipal A-modules coincides with the maximal length of left principal A-modules (but, in general, does not coincide with the maximal length of right principal modules; see Exercise 1 to this chapter).

9.2 Lemma on Separation

In this section we investigate properties of modules which are simultaneously projective and injective. Such modules are called *bijective*[18]. Note that in general, for a given algebra A, there will be no A-bijective modules (see Exercise 2 to this chapter). However, the existence of a bijective module allows to reduce the study of A-modules to the study of modules over some proper quotient algebra of A. This is a property which can be found useful for studying some classes of algebras and their modules.

First we shall establish the following simple but important fact.

Proposition 9.2.1. *If P is a projective and M a faithful module, then for some n, there is a monomorphism $P \to nM$. Similarly, if Q is an injective A-module, then there is an epimorphism $mM \to Q$.*

Proof. Let $E = E_A(M)$. Considering M as a left E-module, we can construct an epimorphism $nE \to M$ for some n. Applying the left exact functor $\operatorname{Hom}_E(-, M)$, we get a monomorphism $\varphi : \operatorname{Hom}_E(M, M) \to \operatorname{Hom}(nE, M) \simeq nM$. In fact, since M is an E-A-bimodule, φ is also an A-module homomorphism. Now assigning to every $a \in A$ the map $f(a) : M \to M$, whose

[18] The usual English terminology for these modules is *projective-injective*

value at m is ma, we get a homomorphism $f : A \to \operatorname{Hom}_E(M, M)$. Clearly, $\operatorname{Ker} f = \operatorname{Ann} M = 0$ because M is faithful. The composition φf is therefore a monomorphism $A \to nM$. In order to obtain the statement for any projective P, it is sufficient to note that P is a submodule (in fact, a direct summand) of the free module kA for some k, and hence the homomorphism $kA \to knM = k(nM)$, coinciding on every component with φf, is obviously also a monomorphism.

Now, the statement on injective modules follows by duality (since M^* is faithful if and only if M is faithful). □

From this result we can derive the following fundamental lemma.

Lemma 9.2.2 (Separation Lemma). *Let W be a bijective A-module. Then there is a non-zero ideal $I \subset A$ such that every A-module M decomposes into a direct sum $M_1 \oplus M_2$, where $\operatorname{Ann} M_1 \supset I$, and every indecomposable direct summand of the module M_2 is isomorphic to a direct summand of the module W.*

Proof. Evidently, it is sufficient to verify that, for every indecomposable module M which is not isomorphic to a direct summand of W, $\operatorname{Ann} M \supset I$. Therefore we may take for I the intersection of the annihilators of all indecomposable A-modules which are not isomorphic to direct summands of W; denote the class of such modules by $\mathcal{M}$. It remains to prove that $I \neq 0$.

Let us assume that $I = 0$. Since A is finite dimensional, A cannot have an infinite chain of subspaces. Consequently, $I = \bigcap_{i=1}^{t} \operatorname{Ann} M_i$, where $M_1, M_2, \ldots, M_t$ is a finite number of modules from $\mathcal{M}$. But then $I = \operatorname{Ann} M$, where $M = M_1 \oplus M_2 \oplus \ldots \oplus M_t$, and thus M is a faithful A-module. By Proposition 9.2.1, there is a monomorphism $W \to nM$, since W is a projective module. However, W is also injective. It follows that $nM \simeq W \oplus X$ and thus, according to the Krull-Schmidt theorem, every indecomposable direct summand of W is isomorphic to one of the M_i's. However, the latter contradicts our assumption, and thus the lemma is proved. □

Hence, if $W = k_1 W_1 \oplus k_2 W_2 \oplus \ldots \oplus k_s W_s$, where W_i are (pairwise non-isomorphic) indecomposable modules, then every A-module has the form $M_1 \oplus \ell_1 W_1 \oplus \ell_2 W_2 \oplus \ldots \oplus \ell_s W_s$, where M_1 is a module over the quotient algebra A/I. We shall denote this quotient algebra by $A^-(W)$.

The Separation lemma has a particularly simple formulation in the case when the module W is indecomposable (such module will be called *biprincipal*): Every A-module M has the form $M \simeq M_1 \oplus kW$, where M_1 is a module over $A^-(W)$. In this case, the converse statement holds, as well.

Lemma 9.2.3. *Let W be an indecomposable A-module such that every A-module M is of the form $M_1 \oplus kW$, where M_1 is a module over a proper quotient algebra B of the algebra A. Then W is a bijective module.*

Proof. Since both the regular and the coregular (right) A-modules are faithful, they are not B-modules. Consequently, they must possess a direct summand isomorphic to W. Therefore W is both projective and injective, as required. □

Now let us fix a biprincipal A-module W. We shall describe the principal and coprincipal modules over the quotient algebra $B = A^{-}(W)$ in more details.

Proposition 9.2.4. *Every principal B-module is either a principal A-module (distinct from W), or a factor module of W by a minimal submodule W_1 (unique by Corollary 9.1.5). Every coprincipal B-module is either a coprincipal A-module or a maximal submodule W_2 of the module W (unique by Corollary 3.2.5). Conversely, if W is not a simple module, then W/W_1 is a principal and W_2 is a coprincipal B-module.*

Proof. Clearly, W/W_1 is a B-module. Write $P = P_B(W/W_1)$. Then the epimorphism $W \to W/W_1$ extends to an epimorphism $\varphi : W \to P$ which is not an isomorphism (because W is not a B-module). Therefore $\operatorname{Ker}\varphi \supset W_1$ and $\ell(P) \leq \ell(W/W_1)$, which implies that $P \simeq W/W_1$. Similarly, $W_2 \simeq Q_B(W_2)$. Besides, W/W_1 possesses a unique maximal submodule and is therefore indecomposable. Also W_2 is indecomposable, since it has a unique minimal submodule.

Now, let P be an arbitrary principal B-module, and $P' = P_A(P)$. If $P' \not\simeq W$, then P' is a projective B-module, and therefore $P' \simeq P$. If $P' \simeq W$, then the epimorphism $W \to P$ is not an isomorphism and can be factored through an epimorphism $W/W_1 \to P$, from where $P \simeq W/W_1$. The statement on the coprincipal modules follows by duality. □

Proposition 9.2.4 and Lemma 9.2.2 yield immediately the following consequence.

Corollary 9.2.5. *Let W be a biprincipal A-module, and let $A = P_1 \oplus P_2$, where $P_1 \simeq nW$ and P_2 has no direct summands isomorphic to W. Then $\operatorname{soc} P_1$ is an ideal of A and $A^{-}(W) = A/\operatorname{soc} P_1$.*

Proposition 9.2.6. *An algebra A has a simple bijective module W if and only if $A \simeq A_1 \times A_2$, where $A_1 \simeq M_n(D)$ with a division algebra D and $A_2 = A^{-}(W)$.*

Proof. Let W be a simple bijective A-module, and let $A \simeq nW \oplus P$, where P has no direct summands isomorphic to W. Every homomorphism $\varphi : W \to P$ is either zero, or a monomorphism. However, in the latter case $P \simeq W \oplus X$ since W is injective, which is impossible. Consequently, $\operatorname{Hom}_A(nW, P) = 0$. Similarly, $\operatorname{Hom}_A(P, nW) = 0$. Therefore $A \simeq A_1 \times A_2$, where $A_1 = E_A(nW) \simeq M_n(D)$ with a division algebra $D = E_A(W)$, and $A_2 = E_A(P)$. Evidently, $A_2 = A^{-}(W)$. The converse statement is trivial. □

9.3 Quasi-Frobenius Algebras

We have already observed that an algebra may not have any bijective modules. An important class of algebras, introduced by T. Nakayama, consists of algebras all of whose projective modules are injective (and thus bijective). Clearly, this is equivalent to the fact that the regular module is injective. Such algebras are called *quasi-Frobenius*.

As a matter of fact, one should have defined right quasi-Frobenius and left quasi-Frobenius algebras. However, the following theorem shows the equivalence of these notions.

Theorem 9.3.1. *The following conditions for an algebra A are equivalent:*

1) *the right regular A-module is injective;*
1a) *the left regular A-module is injective;*
2) *the right coregular A-module is projective;*
2a) *the left coregular A-module is projective.*

Proof. The equivalences 1) $\Leftrightarrow$ 2a) and 1a) $\Leftrightarrow$ 2) follow from duality. It is therefore sufficient to prove, for example, the equivalence 1) $\Leftrightarrow$ 2). Observe that the number of coprincipal right A-modules equals the number of principal left A-modules; this number is, by Corollary 3.2.9, the number of simple left A-modules, i. e. the number of simple components of the semisimple algebra $A/\operatorname{rad} A$. Since the latter is the number of simple right A-modules, it follows that also the number of principal right A-modules is the same. However, the fact that the regular module is injective is clearly equivalent to the fact that every principal module is injective, and thus the number of bijective A-modules coincides with the number of principal ones. In turn, this is equivalent to the statement that the number of bijective modules equals the number of coprincipal modules, and thus that every coprincipal module, or equivalently the coregular module, is projective. The theorem follows. □

Now, quasi-Frobenius algebras can be characterized as follows: If $A \simeq k_1P_1 \oplus k_2P_2 \oplus \ldots \oplus k_sP_s$, where P_i are principal modules, then $A^* \simeq \ell_1P_1 \oplus \ell_2P_2 \oplus \ldots \oplus \ell_sP_s$. Here, in general, $k_i \neq \ell_i$, i. e. $A \not\simeq A^*$. Those algebras which satisfy $A \simeq A^*$ form a proper subclass of the class of quasi-Frobenius algebras. They are called *Frobenius algebras*. Important examples of Frobenius algebras are the group algebras of finite groups (cf. Exercise 8 below).

Arguments similar to those used in the proof of Theorem 9.3.1 produce the following result.

Theorem 9.3.2. *Let A be a quasi-Frobenius algebra. If all multiplicities of the principal A-modules in the regular module are equal, i. e. if $A/\operatorname{rad} A \simeq M_n(D_1) \times M_n(D_2) \times \ldots \times M_n(D_s)$, where D_i are division algebras, then A is Frobenius. In particular, a basic quasi-Frobenius algebra is Frobenius.*

The fact that the definition of quasi-Frobenius algebras has been given in terms of module categories results immediately in the following consequence.

Corollary 9.3.3. *Every algebra isotypic to a quasi-Frobenius algebra is quasi-Frobenius. In particular, every quasi-Frobenius algebra is isotypic to a Frobenius algebra (namely, to its basic algebra).*

Moreover, Proposition 9.2.1 and the Krull-Schmidt theorem yield the following corollary.

Corollary 9.3.4. *Let A be a quasi-Frobenius algebra, $A \simeq k_1 P_1 \oplus k_2 P_2 \oplus \ldots \oplus k_s P_s$, where P_i are (pairwise non-isomorphic) principal modules and $M_0 = P_1 \oplus P_2 \oplus \ldots \oplus P_s$. Then every faithful A-module has a direct summand isomorphic to M_0.*

Using Proposition 9.2.6 and the definition of the diagram of an algebra, we get the following statement.

Corollary 9.3.5. *Let $\mathcal{D}$ be the diagram of a quasi-Frobenius algebra A. If there is a vertex $i \in \mathcal{D}$ which is either a sink (i is not the tail of any arrow) or a source (i is not the head of any arrow), then $A \simeq A_1 \times A_2$, where $A_1 \simeq M_n(D)$ with a division algebra D.*

Proof. In order to prove this statement, it suffices to note that if i is a sink, then the corresponding principal module is simple (and the remaining statement follows by duality). □

By comparing Corollary 9.3.5 and the assertion on diagrams of hereditary algebras (Corollary 3.7.3), we obtain the following corollary.

Corollary 9.3.6. *A hereditary quasi-Frobenius algebra is semisimple.*

If A is a quasi-Frobenius algebra, then every principal A-module P is coprincipal and, by Corollary 9.1.5, its socle is a simple A-module. Moreover, if P' is a principal module which is not isomorphic to P, then $\operatorname{soc} P' \not\simeq \operatorname{soc} P$. It turns out that a converse statement holds, as well. It provides a rather simple and convenient criterion for an algebra to be quasi-Frobenius.

Theorem 9.3.7. *An algebra A is quasi-Frobenius if and only if the socle of each principal A-module is simple and, for any two non-isomorphic principal A-modules P_1 and P_2, $\operatorname{soc} P_1 \not\simeq \operatorname{soc} P_2$.*

Proof. In view of our previous remark, it is enough to prove the sufficiency of the conditions. Without loss of generality, we may assume that A is basic. Then $A = P_1 \oplus P_2 \oplus \ldots \oplus P_s$, where P_i are pairwise non-isomorphic principal modules. Write $Q_i = Q(P_i)$. By Theorem 9.1.6, $\operatorname{soc} Q_i = \operatorname{soc} P_i$

is a simple module, moreover, $\operatorname{soc} Q_i \not\simeq \operatorname{soc} Q_j$ for $i \neq j$. Consequently, Q_i are pairwise non-isomorphic coprincipal modules. It follows that Q_i^* are pairwise non-isomorphic left principal modules, and since A is a basic algebra, $A \simeq Q_1^* \oplus Q_2^* \oplus \ldots \oplus Q_s^*$ (recall that the number of left and right principal modules is the same). We want to show that $P_i \simeq Q_i$. In fact, $\dim P_i \leq \dim Q_i = \dim Q_i^*$ and

$$\sum_{i=1}^{s} \dim P_i = \dim A = \sum_{i=1}^{s} \dim Q_i^* = \sum_{i=1}^{s} \dim Q_i \,,$$

so that $\dim P_i = \dim Q_i$ for all i. Hence, $P_i \simeq Q_i$ and the proof of the theorem is completed. □

Now assume that A is a Frobenius algebra. Then, in view of $A \simeq A^*$ and Corollary 9.1.3, it follows that the lattice of the left ideals and the lattice of the right ideals of the algebra A are anti-isomorphic. It turns out that this holds for an arbitrary quasi-Frobenius algebra. In fact, this condition provides a characterization of quasi-Frobenius algebras.

Theorem 9.3.8. *An algebra A is quasi-Frobenius if and only if the lattices of its left ideals and of its right ideals are anti-isomorphic.*

Proof. Let φ be an anti-isomorphism between the lattices of the left and the right ideals of A. Clearly, $\varphi(0) = A$ and $\varphi(A) = 0$. Decompose the left regular A-module into the direct sum of the principal ones: $A = P_1 \oplus P_2 \oplus \ldots \oplus P_s$. This means that P_i are left ideals of A, $\sum_{i=1}^{s} P_i = A$ and, for every i, $P_i \cap \Big(\sum_{j\neq i} P_j\Big) = 0$.

Applying the lattice anti-isomorphism φ, we obtain the right ideals $\varphi(P_i)$ of A such that $\bigcap_{i=1}^{s} \varphi(P_i) = 0$ and $\varphi(P_i) + \Big(\bigcap_{j\neq i} \varphi(P_j)\Big) = A$ for every i. Write $P_i' = \bigcap_{j\neq i} \varphi(P_j)$. Then $A = \varphi(P_i) \oplus P_i'$ for every i. We are going to show, by induction on k, that

$$A = P_1' \oplus P_2' \oplus \ldots \oplus P_k' \oplus \Big(\bigcap_{j=1}^{k} \varphi(P_j)\Big)\,, \quad k \leq s.$$

For $k = 1$, the statement has been proved. Thus, assume that the formula holds for a certain $k < s$. Then $\bigcap_{j=1}^{k} \varphi(P_j) \supset P_{k+1}'$ and since $A = P_{k+1}' \oplus \varphi(P_{k+1})$, $\bigcap_{j=1}^{k} \varphi(P_j) = P_{k+1}' \oplus \Big(\bigcap_{j=1}^{k+1} \varphi(P_j)\Big)$, as required. In particular, $A = P_1' \oplus P_2' \oplus \ldots \oplus P_s'$, and thus all P_i' are principal right modules. In addition, since $P_i' = A/\varphi(P_i)$, the lattice of the submodules of P_i' is isomorphic to the lattice of the submodules of A containing $\varphi(P_i)$, and thus anti-isomorphic to the lattice of the submodules of P_i (since $\varphi^{-1}(A) = 0$). However, P_i has a unique maximal submodule, and therefore P_i' has a unique minimal submodule, and thus $U_i = \operatorname{soc} P_i'$ is simple.

In order to show that A is quasi-Frobenius, it is sufficient, in view of Theorem 9.3.7, to verify that $U_i \not\simeq U_j$ whenever $P_i' \not\simeq P_j'$.

Observe that $\varphi^{-1}(P_i') = \sum_{j\neq i} P_j$, from where $P_i = \bigcap_{j\neq i} \varphi^{-1}(P_j')$, and thus the correspondence between P_i and P_i' is symmetric.

We are going to show that if $P_i \not\simeq P_j$, then $U_i \not\simeq U_j$, and consequently, $P_i' \not\simeq P_j'$. This will imply immediately that if $P_i' \not\simeq P_j'$, then $P_i \not\simeq P_j$ and thus $U_i \not\simeq U_j$, as required.

Hence, write $P = P_i + P_j$. Then, in view of $P_i \cap P_j = 0$, $P/P_i \simeq P_j$ and $P/P_j \simeq P_i$. Since $P_i/\operatorname{rad} P_i \not\simeq P_j/\operatorname{rad} P_j$, there are just two maximal submodules in P, one containing P_i and the other P_j. However, $\varphi(P) = \varphi(P_i) \cap \varphi(P_j)$. This means that the lattice of the submodules of P is anti-isomorphic to the lattice of the submodules of $A = \varphi(0)$ which contain $\varphi(P_i) \cap \varphi(P_j)$. Since $\varphi(P_i) + \varphi(P_j) = A$, $A/\big(\varphi(P_i) \cap \varphi(P_j)\big) \simeq A/\varphi(P_i) \oplus A/\varphi(P_j) \simeq P_i' \oplus P_j'$. Thus, in $P_i' \oplus P_j'$, there are just two minimal submodules, which is impossible if $U_i \simeq U_j$ (because, in this case, any element $a + b$, where $a \in U_i$, $b \in U_j$, generates a simple submodule of $U_i \oplus U_j$, and therefore there are at least three such submodules).

Now assume that the algebra A is quasi-Frobenius. Then it is possible to display an explicit anti-isomorphism of the lattices of the left and the right ideals of A in the following way. For every right ideal I, put $\ell(I) = \{a \in A \mid aI = 0\}$, and for every left ideal J, put $r(J) = \{b \in A \mid Jb = 0\}$. Evidently, $\ell(I)$ is a left and $r(J)$ is a right ideal and $I \supset I'$ or $J \supset J'$ implies $\ell(I) \subset \ell(I')$ or $r(J) \subset r(J')$, respectively. We are going to show that ℓ and r are mutually inverse maps; it is clear that this means that they realize the required lattice anti-isomorphism.

Consider the exact sequence of (right) A-modules

$$0 \to I \to A \to A/I \to 0.$$

Applying the functor $\operatorname{Hom}_A(-, A)$ and taking into account that A is injective, we obtain the exact sequence

$$0 \to \operatorname{Hom}_A(A/I, A) \to \operatorname{Hom}_A(A, A) \to \operatorname{Hom}_A(I, A) \to 0.$$

Now, every homomorphism $A/I \to A$ is determined uniquely by the image of the class $1 + I$, which, obviously, may be any element $a \in A$ such that $aI = 0$. In other words, $\operatorname{Hom}_A(A/I, A) \simeq \ell(I)$ and the above sequence can be written in the form

$$0 \to \ell(I) \to A \to A/\ell(I) \to 0.$$

Applying the functor $\operatorname{Hom}_A(-, A)$ again and identifying $\operatorname{Hom}_A(A/\ell(I), A)$ with $r\ell(I)$, we get the exact sequence

$$0 \to r\ell(I) \to A \to A/r\ell(I) \to 0,$$

in which $r\ell(I)$ and $A/r\ell(I)$ are isomorphic to $\mathrm{Hom}_A(\mathrm{Hom}_A(I,A),A)$ and $\mathrm{Hom}_A(\mathrm{Hom}_A(A/I,A),A)$, respectively.

Now, observe that there is a unique homomorphism of the modules $M \to \mathrm{Hom}_A(\mathrm{Hom}_A(M,A),A)$ which maps $m \in M$ to the homomorphism $\bar{m} : \mathrm{Hom}_A(M,A) \to A$ such that, for any $f \in \mathrm{Hom}_A(M,A)$, $\bar{m}(f) = f(m)$. It is clear that the set of all such homomorphisms forms a functor morphism $Id_{A\text{-mod}} \to \mathrm{Hom}_A(\mathrm{Hom}_A(-,A),A)$. In particular, we get the following commutative diagram with exact rows

$$\begin{array}{ccccccccc} 0 & \longrightarrow & I & \longrightarrow & A & \longrightarrow & A/I & \longrightarrow & 0 \\ & & \downarrow & & \downarrow & & \downarrow & & \\ 0 & \longrightarrow & r\ell(I) & \longrightarrow & A & \longrightarrow & A/r\ell(I) & \longrightarrow & 0. \end{array} \qquad (9.3.1)$$

Here the middle map is an isomorphism.

We shall show that the map $M \to \mathrm{Hom}_A(\mathrm{Hom}_A(M,A),A)$ is a monomorphism for every M. To this end, embed M into an injective module Q and observe that, since A is quasi-Frobenius, Q is bijective. Consequently, it is a direct summand of a free module. Therefore, there is a monomorphism $M \to nA$ for some n. This means, however, that for any non-zero $m \in M$, there is a homomorphism $f : M \to A$ such that $f(m) \neq 0$. It follows that $\bar{m}(f) = f(m) \neq 0$; hence $\bar{m} \neq 0$ and our map is a monomorphism.

It remains to remark that if in the diagram (9.3.1) the outer maps are monomorphisms, then they are necessarily isomorphisms. Indeed, since the middle map is an isomorphism, the map $A/I \to A/r\ell(I)$ must be an epimorphism, hence it is an isomorphism. Now, one can apply the Five lemma to obtain that the map $I \to r\ell(I)$ is also an isomorphism. The proof of the theorem is completed. □

Corollary 9.3.9. *An algebra A is quasi-Frobenius if and only if $r\ell(I) = I$ for every right ideal I and $r\ell(J) = J$ for every left ideal J.*

9.4 Uniserial Algebras

We have seen that the description of representations of a quasi-Frobenius algebra A can be reduced to the description of representations of some quotient algebra B. However, in general, the algebra B and its representations can have still a rather complex structure. The situation becomes significantly simpler if we assume that all quotient algebras of A are quasi-Frobenius, as well. In that case, the description of the modules can be achieved by successive applications of Corollary 9.3.4. It turns out that the algebras satisfying this condition have themselves a relatively simple description. Moreover, all ideals of such algebras A are principal ideals, i. e. of the form aA for some $a \in A$. Conversely, every algebra of principal ideals has quasi-Frobenius quotients.

Let us formulate these statements in the form of a theorem.

Theorem 9.4.1. *The following properties of an algebra A are equivalent:*

1) *every quotient algebra of A is quasi-Frobenius;*
2) *every ideal of A is a principal right ideal, and thus it is of the form aA;*
2a) *every ideal of A is is a principal left ideal;*
3) *every right ideal of A is principal;*
3a) *every left ideal of A is principal;*
4) $A \simeq A_1 \times A_2 \times \ldots \times A_s$, *where* $A_i \simeq M_{n_i}(B_i)$, *where each* B_i *is a local algebra with a principal right ideal* $\operatorname{rad} B_i$;
4a) $A \simeq A_1 \times A_2 \times \ldots \times A_s$, *where* $A_i \simeq M_{n_i}(B_i)$, *where each* B_i *is a local algebra with a principal left ideal* $\operatorname{rad} B_i$.

If, in addition, the algebra A is indecomposable (into a direct product), then the above properties are equivalent to the following one:
1a) *A and its quotient algebra by some minimal ideal are quasi-Frobenius.*

Proof. Evidently, we may assume that the algebra A cannot be decomposed into a direct product. The implications 1) $\Rightarrow$ 1a), 3) $\Rightarrow$ 2), 3a) $\Rightarrow$ 2a) are trivial.

1a) $\Rightarrow$ 4). Let I be a minimal ideal such that both A and A/I are quasi-Frobenius. There is a principal (and thus also biprincipal) A-module P which is not an A/I-module. But then clearly $A/I = A^-(P)$ (see the Separation lemma, 9.2.2). Let $A \simeq nP \oplus P'$, where P' does not have any direct summands isomorphic to P. Denote by P_1 the maximal submodule of P and by P_2 the factor module P/U_1 , where $U_1 = \operatorname{soc} P$. By Proposition 9.2.4, P_1 is a coprincipal and P_2 a principal A/I-module. Since A/I is a quasi-Frobenius algebra, P_1 is a principal A/I-module. Hence, by Proposition 9.2.4, either $P_1 \simeq P_2$, or P_1 is isomorphic to a direct summand of P'. However, in the latter case, P_1 would be a bijective A-module and thus $P = P_1 \oplus X$, which is impossible.

Consequently, $P_1 \simeq P_2$. Thus P_2 has a minimal submodule $U_2 \simeq \operatorname{soc} P_1 = U_1$ and $P_2/U_2 \simeq P_1/U_1$; it follows that P_2/U_2 also contains a unique minimal submodule U_3 isomorphic to U_2 , and thus to U_1 . Continuing in this process, we construct a composition series of P all of whose factors are isomorphic to U_1. Moreover, this composition series is unique. In view of the Jordan-Hölder theorem, we obtain in particular that $\operatorname{Hom}_A(P', P) = 0$. By duality, P^* is also a principal module with a unique composition series, and $\operatorname{Hom}_A(P, P') \simeq \operatorname{Hom}_A(P'^*, P^*) = 0$. Consequently, $A \simeq A_1 \times A_2$, where $A_1 = E_A(nP) \simeq M_n(B)$ with the local algebra $B = E_A(P)$, and $A_2 = E_A(P')$. Since A is indecomposable, $P' = 0$ and $A \simeq A_1$.

Write $R = \operatorname{rad} B$. Then $\operatorname{rad} M_n(B) = M_n(R)$ (Proposition 3.3.11). Now, $\operatorname{rad} A = nP_1$ and $P(P_1) \simeq P$; thus there is an epimorphism $A \to \operatorname{rad} A$, and $\operatorname{rad} A$, as well as $\operatorname{rad} B$, is a cyclic module, as required.

4) $\Rightarrow$ 3). Let $A = M_n(B)$, where $R = \operatorname{rad} B$ is a principal right ideal. Then there is an epimorphism $\varphi : B \to R$. This implies that R^2 is the unique

maximal submodule of R. Besides, $\varphi(R) = R^2$. Thus $\varphi(R^2) = R^3$ is the unique maximal submodule in R^2, etc. In this way, we see that every right ideal of B is of the form R^m. However, $B \simeq E_A(P)$, where P is a principal A-module. In view of the Morita theorem, the lattice of the B-submodules of B is the same as the lattice of the A-submodules in $B \otimes_B P \simeq P$ and, for every A-submodule $M \subset P$, there is an epimorphism $P \to M$. By induction on k, one can easily deduce that, for any submodule $M \subset kP$, there is an epimorphism $kP \to M$. In particular, for any right ideal $I \subset A$, there is an epimorphism $A \to I$, and thus I is a principal right ideal.

2) $\Rightarrow$ 1). Let $A \simeq nP \oplus P'$, where P is a principal module and P' has no direct summands isomorphic to P. If $M_1 = \operatorname{rad} P$, then $I_1 = nM_1 \oplus P'$ is an ideal in A, and thus a cyclic A-module. Therefore there is an epimorphism $A \to I_1$, and thus an epimorphism $P \to M_1$. Consequently, M_1 also has a unique maximal submodule M_2, and $M_1/M_2 \simeq P/M_1$. But then $I_2 = nM_2 \oplus P'$ is again an ideal, and therefore there is an epimorphism $P \to M_2$. Continuing this process, we construct a unique composition series of P with isomorphic factors. From here, as before, $\operatorname{Hom}_A(P, P') = \operatorname{Hom}_A(P', P) = 0$ and $A = A_1 \times A_2$; since A is indecomposable, $P' = 0$. Now, P contains a unique minimal submodule, and hence, by Theorem 9.3.7, we conclude that A is quasi-Frobenius. It is clear that the condition 2) translates to all quotient algebras of A, and therefore all quotient algebras of A are quasi-Frobenius.

Now, noting that the conditions 1) and 1a) are left-right symmetric, we can similarly deduce that 1a) $\Rightarrow$ 4a) $\Rightarrow$ 3a) $\Rightarrow$ 2a) $\Rightarrow$ 1). The proof of the theorem is completed. □

The algebras which satisfy the conditions of Theorem 9.4.1 are called *uniserial* (or *principal ideal algebras*).

The following corollary is an immediate consequence of Theorem 9.4.1 and Corollary 9.3.4.

Corollary 9.4.2. *Every indecomposable module over a uniserial algebra is isomorphic to a factor module of a principal module. By duality, every indecomposable module over a uniserial algebra is isomorphic to a submodule of a principal module.*

Observe that in the course of the proof of Theorem 9.4.1 we have also established that a principal module over a uniserial algebra has a unique composition series. Consequently, it has just one submodule of a given length, and we can formulate the following corollary.

Corollary 9.4.3. *An indecomposable module over a uniserial algebra is, up to an isomorphism, uniquely determined by its length and projective cover (or injective hull).*

Theorem 9.4.1 provides also a simple criterion for an algebra to be uniserial.

Proposition 9.4.4. *An algebra A is uniserial if and only if its diagram consists of isolated points and loops (one-point cycles).*

Corollary 9.4.5. *An algebra A is uniserial if and only if the algebra A/R^2, where $R = \operatorname{rad} A$, is uniserial.*

Exercises to Chapter 9

1. Let A be a minimal algebra corresponding to the partially ordered set

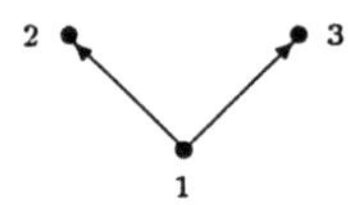

(see Exercise 8 to Chap. 3). Find the lengths of the principal right and left A-modules and see that their maxima are not equal.

2. Show that a three-dimensional algebra A with basis $\{1, a, b\}$ and multiplication table $a^2 = b^2 = ab = ba = 0$ has no bijective modules.

3. Let $A = T_n(K)$ (the algebra of triangular matrices), and let $W = nK$ be the space of n-tuples over the field K viewed as an A-module. Prove that W is the only bijective A-module and find $A^{-}(W)$.

4. Consider the path algebra $K(\mathcal{D})$ of the diagram $\mathcal{D}$:

Let J be the ideal of the paths of non-zero length, $A = K(\mathcal{D})/J^2$, e_1 and e_2 the idémpotents corresponding to the vertices 1 and 2, $P_i = e_i A$ and $P_i' = Ae_i$. Prove that $P_1^* \simeq P_2'$ and $P_2^* \simeq P_1'$. Thus, A is a Frobenius algebra. Deduce that $B = E_A(P_1 \oplus P_2 \oplus P_2)$ is a quasi-Frobenius but not a Frobenius algebra.

5. Prove that a semisimple algebra is a Frobenius algebra.

6. If A is a Frobenius algebra, prove that $M_n(A)$ is a Frobenius algebra.

7. Prove that A is a Frobenius algebra if and only if there is a non-degenerate bilinear form T which is inner A-bilinear (in the sense of Sect. 8.3), i. e. which satisfies $T(ba, c) = T(b, ac)$ for all $a, b, c \in A$.

8. Let $a = \sum_g \alpha_g g$ and $b = \sum_g \beta_g g$ be elements of the group algebra KG of a finite group G. Define

$$T(a, b) = \sum_g \alpha_{g^{-1}} \beta_g .$$

Prove that T is a non-degenerate inner A-bilinear form on KG and thus, KG is a Frobenius algebra.

9. Prove that A is a Frobenius algebra if and only if

$$\dim I + \dim \ell(I) = [A : K] \quad \text{and} \quad \dim J + \dim r(J) = [A : K]$$

for every right ideal I and every left ideal J of A. (Hint: To show that A is a quasi-Frobenius algebra, consider the ideal $I = k(\operatorname{rad} P) \oplus X$, where P is a principal module of A and $A = kP \oplus X$ such that X has no direct summands isomorphic to P.)

10. Let P be a right, or left, principal module over the algebra $T_n(K)$ of triangular matrices. Prove that $\operatorname{soc} P$ is simple. (Note that Exercise 3 implies that $T_n(K)$ is not a quasi-Frobenius algebra.)

11. Let σ be an automorphism of a division algebra D. The (infinite dimensional) algebra $A = D[t, \sigma]$ of "polynomials" $a_n t^n + a_{n-1} t^{n-1} + \ldots + a_0$, where $a_i \in D$ and multiplication is given by the rule $ta = \sigma(a)t$ for every $a \in D$, is called the *skew polynomial algebra* (over D). Verify that $t^n A = At^n$ and prove that $A/t^n A$ is a local uniserial algebra.

12. Prove that every local uniserial algebra of separable type (see Sect. 8.5) is of the form $A/t^n A$, where $A = D[t, \sigma]$ is a skew polynomial algebra over a separable division algebra D constructed in the preceding exercise. Moreover, the division algebra D and the exponent n are determined uniquely, while the automorphism σ is determined up to conjugacy (in the automorphism group) and an inner automorphism of the division algebra D.

13. Making use of Exercise 12, describe all uniserial algebras of separable type.

14. Let $1 = e_1 + e_2 + \ldots + e_n$ be a minimal decomposition of the identity of an algebra A, $n \geq 3$. Prove the following statement: If, for every idempotent e which is a sum of three distinct idempotents of the given decomposition, eAe is a quasi-Frobenius algebra, then A is a quasi-Frobenius algebra. (Hint: Let $P_i = e_i A$ and $\operatorname{soc} P_i$ be not simple; then $P_i \supset U_j \oplus U_k$, where U_j and U_k are simple A-modules, $P(U_j) \simeq P_j = e_j A$, $P(U_k) \simeq P_k = e_k A$; i, j, k are not necessarily distinct. Let P be a direct sum of pairwise distinct modules from P_i, P_j, P_k and $B = E_A(P)$; then the socle of the principal B-module $\bar{P}_i = \operatorname{Hom}_A(P, P_i)$ is not simple, in contradiction to Theorem 9.3.7, since B is quasi-Frobenius. Similarly, if $P_i \not\simeq P_j$, but $\operatorname{soc} P_i \simeq \operatorname{soc} P_j \simeq U_k$, show that $\operatorname{soc} \bar{P}_i \simeq \operatorname{soc} \bar{P}_j$, which contradicts Theorem 9.3.7 again because, in view of Theorem 8.4.4, $\bar{P}_i \not\simeq \bar{P}_j$.)

15. Let $A = K(\mathcal{D})/J^2$, where $K(\mathcal{D})$ is the path algebra of the diagram $\mathcal{D}$ which is a cycle of the form

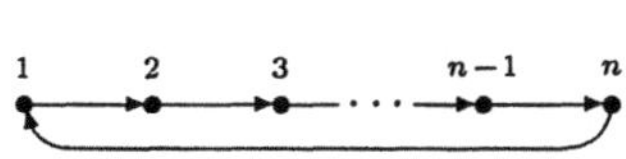

($n \geq 3$) and J is the ideal generated by the paths of non-zero length. Prove that A is a quasi-Frobenius algebra, but eAe, where $e = e_1 + e_2 + \ldots + e_k$, $k < n$ with idempotents e_i corresponding to the vertices i, is not quasi-Frobenius. (Hint: Use Corollary 9.3.5.) Thus, a converse of the statement of Exercise 14 is false.

16. Let $1 = e_1 + e_2 + \ldots + e_n$ be a minimal decomposition of the identity of an algebra A, $n \geq 2$. Let the algebra $(e_i + e_j)A(e_i + e_j)$ be uniserial for every pair of indices i, j. Prove that the algebra A is uniserial.

10. Serial Algebras

Corollaries 9.4.2 and 9.4.3 provide a simple description of modules over uniserial algebras. However, it is easy to see that this description uses not so much the fact that all quotient algebras are quasi-Frobenius as that they all possess bijective modules. In this chapter we shall consider a more general class, the class of serial algebras, also introduced by T. Nakayama, which are characterized by the fact that each of their quotient algebras has a bijective module. We shall show that this is the most general class of algebras for which statements similar to those of Corollaries 9.4.2 and 9.4.3 hold. The structure of serial algebras is substantially more involved than that of uniserial algebras. However, under rather general assumptions, we can obtain a complete description of these algebras (the main results have been obtained by H. Kupisch).

10.1 The Nakayama-Skornjakov Theorem

Let M be a right or left module over an algebra A. The module M is said to be *serial* if the submodule lattice of M is a chain (i. e. a linearly ordered set). This condition is obviously equivalent to the fact that every non-zero submodule N of M has a unique maximal submodule (which is the radical of N), or equivalently, that $N/\operatorname{rad} N$ is simple for every non-zero submodule N of M.

A direct sum of serial modules is called a *semi-serial* module. A trivial example of a serial or semi-serial module is a simple or semisimple module, respectively. In view of the Krull-Schmidt theorem, every direct summand of a semi-serial module is semi-serial. In particular, an indecomposable direct summand is a serial module.

Theorem 10.1.1 (Nakayama-Skornjakov). *Let A be an algebra. Then the following statements are equivalent:*

1) *Every right A-module is semi-serial.*
1a) *Every left A-module is semi-serial.*
2) *The right regular and the left regular A-modules are semi-serial.*
2a) *The right regular and coregular A-modules are semi-serial.*
2b) *The left regular and coregular A-modules are semi-serial.*

3) *Every indecomposable A-module (right and left) is isomorphic to a factor module of a principal module.*

3a) *Every indecomposable A-module (right and left) is isomorphic to a submodule of a coprincipal module.*

4) *Every indecomposable A-module M (right and left) is projective as a module over the quotient algebra* $A/\operatorname{Ann} M$.

4a) *Every indecomposable A-module M (right and left) is injective as a module over the quotient algebra* $A/\operatorname{Ann} M$.

5) *There exists a bijective right module over every quotient algebra of the algebra A.*

5a) *There exists a bijective left module over every quotient algebra of the algebra A.*

The algebras satisfying these equivalent conditions are said to be *serial algebras.*

Proof. The equivalences 1) $\Leftrightarrow$ 1a), 2) $\Leftrightarrow$ 2a) $\Leftrightarrow$ 2b), 3) $\Leftrightarrow$ 3a), 4) $\Leftrightarrow$ 4a) and 5) $\Leftrightarrow$ 5a) follow from duality (note that it is easy to see that $\operatorname{Ann} M^* = \operatorname{Ann} M$ for every module M).

The implications 1) $\Rightarrow$ 2a) and 4) $\Rightarrow$ 3) are obvious.

2) $\Rightarrow$ 5). Observe that condition 2) is preserved for quotient algebras: If $A = P_1 \oplus P_2 \oplus \ldots \oplus P_n$ with serial modules P_i, then $A/I = P_1/P_1I \oplus P_2/P_2I \oplus \ldots \oplus P_n/P_nI$ and all P_i/P_iI are obviously again serial. Therefore it is sufficient to verify that 2) implies the existence of a bijective A-module. It is convenient to consider the condition 2a), equivalent to 2).

Let M be a module of maximal length among all the principal and coprincipal right A-modules. Since M is serial, it contains a unique maximal and a unique minimal submodule. But then Corollaries 3.2.8 and 9.1.7 imply that M is principal and coprincipal, i. e. that the module M is bijective.

5) $\Rightarrow$ 4). Let M be an indecomposable A-module and W a bijective module over the quotient algebra $B = A/\operatorname{Ann} M$. Since M is a faithful indecomposable B-module, the Separation lemma (Lemma 9.2.2) implies that M is isomorphic to a direct summand of W, i. e. the B-module M is bijective and thus, in particular, projective.

3) $\Rightarrow$ 1). Let M be an indecomposable A-module. Since it is isomorphic to a factor module of a principal module, M contains a unique maximal submodule $M_1 = \operatorname{rad} M$ (Corollary 3.2.8). On the other hand, M (and therefore also M_1) is isomorphic to a submodule of a coprincipal module and thus M_1 is indecomposable. We may therefore apply the same argument to M_1: The unique maximal submodule $M_2 = \operatorname{rad} M_1$ of M_1 is indecomposable. Continuing in this manner, we get a chain of submodules $M \supset M_1 \supset M_2 \supset \ldots$ of M in which every submodule is the unique maximal submodule of the preceding one. It turns out that $M_1, M_2, M_3, \ldots$ are the only submodules of M, i. e. that M is serial. The proof of the theorem is completed. □

Remark. It is very essential that the conditions 2), 3) and 4) contain requirements for both right and left modules. An algebra may satisfy one of these conditions for right (or for left) modules, but be not serial. One of the simplest examples of such an algebra is the subalgebra $A \subset M_3(K)$ consisting of all matrices of the form

$$\begin{pmatrix} a_1 & a_2 & a_3 \\ 0 & a_4 & 0 \\ 0 & 0 & a_5 \end{pmatrix}, \quad a_i \in K .$$

Exercise 1 to Chap. 10 shows that A is not a serial algebra; however, A satisfies the conditions 3) and 4) for right modules and 2) for the left regular module.

Since a submodule of a principal module over a serial algebra is uniquely determined by its length, we obtain an analogue of Corollary 9.4.3.

Corollary 10.1.2. *An indecomposable module over a serial algebra is fully determined (up to an isomorphism) by its length and projective cover (or injective hull).*

Since serial algebras can be defined in terms of the module categories (for example, by condition 1) of Theorem 10.1.1), every algebra Morita equivalent to a serial algebra is serial, as well. In particular, A is a serial algebra if and only if its basic algebra is serial.

The condition 2) of Theorem 10.1.1 asserts that an algebra A is serial if and only if all submodules of principal right and left A-modules contain a unique maximal submodule. It turns out that it is sufficient to verify this property only for maximal submodules of principal modules.

Proposition 10.1.3. *Assume that the radical of any principal right (left) A-module contains a unique maximal submodule. Then every principal right (left) A-module is serial.*

Proof. Let P be a principal A-module and $M_1 = \operatorname{rad} P$ its unique maximal submodule. Since M_1 contains a unique maximal submodule $M_2 = \operatorname{rad} M_1$, M_1 is a factor module of some principal A-module P_1 (Corollary 3.2.8) and M_2 is a factor module of $\operatorname{rad} P_1$. In turn, $\operatorname{rad} P_1$ is a factor module of a principal A-module P_2 and thus contains a unique maximal submodule $M_3 = \operatorname{rad} M_2$. Continuing this process, we obtain a chain of submodules $P \supset M_1 \supset M_2 \supset M_3 \supset \dots$ of P in which every module is the unique maximal submodule of the preceding one. From here, it follows immediately that every submodule of P coincides with one of the submodules M_i and thus P is serial. □

Corollary 10.1.4. *Let $R = \operatorname{rad} A$. Then the algebra A/R^2 is serial if and only if A is serial.*

Proof. To prove the statement, recall that $\operatorname{rad} P = PR$, $\operatorname{rad}(\operatorname{rad} P) = PR^2$ and that, by Proposition 10.1.3, it is sufficient to verify that the modules PR/PR^2 (or RQ/R^2Q) are simple for every principal right module P (principal left module Q). □

The latter result, together with Theorem 9.3.7 yield the following consequence.

Corollary 10.1.5. *If the quotient algebra A/R^2 is a quasi-Frobenius algebra, then A is a serial algebra.*

Proof. Let P be a principal right module over the algebra A/R^2. If P is not simple, i. e. if $\operatorname{rad} P \neq 0$, then its unique maximal submodule $\operatorname{rad} P$ coincides with the socle (because $(\operatorname{rad} P)R = PR^2 = 0$). Then, by Theorem 9.3.7, $\operatorname{rad} P$ is simple. This means that P is serial. Similarly, also every principal left A/R^2-module is serial. Hence A/R^2 and therefore also A is a serial algebra. □

Remark. The example of the algebra A of triangular matrices shows that the converse of Corollary 10.1.5 is false: A is a serial algebra but A/R^2 is not quasi-Frobenius.

10.2 Right Serial Algebras

We are going to study the structure of serial algebras. In fact, we shall describe the structure of a wider class of algebras, of the so-called *right serial algebras*, i. e. of algebras whose right regular module is semi-serial. Of course, a similar description holds for *left serial algebras*, i. e. for algebras whose left regular module is semi-serial. The actual formulation of the respective results for left serial algebras is left to the reader.

Observe that Proposition 10.1.3 yields the following corollary.

Corollary 10.2.1. *The algebra A/R^2 is right serial if and only if A is right serial.*

The proof is the same as that of Corollary 10.1.4 (without mentioning left modules).

Moreover, the fact that an algebra is right serial can be fully characterized by its diagram (see Sect. 3.6).

Theorem 10.2.2. *An algebra A is right serial if and only if there is at most one arrow starting at each vertex of its diagram $\mathcal{D}(A)$.*

Proof. In view of Proposition 10.1.3, an algebra A is right serial if and only if the radical $\operatorname{rad} P_i$ of every principal (right) A-module P_i is either zero or isomorphic to a factor module of a principal module P_j. But that means that there is either no arrow, or just one arrow (pointing to the vertex j) starting at the vertex i of the diagram $\mathcal{D}(A)$. □

In order to describe all diagrams of right serial algebras, we introduce the following definitions. A *circuit* of a diagram $\mathcal{D}$ is a sequence of pairwise distinct vertices $\{i_1, i_2, \dots, i_t\}$ and arrows $\{\sigma_1, \sigma_2, \dots, \sigma_t\}$ such that each arrow σ_k points from i_k to i_{k+1} or from i_{k+1} to i_k (assuming that $i_{t+1} = i_1$). Let us remark that it is possible that $t = 1$. Of course, every cycle is a circuit, but the converse is not true in general. A connected diagram without circuits is called a *tree*. A vertex which is not an initial point of any arrow in a given diagram $\mathcal{D}$ is called a *sink*, or a *root* of $\mathcal{D}$. An algebra whose diagram is connected is said to be *connected*; such algebras are indecomposable (cf. Theorem 3.6.2).

Corollary 10.2.3. *A connected algebra A is right serial if and only if its diagram $\mathcal{D}(A)$ is either a tree with a single root, or a diagram with a unique circuit which is a cycle such that when removing all arrows of this cycle, the remaining diagram is a disconnected union of trees whose roots are vertices of the cycle.*

We give two examples for such diagrams.

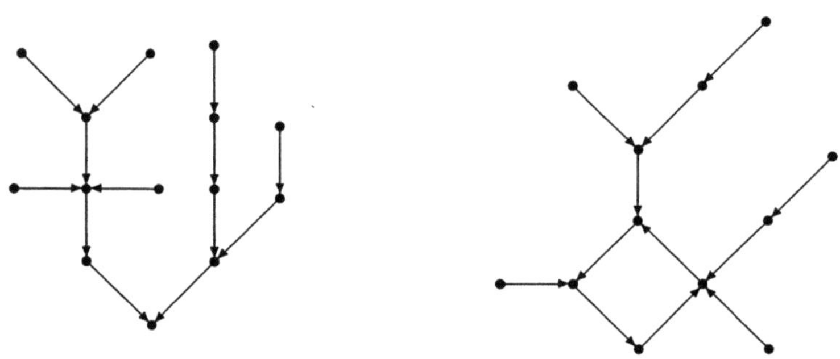

Proof. Assume that A is right serial, i.e. such that each vertex of $\mathcal{D} = \mathcal{D}(A)$ is an initial point of at most one arrow. Let $\{i_1, i_2, \dots, i_t\}$, $\{\sigma_1, \sigma_2, \dots, \sigma_t\}$ be a circuit of $\mathcal{D}$, and assume that $\sigma_1 : i_2 \to i_1$. Then $\sigma_2 : i_2 \to i_3$ is impossible and thus $\sigma_2 : i_3 \to i_2$. Similarly, $\sigma_3 : i_4 \to i_3, \dots, \sigma_t : i_1 \to i_t$, and thus the sequence $\{\sigma_t, \sigma_{t-1}, \dots, \sigma_1\}$ is a cycle (if $\sigma_1 : i_1 \to i_2$, then $\{\sigma_1, \sigma_2, \dots, \sigma_t\}$ is a cycle).

If there is no circuit in $\mathcal{D}$, then $\mathcal{D}$ is a tree. In this case, let i be a root of $\mathcal{D}$ and let $\mathcal{D}'$ be the non-empty set of all vertices of $\mathcal{D}$ from which there is a path to i. Similarly, denote by $\mathcal{D}''$ the set of all vertices from which there is a path to any other root of $\mathcal{D}$. Clearly, $\mathcal{D}' \cup \mathcal{D}'' = \mathcal{D}$ and, in view of our assumption, $\mathcal{D}' \cap \mathcal{D}'' = \emptyset$. Therefore, since A is connected, $\mathcal{D}'' = \emptyset$ by Theorem 3.6.2 and thus $\mathcal{D}$ has a unique root.

Now, let $\{\sigma_1, \sigma_2, \dots, \sigma_t\}$ be a cycle of $\mathcal{D}$. Observe that $\sigma_k : i_k \to i_{k+1}$ (with $i_{t+1} = i_1$) is the only arrow starting at i_k. Denote by $\mathcal{D}_k$ the set of all vertices of $\mathcal{D}$ from which there is a path to i_k which does not contain any arrow of the cycle; in particular, $i_k \in \mathcal{D}_k$. Let $\mathcal{D}_0 = \mathcal{D} \setminus \bigcup_{k=1}^{t} \mathcal{D}_k$. Since every circuit of $\mathcal{D}$ is a cycle, $\mathcal{D}_k$ are pairwise disjoint. Also, since every $i \in \mathcal{D}_k$, $i \neq i_k$ is the initial point for a unique arrow σ whose terminal point is again in $\mathcal{D}_k$, there are no cycles in $\mathcal{D}_k$. Finally, since there is obviously no arrow with initial point in $\mathcal{D}_0$ and terminal point in one of the $\mathcal{D}_k$'s, and since $\mathcal{D}$ is connected, $\mathcal{D}_0 = \emptyset$. Thus, removing all arrows $\sigma_1, \sigma_2, \ldots, \sigma_t$ of the cycle, we obtain a disjoint union of the diagrams $\mathcal{D}_k$ each of which is a tree with a unique root i_k.

To show the converse, let $\mathcal{D} = \mathcal{D}(A)$ be a diagram described in the corollary. First, if $\mathcal{D}$ is a tree with a unique root, then no vertex is an initial point of more than one arrow because there are no circuits in $\mathcal{D}$. It is easy to see that the same conclusion holds if there is a (unique) cycle in $\mathcal{D}$ and its complement is a disjoint union of trees with unique roots which are vertices of the cycle. Thus A is right serial and the proof of the statement is completed. □

Recall that the pair (B, V), where $B = A/R$, $V = R/R^2$ and $R = \operatorname{rad} A$ is called the type of the algebra A (see Sect. 8.5). Proposition 10.1.3 implies the following statement.

Corollary 10.2.4. *Let (B, V) be the type of an algebra A. Then the algebra A is right serial if and only if, for every minimal idempotent $e \in B$, the right B-module eV is simple (or zero).*

A B-bimodule V satisfying this condition will be called *right serial.*

Corollary 10.2.4 and Theorem 8.5.2 yield immediately the following result.

Theorem 10.2.5. *Every right serial algebra of separable type is isomorphic to the quotient algebra of a tensor algebra $T(V)$ by an admissible ideal, where V is a right serial bimodule over a separable algebra B. Conversely, every such quotient algebra is right serial.*

Thus, in order to describe right serial algebras of separable type, it is sufficient to exhibit all admissible ideals of the tensor algebra $T = T(V)$ of a right serial bimodule V over a separable algebra B.

Let $1 = e_1 + e_2 + \ldots + e_n$ be a decomposition of the identity of an algebra B in which all idempotents e_i are minimal, and let I be an admissible ideal of the algebra T (defined by a B-bimodule V). Write $T_i = e_iT$, $J_i = e_iJ$ (where J is the fundamental ideal of the algebra T; see Sect. 8.5) and $I_i = e_iI$. Then $I_i = T_iI$ is a submodule of T_i, and I is admissible if and only if $T_iJ^2 \supset I_i \supset T_iJ^m$ for some m, i. e. if $\ell_i = \ell(T_i/I_i) < \infty$. Moreover, if $J_i \neq 0$, or equivalently $V_i = e_iV \neq 0$, then $\ell_i \geq 2$. Now, Theorem 10.2.5 implies that T_i/T_iJ^m is a serial module and thus ℓ_i uniquely determines the submodule $I_i = T_iJ^{\ell_i}$.

Conversely, given the numbers ℓ_i, one can construct the right ideal $I = \bigoplus_{i=1}^{n} I_i$, where $I_i = T_i J^{\ell_i}$, and it remains to clarify conditions under which I is a two-sided ideal.

First, let us remark that if the simple B-modules e_iB and e_jB are isomorphic, then e_i and e_j are conjugate: $e_j = ae_ia^{-1}$ for an invertible element a from B. Hence, $aI_i = ae_iI = e_jaI = e_jI$ and thus $\ell_i = \ell(T_i/I_i) = \ell(T_j/I_j) = \ell_j$. In other words, the correspondence $i \to \ell_i$ defines a function on the diagram $\mathcal{D}$ of type (B, V).

Now, let $e_iB \not\simeq e_jB$ and let there be an arrow from the vertex corresponding to e_i to the vertex corresponding to e_j in $\mathcal{D}$. This means that $V_i = e_iV \simeq e_jB$. Then, since $J = VT$, $T_iJ = e_iTJ = e_iVT = V_iT$ and $T_iJ^2 = V_iTJ = V_iV_jT$. Continuing in this process, we obtain

$$I_i = TJ^{\ell_i} = V_{i_1}V_{i_2}\ldots V_{i_\ell}T, \quad \ell = \ell_i\,,$$

where $\{i = i_1, i_2, \ldots, i_\ell\}$ is a sequence of vertices in $\mathcal{D}$ such that there is an arrow from i_k to i_{k+1} $(k = 1, 2, \ldots, \ell - 1)$. It follows, in particular, that $BI = I$, and $VI \subset I$ if and only if, for every vertex i_0 from which there is an arrow to i, $V_{i_0}V_{i_1}\ldots V_{i_\ell}T \subset I$. But $V_{i_0}V_{i_1}\ldots V_{i_\ell}T \subset e_{i_0}T$, and therefore $V_{i_0}V_{i_1}\ldots V_{i_\ell}T \subset I_{i_0} = V_{i_0}V_{i_1}\ldots V_{i_m}T$, where $m = \ell_{i_0} - 1$. This is equivalent to the inequality $\ell \geq m$, i.e. $l_i \geq l_{i_0} - 1$.

If i is a sink in the diagram $\mathcal{D}$, then clearly $V_i = 0$ and necessarily $\ell_i = 1$. Consequently, we get the following result.

Proposition 10.2.6. *Let B be a semisimple algebra, V a right serial B-module and $\mathcal{D}$ a diagram of type (B, V). An admissible ideal $I \subset T(V)$ is determined uniquely by assigning a natural number ℓ_i to each vertex i of $\mathcal{D}$ in such a way that $\ell_i = 1$ for every sink i and that $2 \leq \ell_i \leq \ell_j + 1$ for every arrow from i to j.*

Now, having a description of admissible ideals of $T(V)$, it is easy to obtain a complete classification of right serial algebras of separable type.

Theorem 10.2.7. *A right serial algebra A of separable type (B, V) is determined by assigning a natural number ℓ_i to every vertex $i \in \mathcal{D}(A)$ in such a way that $\ell_i = 1$ for every sink i and that $2 \leq \ell_i \leq \ell_j + 1$ for every arrow from i to j. Two algebras, described by $(B, V, \ell_1, \ell_2, \ldots, \ell_s)$ and $(B', V', \ell'_1, \ell'_2, \ldots, \ell'_{s'})$, respectively, are isomorphic if and only if there is an algebra isomorphism $\varphi : B \to B'$, a B-bimodule isomorphism*[19] *$f : V \to V'$ such that $\ell'_{\sigma(i)} = \ell_i$, where σ is the isomorphism of the diagrams of (B, V) and (B', V') induced by the pair (φ, f).*

Proof. Taking into account Theorem 8.5.2 and Proposition 10.2.6, it only remains to verify the isomorphism criterion. Assume that an algebra A de-

[19] The B-bimodule structure of V' is defined by $b_1vb_2 = \varphi(b_1)v\varphi(b_2)$ for every $v \in V'$ and $b_1, b_2 \in B$.

scribed by the data $\{B, V, \ell_1, \ell_2, \ldots, \ell_s\}$ is isomorphic to an algebra A' described by $\{B', V', \ell'_1, \ell'_2, \ldots, \ell'_{s'}\}$. Then an isomorphism $\psi : A \to A'$ induces isomorphisms $\varphi : B \to B'$ and $f : V \to V'$, because $B = A/R$, $B' = A'/R'$, $V = R/R^2$ and $V' = R'/R'^2$, where $R = \operatorname{rad} A$ and $R' = \operatorname{rad} A'$. Moreover, if $A = P_1 \oplus P_2 \oplus \ldots \oplus P_n$ with principal A-modules P_i, then $A' = P'_1 \oplus P'_2 \oplus \ldots \oplus P'_n$, where $P'_i = \psi(P_i)$ are principal A'-modules and $\ell(P_i) = \ell(P'_i)$. Now, if i is a vertex of the diagram $\mathcal{D}(A)$, then, as we have seen, ℓ_i is just the length of the corresponding principal A-module. Consequently, $\ell'_{\sigma(i)} = \ell_i$ for the diagram isomorphism $\sigma : \mathcal{D}(A) \to \mathcal{D}(A')$ induced by the isomorphism ψ, or by the pair (φ, f).

Conversely, given φ and f with the indicated properties, then the isomorphism $T(V) \to T(V')$ induced by the pair (φ, f) clearly carries the ideal I given by the sequence $(\ell_1, \ell_2, \ldots, \ell_s)$ over to the ideal I' given by the sequence $(\ell'_1, \ell'_2, \ldots, \ell'_s)$. Hence, the quotient algebras $T(V)/I$ and $T(V')/I'$ are isomorphic and the theorem is proved. □

If A is a split algebra (for instance, if the field K is algebraically closed), then its type is given by the "multiplicities" n_i assigned to each vertex $i \in \mathcal{D}(A)$: They describe the decomposition $A/R = \prod_{i=1}^{s} M_{n_i}(K)$. Hence, such algebras can be described in the following way.

Corollary 10.2.8. *A split right serial algebra is determined by the data* $\{\mathcal{D}; n_1, n_2, \ldots, n_s; l_1, l_2, \ldots, l_s\}$, *where* $\mathcal{D}$ *is a diagram whose connected components satisfy the condition formulated in Corollary 10.2.3 and* n_i, ℓ_i $(i \in \mathcal{D})$ *are natural numbers such that* $\ell_i = 1$ *for every sink* i *and* $2 \le \ell_i \le \ell_j + 1$ *for every arrow from* i *to* j. *Two algebras described by* $(\mathcal{D}; n_1, n_2, \ldots, n_s; \ell_1, \ell_2, \ldots, \ell_s)$ *and* $(\mathcal{D}'; n'_1, n'_2, \ldots, n'_{s'}; \ell'_1, \ell'_2, \ldots, \ell'_{s'})$ *are isomorphic if and only if there is a diagram isomorphism* $\sigma : \mathcal{D} \to \mathcal{D}'$ *such that* $\ell'_{\sigma(i)} = \ell_i$ *and* $n'_{\sigma(i)} = n_i$ *for all vertices* $i \in \mathcal{D}$.

10.3 The Structure of Serial Algebras

Since serial algebras are right serial, all results of the preceding section apply here. We need only to specify more precisely the form of diagrams and possible types of such algebras.

Theorem 10.3.1. *Let* A *be a connected algebra and* (B, V) *its type; let* $B = \prod_{i=1}^{s} M_{n_i}(D_i)$ *with division algebras* D_i. *The algebra* A *is serial if and only if* $\mathcal{D}(A)$ *is either a cycle or a chain and, moreover,* $D_1 \simeq D_2 \simeq \ldots \simeq D_s$.

Proof. Applying Corollary 10.2.4 and its analogue for left serial algebras, we see that A is serial if and only if both the right B-module eV and the left B-module

Ve are simple for every minimal idempotent $e \in B$. We can, obviously, assume that A is basic.

Let $1 = e_1 + e_2 + \ldots + e_s$ be a decomposition of the identity of the algebra B such that $e_iB \simeq D_i$. There is an arrow from j to i in the diagram $\mathcal{D} = \mathcal{D}(A)$ if and only if $V_{ji} = e_jVe_i \neq 0$. But if $V_{ji} \neq 0$ and $V_{ki} \neq 0$ for $j \neq k$, then the left module Ve_i contains direct summands V_{ji} and V_{ki} and therefore is not simple. Hence, no vertex i of the diagram $\mathcal{D}$ is a terminal point of more than one arrow. Since, by Theorem 10.2.2, no vertex is an initial point of more than one arrow, one can see easily that the diagram $\mathcal{D}$ is either a cycle or an (oriented) chain.

Now, choose in $\mathcal{D}$ a pair of vertices i, j such that there is an arrow j to i. Then $V_{ji} \neq 0$ and thus, evidently, $e_jV = V_{ji} = Ve_i$ is a simple right B-module and a simple left B-module. Since $V_{ji}e_i \neq 0$, it turns out that $V_{ji} \simeq D_i$ as right B-modules. Similarly, $V_{ji} \simeq D_j$ as left B-modules. Thus, we have obtained a D_j-D_i-bimodule $U = V_{ji}$ which is isomorphic to D_i as a right D_i-module and D_j as a left D_j-module. Assigning to every element $a \in D_j$ a D_i-homomorphism $U \to U$ which maps $u \in U$ into au, we get an algebra homomorphism $\varphi : D_j \to \mathrm{End}_{D_i}(U) \simeq D_i$. Since D_j is a division algebra, φ is a monomorphism. Since, moreover, $[D_j : K] = [D_i : K]$, φ is an isomorphism, i. e. $D_j \simeq D_i$. Since A is connected, i. e. the diagram is connected, all division algebras $D_1, D_2, \ldots, D_s$ are isomorphic.

Conversely, let $\mathcal{D}$ be a chain or a cycle and $D_1 \simeq D_2 \simeq \ldots \simeq D_s$. By Theorem 10.2.2, the algebra A is right serial. However, if i is an arbitrary vertex of the diagram $\mathcal{D}$, then Ve_i is either 0 or coincides with V_{ji} for a unique vertex j from which there is an arrow to i. Since A is right serial, V_{ji} is a simple right B-module and thus $V_{ji} \simeq D_i$. On the other hand, since $e_jV_{ji} = V_{ji}$, $V_{ji} \simeq mD_j$ as left B-module. In view of $[D_i : K] = [D_j : K]$, this is possible only for $m = 1$ and then Ve_i is a simple left B-module. Thus A is left serial and, consequently, serial, as required. □

Theorems 10.3.1 and 10.2.7 facilitate a complete description of serial algebras of separable type. Since we may restrict our considerations to connected (indecomposable) basic algebras, it is necessary to describe only those bimodules V over the algebra $B = D^s$, where D is a division algebra, which are simultaneously right and left serial. In this case, the diagram $\mathcal{D}$ of the type (B, V) is either a chain

or a cycle

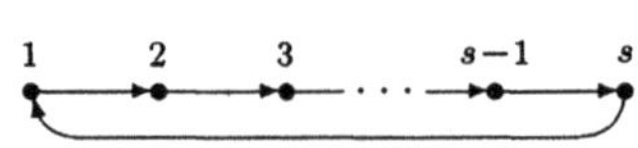

Write $V_i = e_iV$, where e_i is the minimal idempotent of the algebra B corresponding to the vertex i of the diagram $\mathcal{D}$. Then $V_i = Ve_{i+1}$, $i = 1, 2, \ldots, s-1$, and $V_s = 0$ if $\mathcal{D}$ is a chain, or $V_s = Ve_1$ if $\mathcal{D}$ is a cycle. Moreover, V_i is a one-dimensional D-bimodule, i. e. V_i is defined by an automorphism σ_i of the division algebra D (see Sect. 4.1, Example 3). We may assume that $V_i = D$ and that $a \circ v \circ b = \sigma_i(a)vb$ (here, multiplication on the left-hand side of the expression is in the D-bimodule V_i and on the right-hand side, in the division algebra D). Consider an automorphism φ of the algebra B which induces an automorphism τ of the division algebra D in the ith component, and the identity in all other components of B. Then, we may define on a B-bimodule V a new B-bimodule structure V^φ by $a * v * b = \varphi(a) \circ v \circ \varphi(b)$. If $v \in V_j$, where $j \neq i$ and $j \neq i-1$, then clearly $a * v * b = a \circ v \circ b$. If $v \in V_i$, then $a * v * b = \tau(a) \circ v \circ b = \sigma_i\tau(a)vb$. Finally, if $v \in V_{i-1}$, then $a * v * b = a \circ v \circ \tau(b) = \sigma_{i-1}(a)v\tau(b)$. In particular, for $v = 1$, $a * 1 = \sigma_{i-1}(a) = 1 * \tau^{-1}\sigma_{i-1}(a)$, so that the automorphism corresponding to the ith component of the bimodule V^φ is $\sigma_i\tau$ and the one corresponding to the $(i-1)$th component is $\tau^{-1}\sigma_{i-1}$. By Theorem 10.2.7, the pair (B, V) can be replaced by (B, V^φ) (choosing for f the identity map $V \to V^\varphi$).

Taking $i = 2$, $\tau = \sigma_1$, we can replace in this way the original choice of automorphisms by a choice with $\sigma_1 = 1$. Continuing in this manner by taking $i = 3$, $\tau = \sigma_2$ etc., we arrive to the situation when $\sigma_1 = \sigma_2 = \ldots = \sigma_{s-1} = 1$. In the case that $\mathcal{D}$ is a cycle, we are left with an automorphism $\sigma = \sigma_s$. Applying the previous construction to every vertex $i \in \mathcal{D}$ each time with the same automorphism τ, we replace, as we can see easily, the sequence $\{1, 1, \ldots, 1, \sigma\}$ by $\{1, 1, \ldots, 1, \tau^{-1}\sigma\tau\}$. Of course, we can equally replace $\{1, 1, \ldots, \sigma\}$ by $\{1, \ldots, 1, \sigma, 1, \ldots, 1\}$ (σ being at an arbitrary given position).

Let us remark that if U is a one-dimensional D-bimodule defined by an automorphism σ and W a one-dimensional D-bimodule defined by an automorphism τ, then the one-dimensional D-bimodule $U \otimes_D W$ is defined by $\tau\sigma$. Indeed, identifying U and W with D, we have $a \circ (1 \otimes 1) = (a \circ 1) \otimes 1 = \sigma(a) \otimes 1 = \big(1 \circ \sigma(a)\big) \otimes 1 = 1 \otimes \big(\sigma(a) \circ 1\big) = 1 \otimes \tau\big(\sigma(a)\big) = (1 \otimes 1) \circ \tau\sigma(a)$.

It follows that if V_i is a B-bimodule defined by σ_i, then the bimodule $\bar{V} = V_1 \otimes_B V_2 \otimes_B \ldots \otimes_B V_s$ is defined by the automorphism $\bar{\sigma} = \sigma_s \ldots \sigma_2\sigma_1$.

Now, let the diagram $\mathcal{D}$ be a cycle, V' be a B-bimodule defined by the automorphisms $\{\sigma'_1, \sigma'_2, \ldots, \sigma'_s\}$, $\psi : B \to B$ be an automorphism of the algebra B and $f : V \to V'$ an isomorphism such that $f(a \circ v \circ b) = \psi(a) \circ f(v) \circ \psi(b)$. Obviously, $f(V_i) = V'_{k(i)}$, where k is a cyclic permutation of $\{1, 2, \ldots, s\}$. Let, as before, $\bar{V} = V_1 \otimes_B V_2 \otimes_B \ldots \otimes_B V_s$, $\bar{V}' = V'_{k(1)} \otimes_B V'_{k(2)} \otimes_B \ldots \otimes_B V'_{k(s)}$. Then f induces a map $\bar{f} : \bar{V} \to \bar{V}'$ such that $\bar{f}(a \circ \bar{v} \circ b) = \psi(a) \circ \bar{f}(\bar{v}) \circ \psi(b)$. Denoting by τ the automorphism of the division algebra D which coincides with the restriction of ψ to the first component of the algebra B, and taking $\bar{v} = 1 \otimes 1 \otimes \ldots \otimes 1$, we get

$$\bar{f}(a \circ \bar{v}) = \tau(a) \circ \bar{f}(\bar{v}) = \bar{f}\big(\bar{v} \circ \bar{\sigma}(a)\big) = \bar{f}(\bar{v}) \circ \tau\bar{\sigma}(a).$$

Considering $\bar{f}(\bar{v})$ as a basis element of the one-dimensional D-bimodule $\bar{V}'$, we see that $\bar{V}'$ is a bimodule defined by the automorphism $\tau\bar{\sigma}\tau^{-1}$. On the other hand, $\bar{V}'$ is defined by the automorphism $\bar{\sigma}' = \sigma'_{k(s)} \cdots \sigma'_{k(2)}\sigma'_{k(1)}$. It follows that $\tau\bar{\sigma}\tau^{-1}$ and $\bar{\sigma}'$ differ by an inner automorphism of the division algebra D (see Example 3 of Sect. 4.1).

Taking into account that V can always be defined by a sequence $\{1, 1, \ldots, 1, \sigma\}$, we obtain a complete classification of serial algebras of separable type.

Theorem 10.3.2. *A connected serial algebra of separable type is determined by $\{\mathcal{D}, D, \sigma; n_1, n_2, \ldots, n_s; \ell_1, \ell_2, \ldots, \ell_s\}$, where $\mathcal{D}$ is a diagram which is either a chain or a cycle, D a separable division algebra, σ an automorphism of the division algebra D (here $\sigma = 1$ if $\mathcal{D}$ is a chain) and n_i, ℓ_i are natural numbers satisfying $2 \leq \ell_i \leq \ell_{i+1} + 1$ for $i = 1, 2, \ldots, s-1$ and such that $\ell_s = 1$ if $\mathcal{D}$ is a chain, and $2 \leq \ell_s \leq \ell_1 + 1$ if $\mathcal{D}$ is a cycle. Furthermore, the division algebra $\mathcal{D}$ is unique up to an isomorphism, and the automorphism σ up to conjugacy and an inner automorphism; the sequences of natural numbers $\{n_1, n_2, \ldots, n_s\}$ and $\{\ell_1, \ell_2, \ldots, \ell_s\}$ are unique if $\mathcal{D}$ is a chain and unique up to a simultaneous cyclic permutation if $\mathcal{D}$ is a cycle.*

We shall also give a criterion for basic algebras to be serial which will resemble the characterization of uniserial algebras as algebras of principal ideals.

Theorem 10.3.3. *A basic algebra A is serial if and only if its radical R is a principal right and a principal left ideal.*

Proof. Obviously, we may assume that A is connected (indecomposable). If A is serial, then its diagram is either a chain or a cycle and thus the principal right A-modules $P_1, P_2, \ldots, P_s$ can be indexed in such a way that $P(P_iR) \simeq P_{i+1}$, $i = 1, 2, \ldots, s-1$, and that $P_sR = 0$, or $P(P_sR) \simeq P_1$. But then $P(R) \simeq \bigoplus_{i=1}^{s} P(P_iR)$ is isomorphic to a direct summand of A and hence R is a principal right ideal. Similarly, R is a principal left ideal.

Conversely, let R be principal both as a left and as a right ideal. Then there is an epimorphism $A \to R$. Consequently, $P(R)$ is a direct summand of A and thus each of the principal right modules P_i appears in $P(R)$ no more than once. Let $1 = e_1 + e_2 + \ldots + e_s$ be a decomposition of the identity such that $P_i \simeq e_iA$, $V = R/R^2$ and $V_{ij} = e_iVe_j$. Recall that $P = P(R) \simeq P(V)$ and the multiplicity of P_i in P equals the multiplicity of the simple A-module $U_i = P_i/P_iR$ in V (Theorem 3.3.7). Therefore there is at most one index j such that $V_{ji} \neq 0$ and then $V_{ji} \simeq U_i$. In a similar way, the fact that R is a principal left ideal implies that, for each i, there is at most one index j such that $V_{ij} \neq 0$ and then V_{ij} is a simple left A-module.

Now, we are going to show that every vertex i of the diagram $\mathcal{D}(A)$ is an initial point of at most one arrow. Indeed, if there were arrows from i to two distinct points j and k, then $V_{ij} \neq 0$ and $V_{ik} \neq 0$, a contradiction. Also if there were more than one arrow from i to j, the multiplicity of U_j in V_{ij} would be bigger than one, again a contradiction. As a result, the algebra A is right serial (Theorem 10.2.2). Similarly, A is left serial, and thus serial. □

10.4 Quasi-Frobenius and Hereditary Serial Algebras

If a quasi-Frobenius algebra A is right serial, and thus every principal right A-module is serial, then the principal left A-modules, being co-principal, i. e. dual to the principal right ones, are also serial and hence A is serial. This means that for quasi-Frobenius algebras, the properties to be right serial, left serial and serial are all equivalent. The following theorem establishes a criterion for serial algebras to be quasi-Frobenius.

Theorem 10.4.1. *Let A be a connected serial algebra. Then A is a quasi-Frobenius algebra if and only if $\mathcal{D}(A)$ is a cycle and all principal right A-modules have the same length.*

Proof. Assume that A is a quasi-Frobenius algebra. If A is semisimple, the statement is trivial. Otherwise, since A is connected, Corollary 9.3.5 implies that $\mathcal{D}(A)$ has no sinks. Thus $\mathcal{D}(A)$ is a cycle. Moreover, if $\varphi : P_j \to R_i$ is an epimorphism of the principal A-module P_j onto the radical of the principal A-module P_i, then φ is not a monomorphism (in view of the fact that P_j is injective). Thus, $\ell_j = \ell(P_j) > \ell(R_i) = \ell(P_i) - 1$, i. e. $\ell_j \geq \ell_i$. Taking into account that $\mathcal{D}(A)$ is a cycle, we obtain $\ell_1 \leq \ell_2 \leq \ldots \leq \ell_s \leq \ell_1$, so all ℓ_i are equal.

Conversely, let $\mathcal{D}(A)$ be a cycle and $\ell_1 = \ell_2 = \ldots = \ell_s$. Let $P = P_i$ be a principal right A-module, M_k its unique submodule such that $\ell(P/M_k) = k$ (clearly, $M_k = PR^k$, where $R = \operatorname{rad} A$). For convenience, write $P_{s+1} = P_1$, $P_{s+2} = P_2$, etc. Then $P(M_1) \simeq P_{i+1}$ and thus M_2 is an epimorphic image of R_{i+1}; from here, $P(M_2) \simeq P_{i+2}$. In general, $P(M_k) \simeq P_{i+k}$ for $M_k \neq 0$. In particular, $\operatorname{soc} P = M_{\ell-1}$ and thus $P(\operatorname{soc} P_i) = P_{i+\ell-1}$. It is clear that the modules $P_{i+\ell-1}$ are non-isomorphic for $i = 1, 2, \ldots, s$. Therefore the socle of a principal right A-module P_i is simple and for $P_i \not\simeq P_j$, $\operatorname{soc} P_i \not\simeq \operatorname{soc} P_j$. By Theorem 9.3.7, A is a quasi-Frobenius algebra, as required. □

Corollary 10.4.2. *A connected quasi-Frobenius serial algebra A of separable type is determined by $\{s, D, \sigma, \ell; n_1, n_2, \ldots, n_s\}$, where D is a separable division algebra, σ an automorphism of the division algebra D and $s, \ell, n_1, n_2, \ldots, n_s$ are natural numbers. Furthermore, D is unique up to an isomorphism and σ is unique up to conjugacy and an inner automorphism; the sequence $\{n_1, n_2, \ldots, n_s\}$ is unique up to a cyclic permutation (the number s of vertices in the cycle $\mathcal{D}$ and the length ℓ of the principal A-modules are unique).*

Let us describe the hereditary right serial algebras. First, let us remark that, in view of Corollary 3.7.3, there are no cycles in the diagram $\mathcal{D}$ of such algebras. By Corollary 10.2.3, it follows that $\mathcal{D}$ is a disjoint union of trees with unique roots. For algebras of separable type, a complete description can be obtained from Theorem 8.5.4 and 10.2.5: such an algebra is isomorphic to $T(V)$, where V is a right serial bimodule over a separable algebra B such that there are no cycles in the diagram of type (B, V). However, it turns out that the statement holds even without the assumption of separability. This follows from the following result.

Theorem 10.4.3. *Let A be a finite dimensional algebra, $R = \operatorname{rad} A$, $\mathcal{D} = \mathcal{D}(A)$, $B = A/R$ and $V = R/R^2$. Assume that for any arrow $\varphi : i \to j$ in the diagram $\mathcal{D}$, there is no path $\sigma : i \to j$ of length 2 or more (in particular, $\mathcal{D}$ contains no cycles). Then the algebra A is isomorphic to a quotient algebra of the tensor algebra $T(V)$ of the B-bimodule V by an admissible ideal.*

Proof. As in the proof of Theorem 8.5.2, it is sufficient to verify that there is a subalgebra $\bar{A} \simeq B$ of A and a B-subbimodule $\bar{R}$ of R such that $R = \bar{R} \oplus R^2$.

Denote by $P_1, P_2, \ldots, P_s$ the distinct principal A-modules, $R_i = \operatorname{rad} P_i$ and $R_{ij} = \operatorname{Hom}_A(P_j, P_i)$. If $f : P_i \to P_i$ is a homomorphism which is not an isomorphism, then $\operatorname{Im} f \subset R_i$. Consequently, $f = \psi g$ where $\psi : P(R_i) \to R_i$ is an epimorphism and $g : P_i \to P(R_i)$. Since $\mathcal{D}$ has no cycles, it follows that for any direct summand P_j of $P(R_i)$, $j \neq i$ and there is no path from j to i. Then, by Lemma 3.6.1, $R_{ji} = 0$ and hence $g = 0$ and $f = 0$. Therefore, $E_A(P_i) = D_i$ is a division algebra. Let $A \simeq nP_1 \oplus n_2 P_2 \oplus \ldots \oplus n_s P_s$. Applying Theorem 3.5.2 to the algebra $A \simeq \operatorname{End}_A(A)$, we see that $R = \bigoplus_{i \neq j} R_{ij}$, $B = A/R \simeq \prod_{i=1}^{s} M_{n_i}(D_i)$ and A contains a subalgebra $\bar{A} \simeq B$.

Now, let there be an arrow from i to j. Then every path from i to j is an arrow and, according to Lemma 3.6.1, $R_{ij} \simeq V_{ij} = e_i V e_j$, where e_i, e_j are idempotents such that $e_i A \simeq P_i$ and $e_j A \simeq P_j$. Write $\bar{R} = \bigoplus_{i \to j} R_{ij}$ (summation runs over the pairs (i, j) for which there is an arrow from i to j); thus we have a submodule $\bar{R}$ of R such that $R = \bar{R} \oplus R^2$ and the theorem follows. □

Corollary 10.4.4. *A hereditary right serial algebra is isomorphic to $T(V)$, where V is a bimodule over a semisimple algebra B such that the diagram of type (B, V) is a disjoint union of trees with unique roots.*

Corollary 10.4.5. *A hereditary serial algebra is isotypic (i. e. Morita equivalent) to a direct product of algebras of triangular matrices over division algebras.*

Proof. Clearly, it is sufficient to prove that a connected basic hereditary serial algebra A is isomorphic to an algebra of triangular matrices over a division algebra. However, in view of Theorem 10.2.3 and Corollary 10.4.4, such an algebra is isomorphic to $T(V)$, where V is a bimodule over the algebra $B = D \times D \times \ldots \times D = D^s$ (D is a division algebra) such that $V_{ij} = e_i V e_j = 0$ if $j \neq i+1$ and $V_{i(i+1)}$ is a regular D-bimodule for $i = 1, 2, \ldots, s-1$ (here, $1 = e_1 + e_2 + \ldots + e_s$ is a minimal decomposition of the identity of the algebra B). Denote by $e_{i(i+1)}$ the element of $V_{i(i+1)}$ for which $ae_{i(i+1)} = e_{i(i+1)}a$ for $a \in D$.

Let us compute $V^{\otimes 2}$. Clearly, for $j \neq i+1$, $V_{i(i+1)} \otimes_B V_{j(j+1)} = 0$. On the other hand, $V_{i(i+1)} \otimes_B V_{(i+1)(i+2)} \simeq D \otimes_D D \simeq D$ and the element $e_{i(i+2)} = e_{i(i+1)} \otimes e_{(i+1)(i+2)}$ is a basis element of this module. Similarly, we may construct the elements $e_{i(i+3)} = e_{i(i+2)} \otimes e_{(i+2)(i+3)}$, and in general, all elements e_{ij} for $s \geq j > i$. Observe that $V^{\otimes s} = 0$ because there are no paths of length s in the diagram of type (B, V). Therefore every element of the algebra $T(V)$ has a unique form $\sum_{1 \leq i < j \leq s} a_{ij}e_{ij} = \sum_{i,j} e_{ij}a_{ij}$ with $a_{ij} \in D$. Since we can see easily that $e_{ij}e_{k\ell} = 0$ for $j \neq k$ and $e_{ij}e_{j\ell} = e_{i\ell}$, we may associate each element of the algebra with the matrix

$$\begin{pmatrix} a_{11} & a_{12} & \dots & a_{1s} \\ 0 & a_{22} & \dots & a_{2s} \\ \dots & \dots & \dots & \dots \\ 0 & 0 & \dots & a_{ss} \end{pmatrix}$$

and obtain an isomorphism $T(V) \simeq T_s(D)$, as required. □

From the preceding description of $T(V)$ and Theorem 10.4.3 we obtain the following consequence.

Corollary 10.4.6. *Let A be a connected serial algebra whose diagram is a chain. Then A is isotypic (i. e. Morita equivalent) to a quotient algebra of $T_s(D)$, where D is a division algebra.*

Exercises to Chapter 10

1. Let $A \subset M_3(K)$ be a subalgebra consisting of all matrices of the form

$$\begin{pmatrix} a_1 & a_2 & a_3 \\ 0 & a_4 & 0 \\ 0 & 0 & a_5 \end{pmatrix}, \quad a_i \in K.$$

a) Prove that the regular left A-module is semi-serial while the right one is not.

b) Let M be a right A-module, e_{ij} be the matrix units and $M_i = Me_{ii}$. Verify that multiplication by e_{12} (or e_{13}) defines a linear transformation $L_2 : M_1 \to M_2$ (or $L_3 : M_1 \to M_3$, respectively). Conversely, let M_1, M_2, M_3 be three vector spaces and let $L_2 : M_1 \to M_2$ and $L_3 : M_1 \to M_3$ be linear transformations. Put $M = M_1 \oplus M_2 \oplus M_3$ and define multiplication of elements by the basis elements of the algebra by the formulas

$$\begin{gathered}(m_1, m_2, m_3)e_{11} = (m_1, 0, 0); \quad (m_1, m_2, m_3)e_{22} = (0, m_2, 0); \\ (m_1, m_2, m_3)e_{33} = (0, 0, m_3); \quad (m_1, m_2, m_3)e_{12} = (0, m_1 L_2, 0); \\ (m_1, m_2, m_3)e_{13} = (0, 0, m_1 L_3).\end{gathered}$$

Show that in this way M becomes an A-module. Moreover, if N is another A-module obtained in this way by means of transformations L_2' and L_3', then $M \simeq N$ if and only if there are automorphisms φ_i of the spaces M_i, $i = 1, 2, 3$ such that

$$L_i' = \varphi_1 L_i \varphi_i, \quad i = 2, 3.$$

c) Making use of the construction in b), compute all indecomposable A-modules and check that they satisfy the conditions 3) and 4) of Theorem 10.1.1.

d) Verify that $\operatorname{rad} A$ is a principal right ideal but it is not a principal left ideal.

2. Consider the subalgebra $A \subset M_2(\mathbb{C})$ consisting of all matrices of the form

$$\begin{pmatrix} a_1 & a_2 \\ 0 & a_3 \end{pmatrix}, \quad a_1 \in \mathbb{R}, \ a_2, a_3 \in \mathbb{C}.$$

Show that A is right serial, but not left serial, while $\mathcal{D}(A)$ is a chain and $\operatorname{rad} A$ is a principal right ideal.

3. Let $A = T_2(K)$ (the algebra of triangular matrices), $P_i = e_{ii}A$ (where e_{ii} is a matrix unit), $P = 2P_1 \oplus P_2$, $B = \operatorname{End}_A(P)$. Prove that B is a serial algebra and $\operatorname{rad} B$ is not a principal right ideal.

4. Prove that if $R = \operatorname{rad} A$ is principal both as a right and as a left ideal, then A is serial. (Hint: Let $A \simeq \bigoplus_{i=1}^{m} n_i P_i \simeq \bigoplus_{i=1}^{m} n_i P_i'$, where P_i (P_i') are mutually non-isomorphic principal right (left) A-modules so that if $P_i \simeq e_i A$ then $P_i' \simeq A e_i$; let $P(P_i R) = \bigoplus_{j=1}^{m} t_{ij} P_j$, $P(RP_i') = \bigoplus_{j=1}^{m} t_{ij}' P_j'$. From Exercise 5 to Chap. 3, deduce that $\sum_{i=1}^{m} t_{ij} n_i \le n_j$ and $\sum_{i=1}^{m} t_{ij}' n_i \le n_j$ for every j. Using these inequalities and the fact that $t_{ij} = 0$ implies $t_{ji}' = 0$ and vice versa, deduce that $\sum_{j=1}^{m} t_{ij} \le 1$ and $\sum_{j=1}^{m} t_{ij}' \le 1$ for any i.) The converse is, in view of Exercise 3, false.

5. Prove that an algebra A is right serial if and only if any diagram of the form

$$\begin{array}{ccc} P_1 & & \\ & \searrow^{\varphi} & \\ & & P_3\ , \\ & \nearrow_{\psi} & \\ P_2 & & \end{array}$$

where P_1 , P_2 , P_3 are principal A-modules (not necessarily distinct) can be completed to one of the following two commutative diagrams:

$$\begin{array}{ccc} P_1 & & \\ & \searrow^{\varphi} & \\ f\downarrow & & P_3\ ; \\ & \nearrow_{\psi} & \\ P_2 & & \end{array} \qquad\qquad \begin{array}{ccc} P_1 & & \\ & \searrow^{\varphi} & \\ g\uparrow & & P_3\ . \\ & \nearrow_{\psi} & \\ P_2 & & \end{array}$$

6. Let $1 = e_1 + e_2 + \ldots + e_n$ be a minimal decomposition of the identity of an algebra A ($n \geq 3$). Prove that A is right serial (serial) if and only if, for any choice of three indices i, j, k, the algebra eAe, where $e = e_i + e_j + e_k$, is right serial (serial). (Hint: Use Theorem 8.4.4 and the preceding exercise.)

7. Prove a theorem similar to the one formulated in the previous exercise for right serial hereditary algebras.

8. Let $B = D^s$, where D is a finite dimensional division algebra, $1 = e_1 + e_2 + \ldots + e_s$ be a minimal decomposition of the identity of the algebra B, and let V be a B-module satisfying the following conditions: $V_{ij} = e_i V e_j = 0$ for all pairs (i, j), except for $(i, i+1)$ and $(s, 1)$; $V_{i(i+1)}$ is a regular D-bimodule and V_{s1} a D-bimodule defined by an automorphism σ of the division algebra D. Prove that the tensor algebra $T(V)$ is isomorphic to the algebra of matrices of the form

$$\begin{pmatrix} a_{11} & a_{12} & a_{13} & \ldots & a_{1n} \\ ta_{21} & a_{22} & a_{23} & \ldots & a_{2n} \\ ta_{31} & ta_{32} & a_{33} & \ldots & a_{3n} \\ \ldots & \ldots & \ldots & \ldots & \ldots \\ ta_{n1} & ta_{n2} & ta_{n3} & \ldots & a_{nn} \end{pmatrix},$$

where a_{ij} are elements of the skew polynomial algebra $D[t, \sigma]$ (see Exercise 11 to Chap. 9), and that the fundamental ideal J consists of the matrices whose diagonal entries are all multiples of t.

9. From Exercise 8, deduce a description of basic serial algebras of separable type. How should we modify the construction of the respective matrix algebra in order to obtain algebras which are not basic?

10. Prove that a quasi-Frobenius serial algebra A of a separable type (B, V) is isomorphic to $T(V)/J^k$, where J is the fundamental ideal of the tensor algebra of the B-bimodule V.

11. Elements of Homological Algebra

The present chapter has been written for the English edition. The aim of this extension is to present an introduction to homological methods, which play an increasingly important role in the theory of algebras, and in this way to make the book more suitable as a textbook. Besides the fundamental concepts of a complex, resolutions and derived functors, we shall also briefly examine three special topics: homological dimension, almost split sequences and Auslander algebras.

11.1 Complexes and Homology

A *complex* of A-modules $(V_\bullet, d_\bullet)$, or simply $V_\bullet$, is a sequence of A-modules and homomorphisms

$$\cdots \longrightarrow V_2 \xrightarrow{d_2} V_1 \xrightarrow{d_1} V_0 \xrightarrow{d_0} V_{-1} \xrightarrow{d_{-1}} V_{-2} \longrightarrow \cdots$$

such that $d_n d_{n+1} = 0$ for all indices n. Clearly, this means that $\operatorname{Im} d_{n+1} \subset \operatorname{Ker} d_n$. Thus, one can define the *homology modules* $\mathrm{H}_n(V_\bullet) = \operatorname{Ker} d_n / \operatorname{Im} d_{n+1}$.

The set of the maps $d_\bullet = \{d_n\}$ is called the *differential* of the given complex. In what follows, we shall write often dx instead of $d_n x$ for $x \in V_n$ (and use, without mentioning it, other similar simplifications by omitting subscripts). The coset ("homology coset") $x + \operatorname{Im} d_{n+1}$, where $x \in \operatorname{Ker} d_n$, will be denoted by $[x]$.

If $(V'_\bullet, d'_\bullet)$ is another complex, a complex homomorphism $f_\bullet : V_\bullet \to V'_\bullet$ is a family of homomorphisms $f_n : V_n \to V'_n$ "commuting with the differential", i.e. such that $f_{n-1} d_n = d'_n f_n$ for all n. Evidently, such a family induces homology maps

$$\mathrm{H}_n(f_\bullet) : \mathrm{H}_n(V_\bullet) \to \mathrm{H}_n(V'_\bullet)$$

by $\mathrm{H}_n(f_\bullet)[x] = [f_n(x)]$ for all n (it is easy to see that for $dx = 0$, also $d' f(x) = 0$ and $[f(x + dy)] = [f(x)]$). In this way, we can consider the *category of complexes* of A-modules com-A and the family of the functors $\mathrm{H}_n :$ com-$A \to$ mod-A.

Two homomorphisms $f_\bullet$ and $g_\bullet : V_\bullet \to V'_\bullet$ are said to be *homological* if $\mathrm{H}_n(f_\bullet) = \mathrm{H}_n(g_\bullet)$ for all n; we shall denote this fact by $f_\bullet \equiv g_\bullet$. An important example of homological homomorphisms is the case of homotopic homomorphisms in the following sense. Two homomorphisms $f_\bullet$ and $g_\bullet$ are called *homotopic*: $f_\bullet \sim g_\bullet$ if there are homomorphisms $s_n : V_n \to V'_{n+1}$ such that $f_n - g_n = d'_{n+1}s_n + s_{n-1}d_n$ for all n (the sequence $s_\bullet = \{s_n\}$ is called a *homotopy* between $f_\bullet$ and $g_\bullet$).

Proposition 11.1.1. *Homotopic homomorphisms are homological.*

Proof. For every homology class $[x]$,

$$\begin{aligned} \mathrm{H}_n(f_\bullet)[x] &= \big[f(x)\big] = \big[g(x) + d's(x) + s(dx)\big] = \\ &= \big[g(x) + d's(x)\big] = \big[g(x)\big] = \mathrm{H}_n(g_\bullet)[x] \end{aligned}$$

because $dx = 0$. □

Two complexes $V_\bullet$ and $V'_\bullet$ are called *homotopic* if there are homomorphisms $f_\bullet : V_\bullet \to V'_\bullet$ and $f'_\bullet : V'_\bullet \to V_\bullet$ such that $f_\bullet f'_\bullet \sim 1$ and $f'_\bullet f_\bullet \sim 1$. In this case, we shall write $V_\bullet \sim V'_\bullet$.

Corollary 11.1.2. *If $V_\bullet$ and $V'_\bullet$ are homotopic, then $\mathrm{H}_n(V_\bullet) \simeq \mathrm{H}_n(V'_\bullet)$ for all n.*

Remark. The converse of Proposition 11.1.1 and of Corollary 11.1.2 does not hold in general: $f_\bullet \equiv g_\bullet$ does not imply $f_\bullet \sim g_\bullet$ and $\mathrm{H}_n(V_\bullet) \simeq \mathrm{H}_n(V'_\bullet)$ for all n does not imply $V_\bullet \sim V'_\bullet$ (see Exercise 1 and 2).

Along with complexes of the above type ("chain complexes") it is often convenient to consider "cochain complexes" $(V^\bullet, d^\bullet)$ of the form

$$\cdots \longrightarrow V^{-1} \xrightarrow{d^{-1}} V^0 \xrightarrow{d^0} V^1 \xrightarrow{d^1} V^2 \longrightarrow \cdots$$

with the condition $d^n d^{n-1} = 0$. In this case, we obtain the *cohomology modules* $\mathrm{H}^n(V^\bullet) = \operatorname{Ker} d^n / \operatorname{Im} d^{n-1}$. Obviously, one can pass from chain to cochain complexes simply by changing the indices, i.e. putting $V^n = V_{-n}$ and $d^n = d_{-n}$; hereby, H_n becomes H^{-n}. One can usually use the "chain" terminology if the complex is bounded from the right, i.e. there is a number n_0 so that $V_n = 0$ for $n < n_0$ and "cochain" terminology if $V_\bullet$ is bounded from the left, i.e. if there is a number n_0 so that $V_n = 0$ for $n > n_0$.

If F : mod-A $\to$ mod-B is a functor, then F induces a functor $F_\bullet$: com-A $\to$ com-B assigning to a complex $V_\bullet = \{V_n, d_n\}$ the complex $F_\bullet(V_\bullet) = \{F(V_n), F(d_n)\}$. For example, considering the functor h_M : mod-A $\to$ Vect for a fixed A-module M (see Example 1 in Sect. 8.1), we obtain the functor com-A $\to$ com-K assigning to a complex $V_\bullet$ the complex $\mathrm{Hom}_A(M, V_\bullet) = \{\mathrm{Hom}_A(M, V_n)\}$. Similarly, for a left A-module N, we have the functor $- \otimes_A N$ assigning to a complex $V_\bullet$ the complex $V_\bullet \otimes_A N = \{V_n \otimes_A N\}$. A contravariant functor from mod-A to mod-B, i.e. a functor $G : (\text{mod-}A)^o \to \text{mod-}B$ defines a functor $G^\bullet : (\text{com-}A)^o \to \text{com-}B$, but it is more convenient in this case to consider $G^\bullet(V_\bullet)$ as a cochain complex with the nth component equal to $G(V_n)$. For instance, if $G = h^o_M$ (see Example 6 in Sect. 8.1), we obtain a contravariant functor mapping a chain complex $\{V_n\}$ into a cochain complex $\{\mathrm{Hom}_A(V_n, M)\}$.

It is evident that every such functor maps homotopic homomorphisms (and complexes) into homotopic ones; however, again, $f_\bullet \equiv g_\bullet$ does not imply $F_\bullet(f_\bullet) \equiv F_\bullet(g_\bullet)$ (see Exercise 3).

Let $f_\bullet : V_\bullet \to V'_\bullet$ be a complex homomorphism. Then, obviously, $d'_n(\mathrm{Im}\, f_n) \subset \mathrm{Im}\, f_{n-1}$ and $d_n(\mathrm{Ker}\, f_n) \subset \mathrm{Ker}\, f_{n-1}$ for all n, and thus we get the complexes $\mathrm{Im}\, f_\bullet = \{\mathrm{Im}\, f_n\}$ and $\mathrm{Ker}\, f_\bullet = \{\mathrm{Ker}\, f_n\}$. Therefore, one can define *exact sequences* of complexes just the same way as exact sequences of modules in Sect. 8.2. The following theorem seems to play a fundamental role in homological algebra.

Theorem 11.1.3. *Let* $0 \to V'_\bullet \xrightarrow{f_\bullet} V_\bullet \xrightarrow{g_\bullet} V''_\bullet \to 0$ *be an exact sequence of complexes. Then, for each* n*, there is a homomorphism* $\partial_n : \mathrm{H}_n(V''_\bullet) \to \mathrm{H}_{n-1}(V'_\bullet)$ *such that the following sequence is exact:*

$$\cdots \longrightarrow \mathrm{H}_{n+1}(V''_\bullet) \xrightarrow{\partial_{n+1}} \mathrm{H}_n(V'_\bullet) \xrightarrow{\mathrm{H}_n(f_\bullet)} \mathrm{H}_n(V_\bullet) \xrightarrow{\mathrm{H}_n(g_\bullet)}$$
$$\xrightarrow{\mathrm{H}_n(g_\bullet)} \mathrm{H}_n(V''_\bullet) \xrightarrow{\partial_n} \mathrm{H}_{n-1}(V'_\bullet) \xrightarrow{\mathrm{H}_{n-1}(f_\bullet)} \mathrm{H}_{n-1}(V_\bullet) \longrightarrow \cdots$$

Proof. (We shall use the same letter d for differentials in all complexes and omit subscripts.) Let $[x]$ be a homology coset of $\mathrm{H}_n(V''_\bullet)$. Since g_n is an epimorphism, $x = g(y)$ for some $y \in V_n$. Now, $g(dy) = dg(y) = dx = 0$ and thus, in view of the exactness, $dy = f(z)$ for some $z \in V'_{n-1}$. Furthermore, $f(dz) = df(z) = d^2y = 0$ and therefore $dz = 0$ because f is a monomorphism.

Let us verify that the coset $[z] \in \mathrm{H}_{n-1}(V'_\bullet)$ depends neither on the choice of y nor on the choice of x in the homology coset $[x]$. Indeed, if $g(y') = g(y)$, then $g(y'-y) = 0$ and $y'-y = f(u)$ for some u; thus $dy' = dy+df(u) = f(z+du)$ and $[z+du] = [z]$. Furthermore, let $[x'] = [x]$, i. e. $x' = x + dv$ for some $v \in V''_{n+1}$. Then there is $w \in V_{n+1}$ such that $v = g(w)$ and therefore $x' = g(y + dw)$. Since $d(y + dw) = dy$, the choice of x' does not effect the coset $[z]$.

Consequently, setting $\partial_n[x] = [z]$ gives a well-defined homomorphism $\partial_n : \mathrm{H}_n(V''_\bullet) \to \mathrm{H}_{n-1}(V'_\bullet)$. It remains to prove that the long sequence is exact.

We are going to show that $\mathrm{Ker}\, \mathrm{H}_n(f_\bullet) \subset \mathrm{Im}\, \partial_{n+1}$ and $\mathrm{Ker}\, \partial_n \subset \mathrm{Im}\, \mathrm{H}_n(g_\bullet)$ and leave the other (rather easy) verifications to the reader. Let $\mathrm{H}_n(f_\bullet)[x] = 0$. Thus $f(x) = dy$ for some $y \in V_{n+1}$. Put $z = g(y)$. Then $dz = g(dy) = gf(x) = 0$ and we get $[z] \in \mathrm{H}_{n+1}(V''_\bullet)$ satisfying $\partial[z] = [x]$ according to the definition of ∂.

Now, let $\partial_n[x] = 0$. By the definition of ∂, this means that if $x = g(y)$ and $dy = f(z)$, then $z = du$ for some $u \in V'_n$. Hence, $x = g(y - f(u))$ and $d(y - f(u)) = dy - f(du) = 0$, which gives that $[x] = \mathrm{H}_n(g_\bullet)[y - f(u)]$, as required. □

A complex $V_\bullet$ is called *acyclic in dimension* n if $\mathrm{H}_n(V_\bullet) = 0$ and *acyclic* if it is acyclic in all dimensions (trivially, it means that $V_\bullet$ is an exact sequence).

Corollary 11.1.4. *Let* $0 \to V'_\bullet \xrightarrow{f_\bullet} V_\bullet \xrightarrow{g_\bullet} V''_\bullet \to 0$ *be an exact sequence of complexes. Then*

1) $V_\bullet$ *is acyclic in dimension* n *if and only if* ∂_n *is a monomorphism and* ∂_{n+1} *is an epimorphism.*
2) $V'_\bullet$ *is acyclic in dimension* n *if and only if* $\mathrm{H}_n(g_\bullet)$ *is a monomorphism and* $\mathrm{H}_{n+1}(g_\bullet)$ *is an epimorphism.*
3) $V''_\bullet$ *is acyclic in dimension* n *if and only if* $\mathrm{H}_{n-1}(f_\bullet)$ *is a monomorphism and* $\mathrm{H}_n(f_\bullet)$ *an epimorphism.*

Corollary 11.1.5. *Let* $0 \to V'_\bullet \to V_\bullet \to V''_\bullet \to 0$ *be an exact sequence of complexes.*

1) *If* $V'_\bullet$ *and* $V''_\bullet$ *are acyclic in dimension* n*, then* $V_\bullet$ *is acyclic in dimension* n*.*
2) *If* $V_\bullet$ *is acyclic in dimension* n *and* $V'_\bullet$ *in dimension* $n-1$*, then* $V''_\bullet$ *is acyclic in dimension* n*.*
3) *If* $V_\bullet$ *is acyclic in dimension* n *and* $V''_\bullet$ *in dimension* $n+1$*, then* $V'_\bullet$ *is acyclic in dimension* n*.*

The construction of the connecting homomorphisms ∂_n also yields the following statement, whose proof is left to the reader.

Proposition 11.1.6. *Let*

$$\begin{array}{ccccccccc} 0 & \longrightarrow & V'_\bullet & \longrightarrow & V_\bullet & \longrightarrow & V''_\bullet & \longrightarrow & 0 \\ & & {\scriptstyle\alpha_\bullet}\downarrow & & {\scriptstyle\beta_\bullet}\downarrow & & {\scriptstyle\gamma_\bullet}\downarrow & & \\ 0 & \longrightarrow & W'_\bullet & \longrightarrow & W_\bullet & \longrightarrow & W''_\bullet & \longrightarrow & 0 \end{array}$$

be a commutative diagram of complexes with exact rows. Then the following diagram is commutative:

$$\begin{array}{ccc} \mathrm{H}_n(V''_\bullet) & \xrightarrow{\partial_n} & \mathrm{H}_{n-1}(V'_\bullet) \\ {\scriptstyle \mathrm{H}_n(\gamma_\bullet)}\downarrow & & \downarrow{\scriptstyle \mathrm{H}_{n-1}(\alpha_\bullet)} \\ \mathrm{H}_n(W''_\bullet) & \xrightarrow{\partial_n} & \mathrm{H}_{n-1}(W'_\bullet)\,. \end{array}$$

11.2 Resolutions and Derived Functors

Let M be an A-module. A *projective resolution* of M is a complex of A-modules $P_\bullet$ in which $P_n = 0$ for $n < 0$, all P_n are projective, and $P_\bullet$ is acyclic in every dimension $n \neq 0$, while $\mathrm{H}_0(P_\bullet) \simeq M$ is a fixed isomorphism. Observe that $\operatorname{Ker} d_0 = P_0$ and thus $\mathrm{H}_0(P_0) = P_0/\operatorname{Im} d_1$; hence, we have a fixed epimorphism $\pi : P_0 \to M$ whose kernel is $\operatorname{Im} d_1$. Therefore a projective resolution is often considered in the form of an exact sequence

$$\cdots \longrightarrow P_2 \xrightarrow{d_2} P_1 \xrightarrow{d_1} P_0 \xrightarrow{\pi} M \longrightarrow 0 .$$

However, in what follows, we want to underline the fact that M is not included in its projective resolution: the last non-zero term of its resolution is P_0.

In a dual way, one defines an *injective resolution* of an A-module M as a cochain complex $Q^\bullet$ in which $Q^n = 0$ for $n < 0$, all A-modules Q^n are injective and such that $Q^\bullet$ is acyclic in all dimensions $n \neq 0$, while $M \simeq \mathrm{H}^0(Q^\bullet) = \operatorname{Ker} d^0$ is a fixed isomorphism. Such a resolution can be identified with an exact sequence

$$0 \longrightarrow M \xrightarrow{\epsilon} Q^0 \xrightarrow{d^0} Q^1 \xrightarrow{d^1} Q^2 \longrightarrow \cdots$$

Generally speaking, we will deal with projective resolutions, leaving the corresponding formulations (and proofs) for injective resolutions to the reader.

Let $P_\bullet$ be a projective resolution of a module M and $P'_\bullet$ a projective resolution of M'. Then every complex morphism $f_\bullet : P_\bullet \to P'_\bullet$ induces a module homomorphism $\varphi : M \to M'$. The morphism $f_\bullet$ is said to be an *extension of* φ *to the resolutions* $P_\bullet$ and $P'_\bullet$. In other words, an extension of φ to the resolutions is a commutative diagram

$$\begin{array}{ccccccccc}
\cdots \longrightarrow & P_2 & \xrightarrow{d_2} & P_1 & \xrightarrow{d_1} & P_0 & \xrightarrow{\pi} & M & \longrightarrow 0 \\
 & {\scriptstyle f_2}\downarrow & & {\scriptstyle f_1}\downarrow & & {\scriptstyle f_0}\downarrow & & {\scriptstyle \varphi}\downarrow & \\
\cdots \longrightarrow & P'_2 & \xrightarrow{d'_2} & P'_1 & \xrightarrow{d'_1} & P'_0 & \xrightarrow{\pi'} & M' & \longrightarrow 0
\end{array}$$

Theorem 11.2.1. 1) *Every A-module M has a projective resolution.*

2) *Any two projective resolutions of a module M are homotopic.*

3) *Every homomorphism $\varphi : M \to M'$ can be extended to the resolutions $P_\bullet$ and $P'_\bullet$ of the modules M and M', respectively.*

4) *Any two extensions of φ to a given pair of resolutions are homotopic.*

Proof. 1) For every A-module M, there is an epimorphism $\varphi : P_0 \to M$ with a projective module P_0 (Corollary 3.3.4). Write $M_1 = \operatorname{Ker} \pi$ and construct an epimorphism $\pi_1 : P_1 \to M_1$, where P_1 is again projective. This epimorphism can be interpreted as a homomorphism $d_1 : P_1 \to P_0$ with $\operatorname{Im} d_1 = \operatorname{Ker} \pi$. Applying the same construction to $M_2 = \operatorname{Ker} d_1$, we obtain $d_2 : P_2 \to P_1$ with $\operatorname{Im} d_2 = \operatorname{Ker} d_1$. Continuing this process, we get a projective resolution $P_\bullet$ of the module M.

3) Let $P'_\bullet$ be a projective resolution of M'. Consider the homomorphism $\varphi\pi : P_0 \to M'$. Since P_0 is projective and $\pi' : P'_0 \to M'$ is an epimorphism, there is a homomorphism $f_0 : P_0 \to P'_0$ such that $\pi' f_0 = \varphi\pi$. From here, $\pi' f_0 d_1 = \varphi\pi d_1 = 0$ and thus $\operatorname{Im} f_0 d_1 \subset \operatorname{Ker} \pi'$. However, $\operatorname{Im} d'_1 = \operatorname{Ker} \pi'$, and P_1 is projective, so there is $f_1 : P_1 \to P'_1$ such that $f_0 d_1 = d'_1 f_1$. In particular, $d'_1 f_1 d_2 = f_0 d_1 d_2 = 0$ and therefore $\operatorname{Im} f_1 d_2 \subset \operatorname{Ker} d'_1$; hence there is $f_2 : P_2 \to P'_2$ such that $f_1 d_2 = d'_2 f_2$. Continuing this procedure, we construct an extension $f_\bullet : P_\bullet \to P'_\bullet$ of the homomorphism φ.

4) If $g_\bullet : P_\bullet \to P'_\bullet$ is another extension of φ, then $f_\bullet - g_\bullet$ is an extension of the zero homomorphism. Hence, it is sufficient to show that $f_\bullet \sim 0$ for any

extension $f_\bullet$ of the zero homomorphism. In such a case we have a commutative diagram

$$\begin{array}{ccccccccc}
\cdots \longrightarrow & P_3 & \xrightarrow{d_3} & P_2 & \xrightarrow{d_2} & P_1 & \xrightarrow{d_1} & P_0 & \longrightarrow 0 \\
& {\scriptstyle f_3}\downarrow & & {\scriptstyle f_2}\downarrow & & {\scriptstyle f_1}\downarrow & & {\scriptstyle f_0}\downarrow & \\
\cdots \longrightarrow & P'_3 & \xrightarrow{d'_3} & P'_2 & \xrightarrow{d'_2} & P'_1 & \xrightarrow{d'_1} & P'_0 & \longrightarrow 0
\end{array}$$

with $\operatorname{Im} f_0 \subset \operatorname{Im} d'_1$ (since $\mathrm{H}_0(f_\bullet) = 0$).

Since P_0 is projective, $f_0 = d'_1 s_0$ for some $s_0 : P_0 \to P'_1$; thus $f_0 = d'_1 s_0 + s_{-1} d_0$ (because $d_0 = 0$). Consider $\bar{f}_1 = f_1 - s_0 d_1$. Then $d'_1 \bar{f}_1 = d'_1 f_1 - d'_1 s_0 d_1 = d'_1 f_1 - f_0 d_1 = 0$ and therefore $\operatorname{Im} \bar{f}_1 \subset \operatorname{Ker} d'_1 = \operatorname{Im} d'_2$ in view of $\mathrm{H}_1(P'_\bullet) = 0$. Since P_1 is projective, there exists $s_1 : P_1 \to P'_2$ such that $\bar{f}_1 = d'_2 s_1$, i.e. $f_1 = s_0 d_1 + d'_2 s_1$. Now, take $\bar{f}_2 = f_2 - s_1 d_2$; again $d'_2 \bar{f}_2 = d'_2 f_2 - d'_2 s_1 d_2 = d'_2 f_2 - f_1 d_2 + s_0 d_1 d_2 = 0$ and subsequently $\bar{f}_2 = d'_3 s_2$, i.e. $f_2 = s_1 d_2 + d'_3 s_2$ for some $s_2 : P_2 \to P'_3$. Again, by induction, $f_\bullet \sim 0$.

2) Let $P_\bullet$ and $P'_\bullet$ be two projective resolutions of a module M. There are extensions $f_\bullet : P_\bullet \to P'_\bullet$ and $f'_\bullet : P'_\bullet \to P_\bullet$ of the identity homomorphism $1 : M \to M$. But then $f_\bullet f'_\bullet$ and $f'_\bullet f_\bullet$ also extend $1 : M \to M$. Since the identity morphisms $1_\bullet : P_\bullet \to P_\bullet$ and $1_\bullet : P'_\bullet \to P'_\bullet$ extend $1 : M \to M$, as well, 4) implies that $f_\bullet f'_\bullet \sim 1$ and $f'_\bullet f_\bullet \sim 1$. Therefore $P_\bullet \sim P'_\bullet$ and the theorem is proved. □

Taking into account the fact that every functor F : mod-$A \to$ mod-B translates homotopic complexes and homomorphisms into homotopic ones, and applying Proposition 11.1.1 and Corollary 11.1.2, we get the following consequence.

Corollary 11.2.2. 1) *Let F : mod-$A \to$ mod-B be a functor and $P_\bullet$ a projective resolution of an A-module M. Then the homology $\mathrm{H}_n(F(P_\bullet))$ is independent of the choice of the resolution $P_\bullet$.*

2) *If $P'_\bullet$ is a projective resolution of M' and $f_\bullet : P_\bullet \to P'_\bullet$ an extension of a homomorphism $\varphi : M \to M'$, then $\mathrm{H}_n\big(F_\bullet(f_\bullet)\big)$ is independent of the choice of the extension $f_\bullet$.*

In the situation described in Corollary 11.2.2, we shall write $L_nF(M) = \mathrm{H}_n\big(F_\bullet(P_\bullet)\big)$ and $L_nF(\varphi) = \mathrm{H}_n\big(F_\bullet(f_\bullet)\big)$. If $f_\bullet$ is an extension of φ and $g_\bullet$ an extension of $\psi : M' \to M''$, then $g_\bullet f_\bullet$ is an extension of $\psi\varphi$ and thus $L_nF(\psi\varphi) = L_nF(\psi)L_nF(\varphi)$, i.e. L_nF is a functor mod-$A \to$ mod-B, which is called the *n-th left derived functor* of the functor F. Similarly, replacing projective resolutions by injective ones, one can define *right derived functors* R^nF. The definitions of left and right derived functors of a contravariant functor G can be given dually, using injective resolutions for L_nG and projective resolutions for R^nG. All further arguments apply to right derived, as well as contravariant functors.

Proposition 11.2.3. *A right (left) exact functor F satisfies $L_0F \simeq F$ (respectively, $R^0F \simeq F$).*

Proof. If $P_\bullet$ is a projective resolution of M, then $P_1 \xrightarrow{d_1} P_0 \to M \to 0$ is an exact sequence, and thus $F(P_1) \xrightarrow{F(d_1)} F(P_0) \to F(M) \to 0$ is exact, as well. Therefore, $L_0F(M) = H_0\big(F_\bullet(P_\bullet)\big) = F(P_0)/\operatorname{Im} F(d_1) \simeq F(M)$. □

The importance of derived functors stems in many respects from the existence of "long exact sequences". Their construction is based on Theorem 11.1.3 and the following lemmas.

Lemma 11.2.4. *For every exact sequence of modules*

$$0 \longrightarrow M' \xrightarrow{\varphi} M \xrightarrow{\psi} M'' \longrightarrow 0,$$

there are projective resolutions $P'_\bullet$, $P_\bullet$ and $P''_\bullet$ and an exact sequence

$$0 \longrightarrow P'_\bullet \xrightarrow{f_\bullet} P_\bullet \xrightarrow{g_\bullet} P''_\bullet \longrightarrow 0,$$

in which $f_\bullet$ extends φ and $g_\bullet$ extends ψ.

Proof. Let $\pi' : P'_0 \to M'$ and $\pi'' : P''_0 \to M''$ be epimorphisms. Put $P_0 = P'_0 \oplus P''_0$ and consider a homomorphism $\pi = (\pi', \eta) : P_0 \to M$, where η is a homomorphism $P''_0 \to M$ such that $\psi\eta = \pi''$. It is easy to verify that π is also an epimorphism and that we obtain a commutative diagram

$$\begin{array}{ccccccccc}
 & & 0 & & 0 & & 0 & & \\
 & & \downarrow & & \downarrow & & \downarrow & & \\
0 & \longrightarrow & M'_1 & \xrightarrow{\varphi_1} & M_1 & \xrightarrow{\psi_1} & M''_1 & \longrightarrow & 0 \\
 & & \downarrow & & \downarrow & & \downarrow & & \\
0 & \longrightarrow & P'_0 & \xrightarrow{f_0} & P_0 & \xrightarrow{g_0} & P''_0 & \longrightarrow & 0 \\
 & & {\scriptstyle\pi'}\downarrow & & {\scriptstyle\pi}\downarrow & & {\scriptstyle\pi''}\downarrow & & \\
0 & \longrightarrow & M' & \xrightarrow{\varphi} & M & \xrightarrow{\psi} & M'' & \longrightarrow & 0 \\
 & & \downarrow & & \downarrow & & \downarrow & & \\
 & & 0 & & 0 & & 0 & &
\end{array}$$

in which all columns and the two lower rows are exact; here $M'_1 = \operatorname{Ker} \pi'$, $M_1 = \operatorname{Ker} \pi$, $M''_1 = \operatorname{Ker} \pi''$. According to part 3) of Corollary 11.1.5 (see also Exercise 3 to Chapter 8) the first row is also exact, and thus we may apply to it the same construction. By repeating this procedure, we obtain a required exact sequence of resolutions. □

Lemma 11.2.5. *If $0 \to V'_\bullet \to V_\bullet \to V''_\bullet \to 0$ is an exact sequence of complexes, where all modules V''_n are projective, then the sequence $0 \to F_\bullet(V'_\bullet) \to F_\bullet(V_\bullet) \to F_\bullet(V''_\bullet) \to 0$ is exact for every functor F.*

Proof. Since every sequence $0 \to V'_n \to V_n \to V''_n \to 0$ splits, the sequence $0 \to F(V'_n) \to F(V_n) \to F(V''_n) \to 0$ also splits. □

Now we apply the preceding lemmas and Theorem 11.1.3 in order to get a long exact sequence for arbitrary functors.

Corollary 11.2.6. *Let $0 \to M' \xrightarrow{\varphi} M \xrightarrow{\psi} M'' \to 0$ be an exact sequence of modules. Then for any functor F, there exist connecting homomorphisms $\partial_n : L_nF(M'') \to L_{n-1}F(M')$ so that the following sequence is exact*

$$\begin{array}{llllllll} \cdots \to & L_{n+1}F(M'') & \xrightarrow{\partial_{n+1}} & L_nF(M') & \xrightarrow{L_nF(\varphi)} & L_nF(M) & \xrightarrow{L_nF(\psi)} \\ \xrightarrow{L_nF(\psi)} & L_nF(M'') & \xrightarrow{\partial_n} & L_{n-1}F(M') & \xrightarrow{L_{n-1}F(\varphi)} & L_{n-1}F(M) & \to \cdots \end{array}$$

Observe that, by definition, $L_nF = 0$ for $n < 0$ and thus, Corollary 11.2.6 implies that L_0F is always right exact. In particular, if F itself is right exact, then in view of Proposition 11.2.3, the end of the long exact sequence has the following form:

$$\cdots \longrightarrow L_1F(M'') \xrightarrow{\partial_1} F(M') \longrightarrow F(M) \longrightarrow F(M'') \longrightarrow 0\,.$$

Corollary 11.2.7. 1) *A functor F is right (left) exact if and only if $F \simeq L_0F$ (respectively, $F \simeq R^0F$).*

2) *A right (left) exact functor F is exact if and only if $L_1F = 0$ (respectively, $R^1F = 0$).*

Observe that, for an exact F, both $L_nF = 0$ and $R^nF = 0$ for all $n > 0$.

If a module P is projective, then its projective resolution has a very simple form: $P_0 = P$ and $P_n = 0$ for $n > 0$. In particular, $L_nF(P) = 0$ for all $n > 0$.

This trivial observation indicates how to characterize derived functors "axiomatically", in the following way.

Theorem 11.2.8. *Let F be a right exact functor and $\{\Phi_n \mid n \geq 0\}$ a family of functors satisfying the following properties:*

1) *$\Phi_0 \simeq F$ (as functors);*
2) *$\Phi_n(P) = 0$ for all $n > 0$ and all projective P;*
3) *If $0 \to M' \xrightarrow{\varphi} M \xrightarrow{\psi} M'' \to 0$ is an exact sequence of modules, then there are homomorphisms $\Delta_n : \Phi_n(M'') \to \Phi_{n-1}(M')$, $n \geq 0$, so that the following sequence is exact:*

$$\begin{array}{llllllll} \cdots \to & \Phi_{n+1}(M'') & \xrightarrow{\Delta_{n+1}} & \Phi_n(M') & \xrightarrow{\Phi_n(\varphi)} & \Phi_n(M) & \xrightarrow{\Phi_n(\psi)} \\ \xrightarrow{\Phi_n(\psi)} & \Phi_n(M'') & \xrightarrow{\Delta_n} & \Phi_{n-1}(M') & \xrightarrow{\Phi_{n-1}(\varphi)} & \Phi_{n-1}(M) & \to \cdots \end{array}$$

Then $\Phi_n(M) \simeq L_nF(M)$ for all $n \geq 0$ and all modules M.

Proof. The exact sequence $0 \to L \xrightarrow{\alpha} P \to M \to 0$ with a projective module P induces a long exact sequence for the functors Φ_n. For $n = 1$, we get the exact sequence

$$\Phi_1(P) = 0 \longrightarrow \Phi_1(M) \xrightarrow{\Delta_1} \Phi_0(L) \xrightarrow{\Phi_0(\alpha)} \Phi_0(P),$$

from where $\Phi_1(M) \simeq \operatorname{Ker} \Phi_0(\alpha) = \operatorname{Ker} F(\alpha) \simeq L_1F(M)$ by the condition 1). For $n > 1$, the exact sequence has the form

$$\Phi_n(P) = 0 \longrightarrow \Phi_n(M) \xrightarrow{\Delta_n} \Phi_{n-1}(L) \longrightarrow \Phi_{n-1}(P) = 0,$$

thus Δ_n is an isomorphism and the theorem follows by induction. □

Remark. In fact, in Theorem 11.2.8, $\Phi_n \simeq L_nF$ as functors; however, we will not use this result.

From Proposition 11.1.6, we get also the following consequence.

Corollary 11.2.9. *Let*

$$\begin{array}{ccccccccc} 0 & \longrightarrow & M' & \longrightarrow & M & \longrightarrow & M'' & \longrightarrow & 0 \\ & & {\scriptstyle\alpha}\downarrow & & {\scriptstyle\beta}\downarrow & & {\scriptstyle\gamma}\downarrow & & \\ 0 & \longrightarrow & N' & \longrightarrow & N & \longrightarrow & N'' & \longrightarrow & 0 \end{array}$$

be a commutative diagram with exact rows. Then the following diagram is commutative:

$$\begin{array}{ccc} L_nF(M'') & \xrightarrow{\partial_n} & L_{n-1}F(M') \\ {\scriptstyle L_nF(\gamma)}\downarrow & & \downarrow{\scriptstyle L_{n-1}F(\alpha)} \\ L_nF(N'') & \xrightarrow{\partial_n} & L_{n-1}F(N'). \end{array}$$

11.3 Ext and Tor. Extensions

The construction of derived functors applies, in particular, to the functors Hom and $\otimes$ (more precisely, to the functors h_M, h_N°, $X \otimes_A -$ and $- \otimes_A Y$). Since Hom is left exact, it is natural to consider right derived functors $R^n h_M$ (constructed by means of injective resolutions) and $R^n h_N^\circ$ (constructed by means of projective resolutions, since h_N° is contravariant), which coincide for $n = 0$ with h_M and h_N°. It is a remarkable fact that these constructions produce the same result.

Theorem 11.3.1. *For all A-modules M, N and each $n \geq 0$,*

$$R^n h_M(N) \simeq R^n h_N^\circ(M).$$

Proof. Fix a module M and put $\Phi_n(N) = R^n h_N^\circ(M)$. If $\varphi : N \to L$, then φ induces a functor morphism $h_N^\circ \to h_L^\circ$ assigning to a homomorphism $\alpha : M \to N$ the homomorphism $\varphi\alpha : M \to L$, and thus also a derived functor morphism $\Phi_n(\varphi) : \Phi_n(N) \to \Phi_n(L)$. Note that if N is injective, then, in accordance with the definition of injectivity (see Theorem 9.1.4), the functor h_N^o is exact and therefore $\Phi_n(N) = 0$ for $n > 0$. In addition, $\Phi_0(N) = R^0 h_N^o(M) \simeq h_N^o(M) = h_M(N)$ by Proposition 11.2.3. Clearly, this isomorphism is functorial in N, and thus $\Phi_0 \simeq h_M$.

Now, let $0 \to N' \xrightarrow{\varphi} N \xrightarrow{\psi} N'' \to 0$ be an exact sequence. Then, for any complex $P_\bullet$ consisting of projective modules, the sequence of complexes

$$0 \to \mathrm{Hom}_A(P_\bullet, N') \to \mathrm{Hom}_A(P_\bullet, N) \to \mathrm{Hom}_A(P_\bullet, N'') \to 0$$

is exact. Taking for $P_\bullet$ a projective resolution of the module M, we get, according to Theorem 11.1.3, just a long exact cohomology sequence similar to that which appears in the formulation of Theorem 11.2.8 (condition 3)). Thus, all the conditions of this theorem are satisfied, and therefore $\Phi_n(N) \simeq R^n h_M(N)$. The proof of the theorem is completed. □

The common value $R^n h_M(N) \simeq R^n h_N^o(M)$ is denoted by $Ext_A^n(M, N)$.

An analogous result holds for the functors $t_M = M \otimes_A -$ and $t_N = - \otimes_A N$, where M is a right and N is a left A-module.

Theorem 11.3.2. *For any right A-module M and any left A-module N, and each $n \geq 0$,*

$$L_n t_M(N) = L_n t_N(M).$$

The proof is (quite similar to the proof of Theorem 3.1) left to the reader.

The common value of these functors is denoted by $\mathrm{Tor}_n^A(M, N)$. Let us point out that $\mathrm{Ext}_A^0(M, N) \simeq \mathrm{Hom}(M, N)$ and $\mathrm{Tor}_0^A(M, N) \simeq M \otimes_A N$.

The functor $\mathrm{Ext}_A^1(M, N)$ is closely related to the module extensions. Referring to Sect. 1.5, let us reformulate the definition of an extension of a module M with kernel N as an exact sequence ζ of the form

$$\zeta : 0 \to N \xrightarrow{\alpha} X \xrightarrow{\beta} M \to 0.$$

Two extensions ζ and ζ', where

$$\zeta' : 0 \to N \xrightarrow{\alpha'} X' \xrightarrow{\beta'} M \to 0$$

are said to be equivalent (which is denoted by $\zeta \simeq \zeta'$) if there is a homomorphism $\gamma : X \to X'$ such that the following diagram is commutative:

$$\begin{array}{ccccccccc} 0 & \longrightarrow & N & \xrightarrow{\alpha} & X & \xrightarrow{\beta} & M & \longrightarrow & 0 \\ & & {\scriptstyle 1_N}\downarrow & & {\scriptstyle \gamma}\downarrow & & {\scriptstyle 1_M}\downarrow & & \\ 0 & \longrightarrow & N & \xrightarrow{\alpha'} & X' & \xrightarrow{\beta'} & M & \longrightarrow & 0. \end{array}$$

By Lemma 8.2.1 (Five lemma), γ is an isomorphism. Denote by $\mathrm{Ex}(M, N)$ the set of all equivalence classes of extensions of M with kernel N.

By Corollary 11.2.6, an exact sequence ζ induces a connecting homomorphism $\partial_\zeta : \mathrm{Hom}_A(M, M) \to \mathrm{Ext}^1_A(M, N)$. The element $\delta(\zeta) = \partial_\zeta(1_M)$ is called the *characteristic class* of the extension ζ. If $\zeta \simeq \zeta'$, then the diagram

$$\begin{array}{ccc} \mathrm{Hom}_A(M, M) & \xrightarrow{\partial_\zeta} & \mathrm{Ext}^1_A(M, N) \\ \downarrow & & \downarrow \\ \mathrm{Hom}_A(M, M) & \xrightarrow{\partial_{\zeta'}} & \mathrm{Ext}^1_A(M, N) \end{array}$$

is, by Corollary 11.2.9, commutative (with the vertical maps being identity morphisms). From here, $\delta(\zeta) = \delta(\zeta')$, and therefore we get a well defined map $\delta : \mathrm{Ex}(M, N) \to Ext^1_A(M, N)$.

Theorem 11.3.3. *The map δ is one-to-one.*

Proof. We are going to construct an inverse map ω. To this end, fix an exact sequence $0 \to N \xrightarrow{\varepsilon} Q \xrightarrow{\sigma} L \to 0$ with an injective module Q. By Corollary 11.2.6, the sequence

$$\mathrm{Hom}_A(M, Q) \xrightarrow{h_M(\sigma)} \mathrm{Hom}(M, L) \xrightarrow{\partial} \mathrm{Ext}^1_A(M, N) \longrightarrow 0$$

is exact (since $\mathrm{Ext}^1_A(M, Q) = 0$). In particular, every element $u \in \mathrm{Ext}^1_A(M, N)$ is of the form $u = \partial(\varphi)$ for some $\varphi : M \to L$. Consider a lifting of the given exact sequence along φ (see Exercise 5 to Chap. 8), i. e. the exact sequence

$$\xi : 0 \to N \xrightarrow{f} Z \xrightarrow{g} M \to 0,$$

where Z is a submodule of $Q \oplus M$ consisting of the pairs (q, m) such that $\sigma(q) = \varphi(m)$, and f and g are defined by the rules $f(n) = (\varepsilon(n), 0)$ and $g(q, m) = m$. If φ' is another homomorphism satisfying $\partial(\varphi') = u$, then $\varphi' = \varphi + \sigma\eta$ for some $\eta : M \to Q$. Then an equivalence of the extensions ξ and $\xi' : 0 \to N \to Z' \to M \to 0$ constructed as a lifting along φ', is given by a homomorphism $\gamma : Z \to Z'$ sending (q, m) into $(q + \eta(m), m)$ (the simple verification is left to the reader). Consequently, by defining $\omega(u) = \xi$, we get a map $\mathrm{Ext}^1_A(M, N) \to \mathrm{Ex}(M, N)$. The commutative diagram

$$\begin{array}{ccccccccc} 0 & \longrightarrow & N & \xrightarrow{f} & Z & \xrightarrow{g} & M & \longrightarrow & 0 \\ & & {\scriptstyle 1_N}\downarrow & & {\scriptstyle \psi}\downarrow & & {\scriptstyle \varphi}\downarrow & & \\ 0 & \longrightarrow & N & \xrightarrow{\varepsilon} & Q & \xrightarrow{\sigma} & L & \longrightarrow & 0, \end{array}$$

where $\psi(q, m) = q$, yields, in view of Corollary 11.2.9, a commutative square

$$\begin{array}{ccc} \mathrm{Hom}_A(M, M) & \xrightarrow{\partial_\xi} & \mathrm{Ext}^1_A(M, N) \\ {\scriptstyle h_M(\varphi)}\downarrow & & \downarrow{\scriptstyle 1} \\ \mathrm{Hom}_A(M, L) & \xrightarrow{\partial} & \mathrm{Ext}^1_A(M, N), \end{array}$$

and thus $\delta\omega(u) = \partial_\xi(1_M) = \partial(\varphi) = u$.

It remains to show that $\omega\delta(\zeta) \simeq \zeta$ holds for an arbitrary extension $\zeta : 0 \to N \xrightarrow{\alpha} X \xrightarrow{\beta} M \to 0$. Let $\delta(\zeta) = u$. Since Q is injective, the homomorphism $\varepsilon : N \to Q$ extends to $\mu : X \to Q$ such that $\mu\alpha = \varepsilon$, and yields a commutative diagram

$$\begin{array}{ccccccccc} 0 & \longrightarrow & N & \xrightarrow{\alpha} & X & \xrightarrow{\beta} & M & \longrightarrow & 0 \\ & & {\scriptstyle 1_N}\downarrow & & {\scriptstyle \mu}\downarrow & & {\scriptstyle \varphi}\downarrow & & \\ 0 & \longrightarrow & N & \xrightarrow{\varepsilon} & Q & \xrightarrow{\sigma} & L & \longrightarrow & 0\,. \end{array}$$

Therefore the following square is commutative:

$$\begin{array}{ccc} \mathrm{Hom}_A(M,M) & \xrightarrow{\partial_\zeta} & \mathrm{Ext}^1_A(M,N) \\ {\scriptstyle h_M(\varphi)}\downarrow & & \downarrow{\scriptstyle 1} \\ \mathrm{Hom}_A(M,L) & \xrightarrow{\partial} & \mathrm{Ext}^1_A(M,N)\,, \end{array}$$

and $u = \partial(\varphi)$. Using this φ in constructing $\omega(u)$ as above, we get a sequence $\xi : 0 \to N \to Z \to M \to 0$. But then the homomorphism $\gamma : X \to Z$ given by $\gamma(x) = \big(\mu(x), \beta(x)\big)$ establishes the equivalence of ζ and $\xi = \omega(u)$. The theorem is proved. □

In the sequel, we shall identify the elements of $\mathrm{Ext}^1_A(M,N)$ and the respective extensions. Since, for a fixed M, $\mathrm{Ext}^1(M,N)$ is a covariant functor of N (and, for a fixed N, a contravariant functor of M), a homomorphism $\varphi : N \to N'$ (a homomorphism $\psi : M' \to M$) induces a map $\varphi_e : \mathrm{Ext}^1_A(M,N) \to \mathrm{Ext}^1_A(M,N')$ (respectively, a map $\psi^e : \mathrm{Ext}^1_A(M,N) \to Ext^1_A(M',N)$). From the explicit form of the one-to-one correspondence $\omega : \mathrm{Ext}^1_A(M,N) \to \mathrm{Ex}(M,N)$ constructed above, we get immediately the following corollary.

Corollary 11.3.4. 1) *The extension $\omega(\psi^e(u))$ is equivalent to the lifting of $\omega(u)$ along ψ.*
2) *The extension $\omega(\varphi_e(u))$ is equivalent to the descent of $\omega(u)$ along φ.*

(A lifting of an exact sequence has been already defined above. A descent of an extension $0 \to N \xrightarrow{f} Z \xrightarrow{g} M \to 0$ along $\varphi : N \to N'$ is, by definition, the exact sequence $0 \to N' \xrightarrow{f'} Z' \xrightarrow{g'} M \to 0$, where $Z' = (N' \oplus Z)/Y$ with $Y = \big\{\big(-\varphi(n), f(n)\big) \mid n \in N\big\}$ and $f'(n') = [n',0]$, $g'([n',z]) = g(z)$. Here $[n',z]$ denotes the coset $(n',z) + Y$.)

Using the preceding Corollary 11.3.4, we shall write $\psi^e(\zeta) = \omega(\psi^e(u))$ and $\varphi_e(\zeta) = \omega(\varphi_e(u))$ for $\zeta = \omega(u)$.

Corollary 11.3.5. *The following conditions are equivalent:*

1) *The module M is projective (injective).*
2) $\mathrm{Ext}^1_A(M,N) = 0$ *(respectively, $Ext^1_A(N,M) = 0$) for every module N.*

3) $\mathrm{Ext}^1_A(M,N) = 0$ *(respectively,* $Ext^1_A(N,M) = 0$*) for every simple module* N.
4) $\mathrm{Ext}^n_A(M,N) = 0$ *(respectively,* $Ext^n_A(N,M) = 0$*) for each* $n > 0$ *and every module* N.

Proof. The implications 1) $\Rightarrow$ 4) $\Rightarrow$ 2) are trivial and 2) $\Rightarrow$ 1) follows in view of Theorem 11.3.3 and Theorem 3.3.5 (or Theorem 9.1.4 for injectivity). Also, 2) $\Rightarrow$ 3) is trivial, while 3) $\Rightarrow$ 2) can be proved by induction on the length of N, using the long exact sequence. □

It is remarkable that, for modules over finite dimensional algebras, the following statement also holds.

Proposition 11.3.6. *The following conditions are equivalent:*

1) *The module* M *is projective.*
2) $\mathrm{Tor}^A_1(M,N) = 0$ *for every module* N.
3) $\mathrm{Tor}^A_1(M,N) = 0$ *for every simple module* N.
4) $\mathrm{Tor}^A_n(M,N) = 0$ *for every module* N *and each* $n > 0$.

Proof. Again, 1) $\Rightarrow$ 4) $\Rightarrow$ 2) $\Rightarrow$ 3) are trivial. We are going to prove 3) $\Rightarrow$ 1). Consider an exact sequence $0 \to L \to P \xrightarrow{\pi} M \to 0$, where $\pi : P \to M$ is a projective cover of M. Write $\bar{A} = A/R$ with $R = \mathrm{rad}\, A$ and note that $\mathrm{Tor}^A_1(M,\bar{A}) = 0$ because $\bar{A}$ is a direct sum of simple modules. Therefore, in view of Corollary 11.2.6, $0 \longrightarrow L \otimes_A \bar{A} \longrightarrow P \otimes_A \bar{A} \xrightarrow{\pi\otimes 1} M \otimes_A \bar{A} \longrightarrow 0$ is an exact sequence. Now, one can see easily that $M \otimes_A \bar{A} \simeq M/MR$ (an isomorphism can be defined by $x + MR \mapsto x \otimes 1$). Since $\pi : P \to M$ is a projective cover, $\pi \otimes 1$ defines an isomorphism $P/PR \simeq M/MR$. Thus, $L/LR = 0$ and, by Nakayama's lemma, $L = 0$. Hence, $\pi : P \to M$ is an isomorphism and M is projective. □

11.4 Homological Dimensions

The functor mod-$A \to$ Vect assigning to X the space $\mathrm{Ext}^n_A(M,X)$ will be denoted by h^n_M. Notice that if M is a B-A-bimodule then h^n_M can be considered as a functor mod-$A \to$ mod-B. The *projective dimension* of an A-module M is said to be n: $proj.dim_A\, M = n$ if $h^n_M \neq 0$ and $h^m_M = 0$ for all $m > n$; if no such number exists, define $proj.dim_A\, M = \infty$. Dually, considering the functors $h^{o\,n}_M : X \mapsto \mathrm{Ext}^n_A(X,M)$, we define the *injective dimension* $inj.dim_A\, M$ to be n, if $h^{o\,n}_M \neq 0$ but $h^{o\,m}_M = 0$ for all $m > n$, and $inj.dim_A\, M = \infty$ if no such number n exists.

In accordance with Corollary 11.3.5, $proj.dim_A\, M = 0$ means that M is projective and $inj.dim_A = 0$ that M is injective. Furthermore, Corollary 11.2.6 provides an inductive way for computing these dimensions.

Proposition 11.4.1. *Let* $0 \to L \to P \to M \to 0$ *and* $0 \to M \to Q \to N \to 0$ *be exact sequences with a projective module* P *and an injective module* Q. *If* M *is not projective (not injective), then* $proj.dim_A\, M = proj.dim_A\, L + 1$ *(respectively,* $inj.dim_A\, M = inj.dim_A\, N + 1$*).*

Proposition 11.4.2. *Let* $0 \to L \to P_{k-1} \to \cdots \to P_1 \to P_0 \to M \to 0$ *and* $0 \to M \to Q_0 \to Q_1 \to \cdots \to Q_{k-1} \to N \to 0$ *be exact sequences with projective modules* $P_0, P_1, \dots, P_{k-1}$ *and injective modules* $Q_o, Q_1, \dots, Q_{k-1}$. *If* $proj.dim_A\, M \geq k$ *(*$inj.dim_A\, M \geq k$*), then* $proj.dim_A\, M = proj.dim_A\, L + k$ *(respectively,* $inj.dim_A\, M = inj.dim_A\, N + k$*).*

Proposition 11.4.3. *Let* $(P_\bullet, d_\bullet)$ *(respectively,* $(Q^\bullet, d^\bullet)$*) be a projective (injective) resolution of a module* M. *If* M *is not projective (not injective), then* $proj.dim_A\, M = \min\{n \mid \operatorname{Ker} d_{n-1}$ *is projective*$\}$ *(respectively,* $inj.dim_A\, M = \min\{n \mid \operatorname{Coker} d^{n-1}$ *is injective*$\}$*).*

Taking into account Proposition 11.3.6, we obtain also a definition of projective dimension in terms of Tor.

Corollary 11.4.4. $proj.dim_A\, M$ *is equal to* n *if and only if* $\operatorname{Tor}^A_{n+1}(M,N) = 0$ *for all* N *and* $\operatorname{Tor}^A_n(M,N) \neq 0$ *for some module* N *(*$proj.dim_A\, M = \infty$ *if no such* n *exists).*

Let $\bar{A} = A/R$ where $R = \operatorname{rad} A$. In view of condition 3) of Corollary 11.3.5 and Proposition 11.3.6, we get the following result.

Corollary 11.4.5.

$$\begin{aligned} proj.dim_A\, M &= \sup\{n \mid \operatorname{Ext}^n_A(M,\bar{A}) \neq 0\} = \\ &= \sup\{n \mid \operatorname{Tor}^A_n(M,\bar{A}) \neq 0\}\,; \\ inj.dim_A\, M &= \sup\{n \mid \operatorname{Ext}^n_A(\bar{A},M) \neq 0\}\,. \end{aligned}$$

Corollary 11.4.6. *The following values coincide for any finite dimensional algebra* A*:*

$\sup\{proj.dim_A\, M \mid M$ *a right* A*-module*$\}$;
$\sup\{inj.dim_A\, M \mid M$ *a right* A*-module*$\}$;
$\sup\{proj.dim_A\, M \mid M$ *a left* A*-module*$\}$;
$\sup\{inj.dim_A\, M \mid M$ *a left* A*-module*$\}$;
$proj.dim_A\, \bar{A}$;
$inj.dim_A\, \bar{A}$;
$\sup\{n \mid \operatorname{Ext}^n_A(\bar{A},\bar{A}) \neq 0\}$;
$\sup\{n \mid \operatorname{Tor}^n_A(\bar{A},\bar{A}) \neq 0\}$.

(Here, $\bar{A}$ *can always be considered either as a right or as a left* A*-module.)*

This common value is called the *global dimension* of the algebra A and is denoted by $gl.dim\, A$.

Obviously, $gl.dim\, A = 0$ if and only if A is semisimple. In view of Proposition 11.4.1, if A is not semisimple, then $gl.dim\, A = proj.dim_A\, R + 1$. In particular, $gl.dim\, A = 1$ if and only if R is projective, i. e. if and only if A is hereditary (see Theorem 3.7.1). Later we shall also use the following criterion resulting from Proposition 11.4.3.

Corollary 11.4.7. *The following conditions are equivalent:*

1) *$gl.dim\, A \leq 2$;*
2) *the kernel of a homomorphism between projective A-modules is projective;*
3) *the cokernel of a homomorphism between injective A-modules is injective.*

11.5 Duality

Given a complex $(V_\bullet, d_\bullet)$ of right (left) A-modules, one can construct a dual complex $(V_\bullet^*, d_\bullet^*)$:

$$\cdots \longrightarrow V_{-2}^* \xrightarrow{d_{-1}^*} V_{-1}^* \xrightarrow{d_0^*} V_0^* \xrightarrow{d_1^*} V_1^* \xrightarrow{d_2^*} V_2^* \longrightarrow \cdots$$

of left (right) A-modules (in view of indexing, it is natural to consider it as a cochain complex). In order to compute its cohomology, we shall recall (without proofs) some well-known facts from linear algebra.

Proposition 11.5.1. *Let $U \supset W$ be subspaces of a vector space V. Then there is a canonical isomorphism $(U/W)^* \simeq W^\perp/U^\perp$.*

Proposition 11.5.2. *For any linear transformation $f : V \to W$, $(\operatorname{Im} f)^\perp = \operatorname{Ker} f^*$ and $(\operatorname{Ker} f)^\perp = \operatorname{Im} f^*$.*

As a result, we get immediately the following statements.

Corollary 11.5.3. $\mathrm{H}^n(V_\bullet^*) \simeq \mathrm{H}_n(V_\bullet)^*$.

Corollary 11.5.4. *For any right A-module M and any left A-module N, $\operatorname{Ext}_A^n(M, N^*) \simeq \operatorname{Tor}_n^A(M, N)^*$.*

Proof. Consider a projective resolution $P_\bullet$ of the left module N: $\cdots \to P_2 \to P_1 \to P_0 \to N \to 0$. Passing to the dual right modules, we get an injective resolution $P_\bullet^*$ of the module N^*: $0 \to N^* \to P_0^* \to P_1^* \to P_2^* \to \cdots$. It follows from the adjoint isomorphism formula (Proposition 8.3.4) that

$$\begin{aligned}\operatorname{Hom}_A(M, P_\bullet^*) &\simeq \operatorname{Hom}_A\big(M, \operatorname{Hom}_K(P_\bullet, K)\big) \simeq \\ &\simeq \operatorname{Hom}_K(M \otimes_A P_\bullet, K) = (M \otimes_A P_\bullet)^*,\end{aligned}$$

and thus, by Corollary 11.5.3, the cohomology $\mathrm{Ext}_A^n(M,N^*)$ of the complex $\mathrm{H}_A(M,P_\bullet^*)$ is dual to the homology $\mathrm{Tor}_n^A(M,N)$ of the complex $M\otimes_A P_\bullet$. □

In the sequel, we shall find useful another kind of duality defined by the functor $M \mapsto M^{\wedge} = \mathrm{Hom}_A(M,A)$. As the "usual" duality, this is a contravariant functor, or more precisely, a pair of contravariant functors mod-A $\to$ A-mod and A-mod $\to$ mod-A. However, these functors are not exact (in fact, they are only left exact) and not reciprocal. Nevertheless, there is a canonical map $\sigma_M : M \to M^{\wedge\wedge}$, sending $m\in M$ into $\sigma_M(m) : M^{\wedge} \to A$ such that $\sigma_M(m)(f) = f(m)$ for all $f : M\to A$.

If M,N are two right modules, then there is a unique map $\lambda = \lambda(M,N)$: $N\otimes_A M^{\wedge} \to \mathrm{Hom}_A(M,N)$ such that $\lambda(n\otimes f)(m) = nf(m)$ for all $m\in M$, $n\in N$ and $f\in M^{\wedge}$.

Proposition 11.5.5. 1) *If M is a projective module, then σ_M is an isomorphism.*

2) *A homomorphism $\varphi : M\to N$ belongs to the image of $\lambda(M,N)$ if and only if it can be factored into a product $\varphi = \beta\alpha$, where $\alpha : M\to P$ and $\beta : P\to N$ with a projective module P.*

Proof. 1) Obviously, σ_A is an isomorphism and therefore also σ_{nA} is an isomorphism. Thus, in view of Theorem 3.3.5, the statement follows.

2) Similarly to 1), if P is a projective module, we can immediately see that $\lambda(P,N)$ is an isomorphism. Now, let $\alpha : M\to P$ with a projective P. Then the following diagram commutes:

$$\begin{array}{ccc} N\otimes_A P^{\wedge} & \xrightarrow{1\otimes\alpha^{\wedge}} & N\otimes_A M^{\wedge} \\ {\scriptstyle\lambda(P,N)}\downarrow & & {\scriptstyle\lambda(M,N)}\downarrow \\ \mathrm{Hom}_A(P,N) & \xrightarrow{h_N^{o}(\alpha)} & \mathrm{Hom}_A(M,N)\,, \end{array} \tag{11.5.1}$$

and we get that $\mathrm{Im}\, h_N^{o}(\alpha) = \{\beta\alpha \mid \beta : P\to N\} \subset \mathrm{Im}\,\lambda(M,N)$.

In order to complete the proof, we shall need the following obvious lemma.

Lemma 11.5.6. *For a right B-module M, a left A-module N and an A-B-bimodule L, there is an isomorphism*

$$\mathrm{Hom}_B\big(M,\mathrm{Hom}_A(N,L)\big) \simeq \mathrm{Hom}_A\big(N,\mathrm{Hom}_B(M,L)\big)$$

assigning to a homomorphism $f : M\to \mathrm{Hom}_A(N,L)$ the homomorphism $f' : N\to \mathrm{Hom}_B(M,L)$ such that $f'(n)(m) = f(m)(n)$ for all $m\in M$ and $n\in N$.

If, in particular, P is a projective module, then

$$\begin{aligned}\mathrm{Hom}_A(P^{\wedge},M^{\wedge}) &= \mathrm{Hom}_A\big(P^{\wedge},\mathrm{Hom}_A(M,A)\big) \simeq \mathrm{Hom}_A\big(M,\mathrm{Hom}_A(P^{\wedge},A)\big) = \\ &= \mathrm{Hom}_A(M,P^{\wedge\wedge}) \simeq \mathrm{Hom}_A(M,P)\,.\end{aligned}$$

Consider now an epimorphism $\psi : P' \to M\hat{}$, where P' is projective. According to 1), we may assume that $P' = P\hat{}$ and $\psi = \alpha\hat{}$ for a projective module P and $\alpha : M \to P$. Then the homomorphism $1 \otimes \alpha\hat{}$ of (11.5.1) is an epimorphism by Proposition 8.3.6. Consequently $\operatorname{Im}\lambda(M,N) = \operatorname{Im} h_N^o(\alpha)$ and the proof of 2) is completed. □

In what follows, we shall write $\operatorname{Pr}_A(M,N) = \operatorname{Im}\lambda(M,N)$ and call the homomorphisms from $\operatorname{Pr}_A(M,N)$ the *projective homomorphisms.* Let us also introduce the following notation: $\underline{\operatorname{Hom}}_A(M,N) = \operatorname{Hom}_A(M,N)/\operatorname{Pr}_A(M,N)$.

11.6 Almost Split Sequences

In this section, we are going to prove a theorem which plays a fundamental role in the contemporary investigations of representations and structure of finite dimensional algebras. It is related to the concept of almost split sequences, often called Auslander-Reiten sequences.

Proposition 11.6.1. *Let $\zeta : 0 \to N \xrightarrow{f} X \xrightarrow{g} M \to 0$ be a non-split exact sequence with indecomposable modules M and N. Then the following conditions are equivalent:*

1) *For every $\varphi : M' \to M$, where M' is indecomposable and φ is not an isomorphism, the lifting $\varphi^e(\zeta)$ splits.*
1') *For every $\varphi : M' \to M$, where M' is indecomposable and φ is not an isomorphism, there is a factorization $\varphi = g\alpha$ for some $\alpha : M' \to X$.*
2) *For every $\psi : N \to N'$, where N' is indecomposable and ψ is not an isomorphism, the descent $\psi_e(\zeta)$ splits.*
2') *For every $\psi : N \to N'$, where N' is indecomposable and ψ is not an isomorphism, there is a factorization $\psi = \beta f$ for some $\beta : X \to N'$.*

Proof. 1) ⇒ 1'). Consider the commutative diagram involving the lifting $\varphi^e(\zeta)$:

$$\begin{array}{ccccccccc}
\varphi^e(\zeta): & 0 & \longrightarrow & N & \xrightarrow{f'} & X' & \xrightarrow{g'} & M' & \longrightarrow & 0 \\
& & & \downarrow{\scriptstyle 1_N} & & \downarrow{\scriptstyle \varphi'} & & \downarrow{\scriptstyle \varphi} & & \\
\zeta: & 0 & \longrightarrow & N & \xrightarrow{f} & X & \xrightarrow{g} & M & \longrightarrow & 0.
\end{array}$$

Since $\varphi^e(\zeta)$ is split, there is a homomorphism $\gamma : M' \to X'$ for which $g'\gamma = 1$. But then $\varphi = \varphi g'\gamma = g\varphi'\gamma$, as required.

1') ⇒ 1). If $\varphi = g\alpha$, then the homomorphism $\gamma : M' \to X'$ given by the formula $\gamma(m') = (\alpha(m'), m')$ defines a splitting of $\varphi^e(\zeta)$. (Recall that, in the construction of lifting, $X' = \{(x,m') \mid g(x) = \varphi(m')\} \subset X \oplus M'$, and $g'(x,m') = m'$.)

1′) ⇒ 2). Consider the commutative diagram involving the descent $\psi_e(\zeta)$:

$$\begin{array}{ccccccccccc} \zeta: & 0 & \longrightarrow & N & \xrightarrow{f} & X & \xrightarrow{g} & M & \longrightarrow & 0 \\ & & & \psi\big\downarrow & & \psi'\big\downarrow & & 1_M\big\downarrow & & \\ \psi_e(\zeta): & 0 & \longrightarrow & N' & \xrightarrow{f'} & X' & \xrightarrow{g'} & M & \longrightarrow & 0\,. \end{array}$$

Let $X' = X_1 \oplus X_2 \oplus \ldots \oplus X_m$ be a direct decomposition into indecomposable summands X_i and g_i the restrictions of g' to X_i. If any of g_i is invertible, i.e. $g_i h = 1_M$ for some $h : M \to X_i$, then the sequence $\psi_e(\zeta)$ splits due to the homomorphism $\gamma : M \to X'$ defined by $\gamma(m) = (0, \ldots, 0, h(m), 0, \ldots, 0)$ with $h(m)$ at the ith position. Thus, assume that none of g_i is invertible. Then in view of the condition 1′), $g_i = g\alpha_i$ for some $\alpha_i : X_i \to X$ and hence $g' = g\eta$, where $\eta(x_1, x_2, \ldots, x_m) = \sum_i \alpha_i(x_i)$.

Since $g\eta f' = g'f' = 0$, $\operatorname{Im} \eta f' \subset \operatorname{Ker} g = \operatorname{Im} f$, and thus $\eta f' = f\theta$ for some $\theta : N' \to N$. Similarly, since $g(1 - \eta\psi') = g - g'\psi' = 0$, we have a factorization $1 - \eta\psi' = fu$ for some $u : X \to N$. Furthermore, multiplying the equality $1 = \eta\psi' + fu$ by f we get $f = \eta\psi' f + fuf = \eta f'\psi + fuf = f\theta\psi + fuf$. Since f is a monomorphism, this equality yields $1_N = \theta\psi + uf$. Now, N is indecomposable and thus the algebra $E_A(N)$ is local. Consequently, $\theta\psi$ or uf is invertible. However, if $\theta\psi$ is invertible, so is ψ (since N' is also indecomposable) and if uf is invertible, then ζ is split. This contradiction completes the proof.

The assertions 2) ⇔ 2′) and 2′) ⇒ 1) can be proved similarly, or follow by duality. □

A sequence ζ possessing the properties listed in Proposition 11.6.1 is called an *almost split sequence with end M and beginning N.*

It is clear that in order that such an almost split sequence exists, it is necessary that M is not projective and N is not injective. It is rather remarkable that this condition is also sufficient.

Theorem 11.6.2 (Auslander-Reiten). 1) *For any indecomposable module M which is not projective, there is an almost split sequence with end M.*
2) *For any indecomposable module N which is not injective, there is an almost split sequence with beginning N.*

Proof. 1) Theorem 3.3.7 implies that there is an epimorphism $\pi : P_0 \to M$ such that P_0 is projective and $\operatorname{Ker} \pi \subset \operatorname{rad} P_0$. Repeating the same procedure for $\operatorname{Ker} \pi$, we get an exact sequence $P_1 \xrightarrow{\theta} P_0 \xrightarrow{\pi} M \to 0$ for which $\operatorname{Im} \theta = \operatorname{Ker} \pi \subset \operatorname{rad} P_0$ and $\operatorname{Ker} \theta \subset \operatorname{rad} P_1$. Now, apply the functor $\hat{} = h_A^o$ (see Sect. 11.5) and put $T = \operatorname{Tr} M = \operatorname{Coker}(\theta\hat{})$. We obtain the following exact sequence:

$$0 \longrightarrow M\hat{} \xrightarrow{\pi\hat{}} P_0\hat{} \xrightarrow{\theta\hat{}} P_1\hat{} \xrightarrow{\sigma} T \longrightarrow 0\,. \tag{11.6.1}$$

We are going to show that T is indecomposable. Indeed, assuming that T is decomposable, we get from Corollary 3.3.8 that $P_1\hat{} = Y_1 \oplus Y_2$ and

$P_0\hat{} = Z_1 \oplus Z_2$ such that $\theta\hat{}(Z_1) \subset Y_1$ and $\theta\hat{}(Z_2) \subset Y_2$. But then, taking into account part 1) of Proposition 11.5.5, we see that $P_1 = Y_1\hat{} \oplus Y_2\hat{}$, $P_0 = Z_1\hat{} \oplus Z_2\hat{}$ with $\theta(Y_1\hat{}) \subset Z_1\hat{}$ and $\theta(Y_2\hat{}) \subset Z_2\hat{}$. From here, $M \simeq Z_1\hat{}/\theta(Y_1\hat{}) \oplus Z_2\hat{}/\theta(Y_2\hat{})$ and, in view of the fact that $\operatorname{Im}\theta \subset \operatorname{rad} P_0$, both summands are non-zero. This contradiction shows that T is indecomposable. Put $N = T^*$.

According to Corollary 11.5.4, for any module L, there is an isomorphism $\operatorname{Ext}_A^1(L,N) \simeq \operatorname{Tor}_1^A(L,T)^*$. To compute $\operatorname{Tor}_1^A(L,T)$, we will use the exact sequence (11.6.1): It turns out that $\operatorname{Tor}_1^A(L,T)$ is isomorphic to the factor space $\operatorname{Ker} t_L(\theta\hat{})/\operatorname{Im} t_L(\pi\hat{})$ (here t_L is the functor $L \otimes_A -$). Making use of part 2) of Proposition 11.5.5 we obtain $L \otimes_A P_i\hat{} \simeq \operatorname{Hom}_A(P_i, L)$, and hence $\operatorname{Ker} t_L(\theta\hat{}) \simeq \operatorname{Ker} h_L^o(\theta) \simeq \operatorname{Hom}_A(M,L)$, since the sequence $0 \to \operatorname{Hom}_A(M,L) \to \operatorname{Hom}_A(P_0,L) \to \operatorname{Hom}_A(P_1,L)$ is exact. Moreover, $\operatorname{Im} t_L(\pi\hat{})$ is mapped in this isomorphism into $\operatorname{Im}\lambda(M,L) = \operatorname{Pr}_A(M,L)$. Consequently, $\operatorname{Tor}_1^A(L,T) \simeq \underline{\operatorname{Hom}}_A(M,L)$ and $\operatorname{Ext}_A^1(L,N) \simeq \underline{\operatorname{Hom}}_A(M,L)^*$. In particular, $\operatorname{Ext}_A^1(M,N) \simeq \underline{\operatorname{Hom}}_A(M,M)^*$. However, $H = \underline{\operatorname{Hom}}_A(M,M)$ is a quotient algebra of $E_A(M)$ and thus it is a local algebra. Denote by R its radical and consider a non-zero linear functional $\zeta \in H^*$ such that $\zeta(R) = 0$. Let M' be an indecomposable A-module. For any $\varphi : M' \to M$ which is not an isomorphism, the induced map $\operatorname{Hom}_A(M,M') \to \operatorname{Hom}_A(M,M)$ assigns to a homomorphism $f : M \to M'$ the non-invertible endomorphism φf. Thus, denoting by $\bar{f}$ the coset of f in $\underline{\operatorname{Hom}}_A(M,M')$, we get that $\varphi^e(\zeta)(\bar{f}) = \zeta(\varphi\bar{f}) = 0$, which means that the extension of M by kernel N corresponding to the element ζ is an almost split sequence.

The assertion 2) follows from 1) by duality (or can be proved similarly). Let us point out that our computations yield also isomorphisms $M \simeq \operatorname{Tr} N^*$ and $\operatorname{Ext}_A^1(M,L) \simeq \overline{\operatorname{Hom}}_A(L,N)^*$ for every module L; here $\overline{\operatorname{Hom}}_A(L,N)$ denotes the factor space of $\operatorname{Hom}_A(L,N)$ by the subspace $\operatorname{In}_A(L,N)$ consisting of those homomorphisms which factor through an injective module. □

11.7 Auslander Algebras

In conclusion, we will give a homological characterization of an important class of algebras. We call an algebra A an *Auslander algebra* if there is an algebra B possessing only a finite number of non-isomorphic indecomposable modules $M_1, M_2, \dots, M_n$, so that $A \simeq E_B(M)$, where $M = M_1 \oplus M_2 \oplus \dots \oplus M_n$ (more precisely, A is called the Auslander algebra of the algebra B). By definition, such an algebra is always basic. Obviously, a basic semisimple algebra is always an Auslander algebra.

Theorem 11.7.1 (Auslander). *A basic algebra A is an Auslander algebra if and only if $gl.dim\, A \le 2$ and there is an exact sequence $0 \to A \to I_0 \to I_1$ in which the A-modules I_0 and I_1 are bijective.*

The necessity of the statement will be based on the following lemma.

Lemma 11.7.2. *Let $A = E_B(M)$ be an Auslander algebra. Then:*

1) *M is a projective left A-module.*
2) *The functors $F : N \mapsto \mathrm{Hom}_B(M, N)$ and $G : P \mapsto P \otimes_A M$ establish an equivalence between the category* mod-B *and the category* pr-A *of the projective A-modules.*

Proof. 1) Since M is a direct sum of all indecomposable B-modules, $mM \simeq B \oplus L$ for some L, and thus $mA \simeq \mathrm{Hom}_B(mM, M) \simeq \mathrm{Hom}_B(B, M) \oplus \mathrm{Hom}_B(L, M)$. Therefore, $M \simeq \mathrm{Hom}_B(B, M)$ is a projective A-module.

2) The fact that $F(N)$ is always projective can be verified the same way as the first statement 1). The natural transformation of functors (see Sect. 8.4) $\varphi : 1_{\text{pr-}A} \to FG$ and $\psi : GF \to 1_{\text{mod-}B}$ are isomorphisms on A_A and M_B, respectively, and therefore on all their direct summands. Hence φ and ψ are isomorphisms, respectively, on all projective A-modules and all B-modules, as required. □

Proof of necessity in Theorem 11.7.1. Let $A = E_B(M)$ be the Auslander algebra of an algebra B and $g : P_0 \to P_1$ a homomorphism of projective A-modules. In view of Lemma 11.7.2, we may assume that $P_i = F(N_i)$ and $g = F(f)$ for some B-module homomorphism $f : N_0 \to N_1$. Since F is left exact, $\mathrm{Ker}\, g \simeq F(\mathrm{Ker}\, f)$ is a projective A-module and $gl.dim\, A \leq 2$ by Corollary 11.4.7.

Now, construct an exact sequence $0 \to M \to Q_0 \to Q_1$ with injective B-modules Q_0, Q_1. Applying the functor F, we obtain an exact sequence $0 \to A \to F(Q_0) \to F(Q_1)$. It remains to show that $F(Q_i)$ are injective A-modules. In view of Theorem 11.1.4, it is sufficient to know that $F(B^*)$ is an injective A-module. However, $F(B^*) = \mathrm{Hom}_B\big(M, \mathrm{Hom}_K(B, K)\big) \simeq \mathrm{Hom}_K(M \otimes_B B, K) \simeq M^*$ is injective by part 1) of Lemma 11.7.2.

Proof of sufficiency. Assume that $gl.dim\, A \leq 2$ and that there is an exact sequence $0 \to A \to I_0 \to I_1$ with bijective A-modules I_0 and I_1. Denote by I the direct sum of all indecomposable bijective A-modules, $B = E_A(I)$ and consider the contravariant functors $F' : N \mapsto \mathrm{Hom}_B(N, I)$ and $G' : P \mapsto \mathrm{Hom}_A(P, I)$. For a left B module N, a projective resolution $P_1 \to P_0 \to N \to 0$ translates to the exact sequence $0 \to F'(N) \to F'(P_0) \to F'(P_1)$. However $F'(B) \simeq I$ and therefore $F'(P_i)$ are projective (even bijective) A-modules. By Corollary 11.4.7, $F'(N)$ is also projective, and thus F' can be viewed as a functor $(B\text{-mod})^o \to \text{pr-}A$.

Consider the natural transformations $\varphi' : 1_{\text{pr-}A} \to F'G'$ and $\psi' : 1_{\text{mod-}B} \to G'F'$ (they act the same way: $\varphi'(P)$ assigns to an element $x \in P$ the B-homomorphism $\mathrm{Hom}_A(P, I) \to I$ sending f into $f(x)$; $\psi'(N)$ acts similarly). Clearly, $\varphi'(I)$ and $\psi'(B)$ are isomorphisms. Thus, if P is bijective and N is projective, also $\varphi'(P)$ and $\psi'(N)$ are isomorphisms. Besides, the functor $F'G'$ is left exact and $G'F'$ is right exact, since I is an injective A-module and thus G' is exact. Therefore the exact sequence $0 \to A \to I_0 \to I_1$ can be extended to the following commutative diagram with exact rows:

$$\begin{array}{ccccccc}
0 & \longrightarrow & A & \longrightarrow & I_0 & \longrightarrow & I_1 \\
 & & {\scriptstyle \varphi'(A)}\big\downarrow & & {\scriptstyle \varphi'(I_0)}\big\downarrow & & {\scriptstyle \varphi'(I_1)}\big\downarrow \\
0 & \longrightarrow & F'G'(A) & \longrightarrow & F'G'(I_0) & \longrightarrow & F'G'(I_1).
\end{array}$$

As a consequence, $\varphi'(A)$ is an isomorphism and thus $\varphi'(P)$ is an isomorphism for every projective P. Similarly, $\psi'(N)$ is an isomorphism for every N and we conclude that F' and G' establish an equivalence of the categories $(B\text{-mod})^o$ and pr-A. In particular, since $G'(A) = I$, the algebra A is anti-isomorphic to $\mathrm{End}_B(I)$. Furthermore, A is basic, and thus is a direct sum of non-isomorphic principal A-modules; therefore I is a direct sum of all non-isomorphic indecomposable left B-modules. It follows that I^* is a direct sum of all non-isomorphic indecomposable right B-modules and $E_B(I^*) \simeq E_B(I)^o \simeq A$, so A is an Auslander algebra. □

Exercises to Chapter 11

1. Verify that for a complex $V_\bullet$ which is a short exact sequence $0 \to M \to N \to L \to 0$, $V_\bullet \sim 0$ if and only if the sequence splits. (Clearly, $\mathrm{H}_n(V_\bullet) = 0$ for all n.)

2. Let $A = K[a]$, where $a^2 = 0$, $M = A/aA$ and $\pi : A \to M$ the canonical projection. Furthermore, let $\varepsilon : M \to A$ be the embedding sending $x + aA$ into ax and $f_\bullet : V_\bullet \to V'_\bullet$ the complex homomorphism defined by the following diagram:

$$\begin{array}{ccccccc}
0 & \longrightarrow & M \oplus M & \xrightarrow{\left(\begin{smallmatrix}1&0\\0&0\end{smallmatrix}\right)} & M \oplus M & \longrightarrow & 0 \\
 & & {\scriptstyle (\varepsilon\ 0)}\big\downarrow & & \big\downarrow{\scriptstyle \left(\begin{smallmatrix}0&1\\0&0\end{smallmatrix}\right)} & & \\
0 & \longrightarrow & A & \xrightarrow{\left(\begin{smallmatrix}\pi\\0\end{smallmatrix}\right)} & M \oplus M & \longrightarrow & 0.
\end{array}$$

Show that $f_\bullet \equiv 0$, but $f_\bullet \not\sim 0$.

3. Give an example of a complex $V_\bullet$ and a functor F such that $\mathrm{H}_n(V_\bullet) = 0$ for all n, but $\mathrm{H}_n\big(F(V_\bullet)\big) \neq 0$ for some n.

4. Let $V_\bullet$ and $V'_\bullet$ be complexes of projective modules over a hereditary algebra, bounded from the right, and $f_\bullet$ and $g_\bullet$ two homomorphisms $V_\bullet \to V'_\bullet$. Prove that $f_\bullet \equiv g_\bullet$ implies $f_\bullet \sim g_\bullet$.

5. Prove that for every module M there exists a projective resolution $(P_\bullet, d_\bullet)$ satisfying $\mathrm{Im}\, d^n \subset \mathrm{rad}\, P_{n-1}$ for all n, and that any two such resolutions are isomorphic. (Resolutions satisfying this property are called minimal projective resolutions of the module M and are denoted by $P_\bullet(M)$.) Formulate and prove an analogous result for injective resolutions.

6. Let $0 \to N \xrightarrow{\varphi} P_{k-1} \to \cdots \to P_0 \to M \to 0$ be an exact sequence with projective modules $P_0, P_1, \ldots, P_{k-1}$. Let F be a right exact functor. Prove that $L^n F(M) \simeq L^{n-k} F(N)$ for $n > k$ and $L^k F(M) \simeq \mathrm{Ker}\, F(\varphi)$. Formulate and prove similar statements for right derived functors and contravariant functors.

7. Let $P_\bullet(M) = (P_\bullet, d_\bullet)$ be a minimal projective resolution of a right A-module M (see Exercise 5). Prove that, for any simple right A-module V (simple left A-module W), $\mathrm{Ext}^n_A(M,V) \simeq \mathrm{Hom}_A(P_n,V)$ and $\mathrm{Tor}^A_n(M,W) \simeq P_n \otimes_A W$.

8. Let A be a split algebra, $\mathcal{D} = \mathcal{D}(A)$ its diagram and V_i the simple A-module corresponding to the vertex $i \in \mathcal{D}$. Prove that $\mathrm{Ext}^1_A(V_i, V_j) \simeq t_{ij}K$, where (t_{ij}) is the incidence matrix of the diagram $\mathcal{D}$.

9. Construct a one-to-one map $\delta' : \mathrm{Ex}(M,N) \to \mathrm{Ext}^1_A(M,N)$ using the connecting homomorphism with respect to the first variable (and projective resolutions).

10. Prove that $\mathit{proj.dim}_A(\bigoplus_i M_i) = \max_i(\mathit{proj.dim}_A M_i)$ and $\mathit{inj.dim}_A(\bigoplus_i M_i) = \max_i(\mathit{inj.dim}_A M_i)$.

11. Prove that $\mathit{gl.dim}(\prod_i A_i) = \max_i(\mathit{gl.dim}\, A_i)$.

12. Assume that there are no cycles in the diagram $\mathcal{D}(A)$ of an algebra A.
    a) Prove that $\mathit{gl.dim}\, A \leq \ell$, where ℓ is the maximal length of paths in $\mathcal{D}(A)$.
    b) If $(\mathrm{rad}\, A)^2 = 0$, prove that $\mathit{gl.dim}\, A = \ell$.

13. Let L be an extension of the field K. Prove that $\mathit{gl.dim}\, A_L \geq \mathit{gl.dim}\, A$. Prove that the inequality becomes equality if L is a separable extension or if the quotient algebra $A/\mathrm{rad}\, A$ is separable over K.

14. Prove that $\mathit{gl.dim}\, A \leq \mathit{proj.dim}_{A \otimes A^\circ} A$ and that equality holds if $A/\mathrm{rad}\, A$ is separable.

15. Prove that any two almost split sequences with a common beginning (or end) are isomorphic.

16. Prove that a hereditary Auslander algebra is semisimple.

References

1. Cohn, P.M., *Free Rings and Their Relations.* Academic Press, London, 1971.
2. Curtis, C.W., Reiner, I., *The Representation Theory of Finite Groups and Associative Algebras.* Wiley-Interscience, New York-London-Sydney, 1962.
3. Deuring, M., *Algebren.* Ergebnisse der Mathematik, Band 4, Springer-Verlag, Berlin, 1935.
4. Faith, C., *Algebra: Rings, Modules and Categories I.* Grundlehren der Mathematischen Wissenschaften, Band 190, Springer-Verlag, Berlin-Heidelberg-New York, 1973.
5. Herstein, I.N., *Noncommutative Rings.* Carus Mathematical Monographs, No. 15, Mathematical Association of America, 1968.
6. Jacobson, N., *The Theory of Rings.* American Mathematical Society Surveys, Vol. 2, American Mathematical Society, Providence, 1943.
7. Jacobson, N., *Structure of Rings.* American Mathematical Society Colloquium Publications, Vol. 37, American Mathematical Society, Providence, 1956.
8. Lambek, J., *Lectures on Rings and Modules.* Blaisdell-Ginn, Waltham-Toronto-London, 1966.
9. Van der Waerden, B.L., *Algebra I, II.* Springer-Verlag, Berlin-Heidelberg-New York, 1971, 1967.
10. Weil, A., *Basic Number Theory.* Springer-Verlag, New York-Heidelberg-Berlin, 1967.

Quasi-hereditary Algebras

Appendix by Vlastimil Dlab

I wish to express my gratitude to Yu.A. Drozd and V.V. Kirichenko for this opportunity to append a brief exposition on a recently introduced class of algebras. The class of quasi-hereditary algebras has been introduced by Cline, Parshall and Scott ([CPS1],[PS]) in connection with their study of highest weight categories arising in the representation theory of semi-simple complex Lie algebras and algebraic groups.

This presentation is intended for readers who may be interested in getting basic information on some of the developments in this field. It is by no means exhaustive, nor is it homogeneous; ring and module theoretical methods mix in order to provide as broad an introduction to the existing literature as possible. Although the concept of a quasi-hereditary algebra relates naturally to a partial order (of the set of all simple modules), there is no substantial loss of generality to restrict ourselves to a total (refinement) order. This, together with a restriction to basic algebras, may in our view help to make this introductory text more accessible. The text is not entirely self-contained; a few fundamental concepts, notably from category theory, are used without a formal definition; moreover, due to space limitations, some results are presented without proofs. I apologize for an unavoidable bias in the selection of the material and its presentation; references to the literature are kept to a minimum.

Finally, I wish to thank whole-heartedly my friends and colleagues István Ágoston and Erzsébet Lukács for their valuable comments, suggestions and corrections in the preliminary manuscript. Of course the responsibility for any inaccuracies in the text remains my own.

Ottawa, December 1992

A.1 Preliminaries. Standard and Costandard Modules

Throughout this appendix, A will always denote a finite dimensional K-algebra which will be, unless stated otherwise, basic and connected; we put $\bar{A} = A/\mathrm{rad}\, A$. Furthermore, $\mathbf{e} = (e_1, e_2, \ldots, e_n)$ will always denote an (ordered) complete set of primitive orthogonal idempotents; write $\varepsilon_i = e_i + e_{i+1} + \ldots + e_n$ for $1 \leq i \leq n$ and $\varepsilon_{n+1} = 0$. Considering the sequence $\mathbf{e}$ of primitive orthogonal idempotents is equivalent to ordering the set of all non-isomorphic simple A-modules $S(i) \simeq e_i\bar{A}$, or the set of their projective covers $P(i) \simeq e_iA$, $1 \le i \le n$. Of course, we may also consider the ordered set of all simple left A-modules $S^\circ_{(i)} \simeq \bar{A}e_i$, or their projective covers $P^\circ(i) \simeq Ae_i$. Note that $\overset{n}{\underset{j=i}{\oplus}} P(j) \simeq \varepsilon_iA$, $\overset{n}{\underset{j=i}{\oplus}} P^\circ(j) \simeq A\varepsilon_i$ and that the endomorphism algebras $E_A(\varepsilon_iA) \simeq E_A(A\varepsilon_i) \simeq \varepsilon_iA\varepsilon_i$ for all $1 \le i \le n$. Finally, the division algebra $E_A(S(i)) \simeq e_i\bar{A}e_i$ will be denoted by D_i and $\dim_K D_i = d_i$ for $1 \le i \le n$.

If X is an A-module, denote by $[X : S(i)]$ the number (*multiplicity*) of the factors isomorphic to $S(i)$ in a composition series of X, and by $\underline{\dim}\, X$ its *dimension vector*, i. e. the n-tuple whose coordinates are $[X : S(i)]$, $1 \le i \le n$. Obviously, $[X : S(i)] = \dim_{D_i} \mathrm{Hom}(P(i), X)$.

Given A-modules X and Y, define the *trace* $\tau_Y(X)$ of Y in X as the submodule of X generated by all homomorphic images of Y in X:

$$\tau_Y(X) = \langle \mathrm{Im}\, \varphi \mid \varphi \in \mathrm{Hom}_A(Y, X)\rangle_A .$$

Thus, $\tau_{\varepsilon_iA} X = X\varepsilon_iA$; in particular, $\tau_{P(i)}X = Xe_iA$.

Of course, we can also define the *"reject"* $\rho_Z(X)$ of Z in X by

$$\rho_Z(X) = \bigcap \{\mathrm{Ker}\, \varphi \mid \varphi \in \mathrm{Hom}_A(X, Z)\} .$$

The following definition, depending on the order $\mathbf{e}$ (!), is crucial for the subject.

Definition A.1.1. The sequence

$$\Delta = \Delta_A = (\Delta(i) \mid 1 \le i \le n)$$

of the (right) *standard modules* with respect to a given order $\mathbf{e}$ is given by

$$\Delta(i) = \Delta_A(i) = P(i)/\tau_{\varepsilon_{(i+1)}A}P(i) \simeq e_iA/e_iA\varepsilon_{i+1}A.$$

Similarly, there is a sequence $\Delta^\circ = \Delta^\circ_A$ of the left standard A-modules $\Delta^\circ(i) = \Delta^\circ_A(i) \simeq Ae_i/A\varepsilon_{i+1}Ae_i$, or the sequence $\nabla = \nabla_A$ of its duals, the (right) *costandard A-modules*

$$\nabla(i) = \nabla_A(i) = \mathrm{Hom}_K(\Delta^\circ(i), K) .$$

Observe that $\Delta(i)$ is the maximal factor module of $P(i)$ whose composition factors are isomorphic to $S(j)$ for $j \le i$. Dually, $\nabla(i)$ is the maximal submodule of the injective hull $Q(i)$ of $S(i)$ whose composition factors are isomorphic to $S(j)$ for $j \le i$.

Let us summarize some of the basic properties of the standard and costandard modules. As a rule, formulations of the dual statements as well as simple verifications will be left to the reader.

Lemma A.1.2. *An A-module X satisfies $X \simeq \Delta(i)$ if and only if*

1) $X/\operatorname{rad} X \simeq S(i)$;
2) $[X : S(j)] \neq 0$ *implies* $j \leq i$ *and*
3) $\operatorname{Ext}^1(X, S(j)) \neq 0$ *implies* $j > i$.

Thus, $\operatorname{Hom}(\Delta(i), X) \neq 0$ implies $[X : S(i)] \neq 0$ and $\operatorname{Ext}^1(\Delta(i), X) \neq 0$ implies $[X : S(j)] \neq 0$ for some $j > i$. Consequently, we obtain the following implications.

Lemma A.1.3. 1) $\operatorname{Hom}(\Delta(i), \Delta(j)) \neq 0$ *implies* $i \leq j$.
2) $\operatorname{Ext}^1(\Delta(i), \Delta(j)) \neq 0$ *implies* $i < j$.

In combination with their dual versions, the previous statements yield also the following lemma.

Lemma A.1.4. 1) *If* $\operatorname{Hom}(\Delta(i), \nabla(j)) \neq 0$, *then* $i = j$.
2) $\operatorname{Ext}^1(\Delta(i), \nabla(j)) = 0$ *for all* i, j.
2′) $\operatorname{Tor}_1(\Delta(i), \Delta^\circ(j)) = 0$ *for all* i, j.

Writing $B_i = A/A\varepsilon_{i+1}A$ $(1 \leq i \leq n)$, notice that, as a module,

$$B_i \simeq \bigoplus_{j=1}^{i} e_jA/e_jA\varepsilon_{i+1}A\,,$$

and thus $\Delta(i)$ is a projective B_i-module.

Clearly,

$$E_A(\Delta(i)) \simeq E_A(\Delta^\circ(i)) \simeq E_A(\nabla(i)) \simeq e_iAe_i/e_iA\varepsilon_{i+1}Ae_i\,, \quad 1 \leq i \leq n\,.$$

Call the sequence Δ *Schurian* if every $\Delta(i)$ is Schurian, i.e. $E_A(\Delta(i))$ is a division algebra for all $1 \leq i \leq n$. Let us mention some immediate reformulations.

Proposition A.1.5. *The following properties are equivalent:*

1) $\Delta(i)$ *is Schurian;*
1°) $\Delta^\circ(i)$ *(and thus* $\nabla(i)$*) is Schurian;*
2) $E_A(\Delta(i)) \simeq E_A(S(i))$;
2°) $E_A(\nabla(i)) \simeq E_A(S(i))$;
3) $[\Delta(i) : S(i)] = 1$;
3°) $[\Delta^\circ(i) : S^\circ(i)] = ([\nabla(i) : S(i)] =)1$;
4) $e_iA\varepsilon_{i+1}Ae_i = e_i \operatorname{rad} A\, e_i$.

In the sequel, we shall, as a rule, refrain from formulating dual statements. Let us point out that if Δ is Schurian, always

$$\Delta(1) \simeq S(1)\,.$$

In fact, in this case, we obtain a bound on the Loewy length of the regular representation of A.

Proposition A.1.6. *If A has a sequence $\Delta = \{\Delta(i) \mid 1 \le i \le n\}$ which is Schurian, then*

$$\operatorname{rad}^d A = (\operatorname{rad} A)^d = 0 \quad \textit{for} \quad d = 2^n - 1\,.$$

The exponent $2^n - 1$ is optimal.

Proof. Let us write, as before, $B_{n-1} = A/A\varepsilon_n A$. If $\operatorname{rad}^{d'} B_{n-1} = 0$, then $\operatorname{rad}^{d'+1} P(n) = 0$, since $\Delta(n) = P(n)$ is Schurian. Moreover,

$$\operatorname{rad}^{2d'+1}\big(\bigoplus_{i=1}^{n-1} P(i)\big) = 0\,.$$

Thus, if by induction $d' = 2^{n-1} - 1$, then $2d' + 1 = 2^n - 1$, as required.

In order to show that the exponent is optimal, consider the path K-algebra A of the complete graph with n vertices without loops, modulo the ideal generated by all paths

$$\alpha_{ii_1}\alpha_{i_1i_2}\cdots\alpha_{i_ti} \quad \text{with} \quad i_r < i \quad \text{for} \quad 1 \le r \le t$$

i.e. the canonical deep algebra over that graph (see Sect. A.4). Clearly, $(\operatorname{rad} A)^{2(2^{n-1}-1)} \neq 0$. □

Let us conclude this introductory section by a remark concerning the centralizer algebras $C_i = \varepsilon_i A \varepsilon_i$ $(1 \le i \le n)$ of A. We have seen that for the algebras $B_i = A/A\varepsilon_{i+1}A$, $1 \le i \le n$,

$$\Delta_{B_i} = (\Delta_A(j) \mid 1 \le j \le i)\,.$$

For the algebras C_i, we can verify readily that

$$\Delta_{C_i} = (\varepsilon_i \Delta_A(j)\varepsilon_i \mid i \le j \le n)\,.$$

A.2 Trace Filtrations. The Categories $\mathcal{F}(\Delta)$ and $\mathcal{F}(\nabla)$

We are dealing again with an algebra A together with a (complete) sequence $\mathbf{e} = (e_1, e_2, \ldots, e_n)$ of primitive orthogonal idempotents.

Definition A.2.1. Given an A-module X, define its *trace filtration* (with respect to $\mathbf{e}$) by

$$X = X^{(1)} \supseteq X^{(2)} \supseteq \ldots \supseteq X^{(i)} \supseteq X^{(i+1)} \supseteq \ldots \supseteq X^{(n)} \supseteq X^{(n+1)} = 0, \tag{A.2.1}$$

where $X^{(i)} = \tau_{\varepsilon_i A} X$ for $1 \le i \le n$.

Alternatively, (A.2.1) can be rewritten as follows:

$$X = X\varepsilon_1 A \supseteq X\varepsilon_2 A \supseteq \ldots \supseteq X\varepsilon_i A \supseteq X\varepsilon_{i+1} A \supseteq \ldots \supseteq X\varepsilon_n A \supseteq X\varepsilon_{n+1} A = 0\,.$$

Obviously, trace filtrations are compatible with direct sums. Applied to the regular representation of A, we obtain a filtration of the algebra A by the idempotent ideals $I_i = A\varepsilon_i A$:

$$A = I_1 \supset I_2 \supset \ldots \supset I_i \supset I_{i+1} \supset \ldots \supset I_n \supset I_{n+1} = 0\,.$$

Thus, a choice $\mathbf{e}$ of order of idempotents amounts to a choice of a (saturated) chain in the Boolean lattice of all idempotent ideals of A.

Observe that the right module

$$I_i/I_{i+1} \simeq \bigoplus_{j=1}^{i} e_j A\varepsilon_i A/e_j A\varepsilon_{i+1} A$$

with the last summand isomorphic to $\Delta(i)$. In general, $X^{(i)}/X^{(i+1)}$ is a module over $B_i = A/I_{i+1}$, whose projective B_i-cover is a (finite) direct sum of $\Delta(i)$'s.

Let us point out that we can also define the *"reject" filtration* of an A-module X (with respect to $\mathbf{e}$) by

$$X = X^{[n+1]} \supseteq X^{[n]} \supseteq \ldots \supseteq X^{[i+1]} \supseteq X^{[i]} \supseteq \ldots \supseteq X^{[2]} \supseteq X^{[1]} = 0,$$

where $X^{[i]} = \rho_{Q_i} X$ with $Q_i = \bigoplus_{j=i}^{n} Q(j)$.

We shall turn our attention to the modules X whose trace filtrations satisfy the condition that $X^{(i)}/X^{(i+1)}$ equals its projective B_i-cover (or equivalently, is a direct sum of $\Delta(i)$'s) for every $1 \leq i \leq n$. In view of Lemma A.1.3.2), these are just the modules X possessing a Δ-filtration, i. e. a chain of submodules with factors isomorphic to standard modules $\Delta(i)$ for various i's. Denote the full subcategory of all A-modules with Δ-filtration by $\mathcal{F}(\Delta)$. Similarly, denote by $\mathcal{F}(\Delta^\circ)$ the full subcategory of all left A-modules with Δ°-filtration, and by $\mathcal{F}(\nabla) \simeq \mathcal{F}(\Delta^\circ)^o$ the category of all (right) ∇-filtered A-modules. Clearly, these categories are closed under directs summands, and trivially, under extensions.

Now, if $f : X \to Y$ is a homomorphism and $X = (X^{(i)} \mid 1 \leq i \leq n)$, $Y = (Y^{(i)} \mid 1 \leq i \leq n)$ the trace filtrations, then $f(X^{(i)}) \subseteq Y^{(i)}$ for all $1 \leq i \leq n$. In fact, if f is an epimorphism, then $f(X^{(i)}) = Y^{(i)}$ for all $1 \leq i \leq n$. This follows from the first part of the following lemma.

Lemma A.2.2. *Let $f : X \to Y$ be an epimorphism of A-modules. Let P be a projective A-module. Then f induces an epimorphism $f_P : \tau_P(X) \to \tau_P(Y)$ and the following short exact sequence of $A/\tau_P(A)$-modules*

$$0 \longrightarrow \operatorname{Ker} f/\tau_P(\operatorname{Ker} f) \longrightarrow X/\tau_P(X) \longrightarrow Y/\tau_P(Y) \longrightarrow 0\,. \qquad \text{(A.2.2)}$$

Proof. This follows immediately from the commutative diagram

$$\begin{array}{ccccccccc} & & 0 & & 0 & & 0 & & \\ & & \downarrow & & \downarrow & & \downarrow & & \\ 0 & \longrightarrow & \tau_P(\operatorname{Ker} f) & \longrightarrow & \tau_P(X) & \longrightarrow & \tau_P(Y) & \longrightarrow & 0 \\ & & \downarrow & & \downarrow & & \downarrow & & \\ 0 & \longrightarrow & \operatorname{Ker} f & \longrightarrow & X & \longrightarrow & Y & \longrightarrow & 0; \end{array}$$

indeed, the homomorphism $f_P : \tau_P(X) \to \tau_P(Y)$ induced by f is surjective due to the fact that every map from P to Y lifts to X, and $\tau_P(\operatorname{Ker} f) = \operatorname{Ker} f \cap \tau_P(X) = \operatorname{Ker} f_P$. Moreover, all modules in (A.2.2) are annihilated by the (two-sided) ideal $\tau_P(A)$. □

As a consequence, we can formulate the following statement.

Proposition A.2.3. *The category $\mathcal{F}(\Delta)$ is closed under kernels of epimorphisms. Hence, if $A_A \in \mathcal{F}(\Delta)$, then*

$$\operatorname{Ext}^t(\Delta, \nabla) = \operatorname{Tor}_t(\Delta, \Delta^\circ) = 0 \quad \textit{for all} \quad t \geq 1\,.$$

Here, and in what follows, $\operatorname{Ext}^t(\Delta, \nabla) = 0$ means $\operatorname{Ext}^t(\Delta(i), \nabla(j)) = 0$ for all i, j, or equivalently, $\operatorname{Ext}^t(X, Y) = 0$ for all $X \in \mathcal{F}(\Delta)$ and $Y \in \mathcal{F}(\nabla)$.

Proof. For each $1 \leq i \leq n$, Lemma A.2.2 gives the (split) exact sequence

$$0 \longrightarrow (\operatorname{Ker} f)^{(i)}/(\operatorname{Ker} f)^{(i+1)} \longrightarrow X^{(i)}/X^{(i+1)} \longrightarrow Y^{(i)}/Y^{(i+1)} \longrightarrow 0$$

of projective B_i-modules. Hence, $(\operatorname{Ker} f)^{(i)}/(\operatorname{Ker} f)^{(i+1)} \simeq \oplus \Delta(i)$ and $\operatorname{Ker} f \in \mathcal{F}(\Delta)$.

Now, $\operatorname{Ext}^1(\Delta, \nabla) = 0$ by Lemma A.1.4. Given $X \in \mathcal{F}(\Delta)$ and an exact sequence $0 \to X' \to P \to X \to 0$ with a free module P, we have $X' \in \mathcal{F}(\Delta)$. Since $\operatorname{Ext}^{t+1}(X, \nabla(j)) \simeq \operatorname{Ext}^t(X', \nabla(j))$ for all $t \geq 1$, we complete the proof by induction. □

There is a converse to the last statement of Proposition A.2.3.

Proposition A.2.4. *Let* $\operatorname{Ext}^2(\Delta, \nabla) = 0$ *and Δ be Schurian. Then*

$$\mathcal{F}(\Delta) = \{X \mid \operatorname{Ext}^1(X, \nabla) = 0\}\,.$$

In particular, $A_A \in \mathcal{F}(\Delta)$.

Proof. By Lemma A.1.4, $\mathcal{F}(\Delta) \subseteq \{X \mid \operatorname{Ext}^1(X, \nabla) = 0\}$. We are going to show the opposite inclusion by induction on the "trace length" of X. Assume that

$$\{Y \mid \operatorname{Ext}^1(Y, \nabla) = 0 \text{ and } Y^{(i)} \big(= \tau_{\varepsilon_i A} Y\big) = 0\} \subseteq \mathcal{F}(\Delta)$$

and consider X with $X^{(i+1)} = 0$.

We have two exact sequences:

$$0 \longrightarrow X^{(i)} \longrightarrow X \longrightarrow Y \longrightarrow 0 \quad \text{with } Y^{(i)} = 0 \tag{A.2.3}$$

and, since $\Delta(i)$ is Schurian,

$$0 \longrightarrow Z \longrightarrow \oplus \Delta(i) \longrightarrow X^{(i)} \longrightarrow 0 \quad \text{with } Z^{(i)} = 0 \tag{A.2.4}$$

Now, $[Y : S(j)] = 0$ for $j \geq i$ and thus (in view of the statement dual to Lemma A.1.2) $\mathrm{Ext}^1(Y, \nabla(j)) = 0$ for all $j \geq i$. Moreover, for $j < i$ we have $\mathrm{Hom}(X^{(i)}, \nabla(j)) = 0$ and thus the exact sequence

$$\mathrm{Hom}(X^{(i)}, \nabla(j)) \longrightarrow \mathrm{Ext}^1(Y, \nabla(j)) \longrightarrow \mathrm{Ext}^1(X, \nabla(j)) = 0,$$

derived from (A.2.3), yields $\mathrm{Ext}^1(Y, \nabla(j)) = 0$. By induction, we get $Y \in \mathcal{F}(\Delta)$.

In view of our assumption, the last term of the exact sequence

$$\mathrm{Ext}^1(X, \nabla(j)) \longrightarrow \mathrm{Ext}^1(X^{(i)}, \nabla(j)) \longrightarrow \mathrm{Ext}^2(Y, \nabla(j)) = 0,$$

derived again from (A.2.3), is zero and therefore $\mathrm{Ext}^1(X^{(i)}, \nabla) = 0$.

Now, since $Z^{(i)} = 0$, $\mathrm{Hom}(Z, \nabla(j)) = 0$ for $j \geq i$; for $j < i$, the first term of the exact sequence

$$\mathrm{Hom}(\oplus \Delta(i), \nabla(j)) \longrightarrow \mathrm{Hom}(Z, \nabla(j)) \longrightarrow Ext^1(X^{(i)}, \nabla(j)) = 0,$$

derived from (A.2.4), is zero. Hence $\mathrm{Hom}(Z, \nabla) = 0$. However, this means that $Z = 0$ and so $X^{(i)} \simeq \oplus \Delta(i)$ and $X \in \mathcal{F}(\Delta)$, as required. □

Here is the central definition.

Definition A.2.5. A K-algebra A is said to be *quasi-hereditary* (with respect to $\mathbf{e}$, or equivalently, with respect to Δ) if Δ is Schurian and $A_A \in \mathcal{F}(\Delta)$.

Let us point out that this is a version of the original definition of Cline, Parshall and Scott, rephrasing properties of the so-called *heredity chain* in terms of the *heredity ideals* by conditions for the trace filtration of A_A. Let us call in this case $\mathbf{e}$ a *heredity sequence.* Observe that if $\mathbf{e}$ is a heredity sequence of A, then $(\bar{e}_1, \bar{e}_2, \ldots, \bar{e}_i)$ is a heredity sequence of $B_i = A/A\varepsilon_{i+1}A$ and $(e_i, e_{i+1}, \ldots, e_n)$ is a heredity sequence of $C_i = \varepsilon_i A \varepsilon_i$. Also, $A_A \in \mathcal{F}(\Delta)$ is clearly equivalent to saying that all projective A-modules possess Δ-filtrations.

Propositions A.2.3 and A.2.4 yield immediately the following theorem.

Theorem A.2.6. *Let A be a K-algebra with a Schurian sequence Δ. Then the following conditions are equivalent:*

1) *A is quasi-hereditary (with respect to Δ).*
2) $\mathrm{Ext}^t(\Delta, \nabla) = 0$ *for all* $t \geq 1$.
3) $\mathrm{Ext}^2(\Delta, \nabla) = 0$.
4) $\mathcal{F}(\Delta) = \{X \mid \mathrm{Ext}^1(X, \nabla) = 0\}$.
5) $\mathcal{F}(\Delta) = \{X \mid \mathrm{Ext}^t(X, \nabla) = 0\}$ *for all* $t \geq 1$.

Clearly, each of the above conditions 2)–5) can also be formulated in terms of (right and left) standard modules only, using the functors Tor_t (as in Lemma A.1.4). Moreover, since, in view of 2) or 3), the quasi-hereditary algebra is a two-sided concept, we can formulate also dual equivalences (such as $\mathcal{F}(\nabla) = \{Y \mid \mathrm{Ext}^1(\Delta, Y) = 0\}$) in terms of costandard modules.

Let us mention that C. M. Ringel has shown in [R] that both $\mathcal{F}(\Delta)$ and $\mathcal{F}(\nabla)$ are functorially finite subcategories of mod-A and thus both have (relative) almost split sequences.

Observe that the Ext-projective objects in $\mathcal{F}(\Delta)$ are just the projective A-modules and that the Ext-injective objects in $\mathcal{F}(\nabla)$ are the injective modules. The category $\mathcal{F} = \mathcal{F}(\Delta) \cap \mathcal{F}(\nabla)$ consists of the Ext-injective objects in $\mathcal{F}(\Delta)$ (which coincide with the Ext-projective objects in $\mathcal{F}(\nabla)$). In [R], Ringel identified the indecomposable objects $\{T(i) \mid 1 \le i \le n\}$ of $\mathcal{F}$; furthermore, he showed that the *characteristic module* $T(A) = \bigoplus_{i=1}^{n} T(i)$ is both *tilting* and *cotilting* and that the endomorphism algebra $B = E_A(T(A))$ is again quasi-hereditary (with respect to the opposite order of the idempotents). In fact, $\mathcal{F}(\nabla_A) \simeq \mathcal{F}(\Delta_B)$. Moreover, this procedure is involutory: $E_A(T(B)) \simeq A$ (if A is basic), and thus $\mathcal{F}(\Delta_A) \simeq \mathcal{F}(\nabla_B)$. Let us formulate, without proofs some of the results of [R] to which we shall refer later.

Theorem A.2.7. *Let A be a quasi-hereditary algebra with respect to a sequence Δ. Then $T(i)$, $1 \le i \le n$, are the indecomposable modules defined by the exact sequences*

$$0 \longrightarrow \Delta(i) \longrightarrow T(i) \longrightarrow X(i) \longrightarrow 0$$

and

$$0 \longrightarrow Y(i) \longrightarrow T(i) \longrightarrow \nabla(i) \longrightarrow 0\,,$$

where $X(i) \in \mathcal{F}(\Delta)$ with $\big(X(i)\big)^{(i)} = 0$ and $Y(i) \in \mathcal{F}(\nabla)$ with $\big(Y(i)\big)^{[i]} = Y(i)$. The category $\mathcal{F} = \mathcal{F}(\Delta) \cap \mathcal{F}(\nabla)$ consists of all direct sums of modules $T(i)$, i. e. $\mathcal{F} = \operatorname{add} T$, where $T = \bigoplus_{i=1}^{n} T(i)$ is the characteristic module.

Theorem A.2.8. *Let T be the characteristic module of a quasi-hereditary algebra A. Then*

$$\mathcal{F}(\Delta) = \{X \in \text{mod-}A \mid \mathrm{Ext}^t_A(X, T) = 0 \textit{ for all } t \ge 1\}$$

and

$$\mathcal{F}(\nabla) = \{Y \in \text{mod-}A \mid \mathrm{Ext}^t_A(T, Y) = 0 \textit{ for all } t \ge 1\}\,.$$

Thus T determines both Δ and ∇.

Let us conclude this section with a result providing some justification for the choice of terminology. It is clear that every hereditary algebra is quasi-hereditary with respect to an arbitrary order $\mathbf{e}$ of primitive orthogonal idempotents. Here, we have a converse of that statement.

Theorem A.2.9. *Let A be an algebra which is quasi-hereditary with respect to any order* $\mathbf{e}$ *of its complete set of primitive orthogonal idempotents. Then A is a hereditary algebra.*

Proof. Let A be a basic algebra and $\mathbf{e} = (e_1, e_2, \ldots, e_n)$ a complete sequence of primitive orthogonal idempotents such that $\ell(e_iA) \leq \ell(e_{i+1}A)$ for all $1 \leq i \leq n-1$. Thus, $e_iAe_j = 0$ for all $i < j$.

We shall proceed by induction on n, noting that the case $n = 1$ is trivial.

Now, since $(Ae_nA)_A = \bigoplus_{i=1}^{n} e_iAe_nA$ is projective, evidently $Ae_nA = e_nA$. By induction, A/Ae_nA is hereditary. Hence, to establish our claim that A is hereditary, it is sufficient to verify that $e_n \operatorname{rad} A$ is a projective A-module.

The quotient algebra $\bar{A} = A/Ae_{n-1}A$ is hereditary and thus

$$P = (Ae_{n-1}A + e_n \operatorname{rad} A)/Ae_{n-1}A \simeq e_n \operatorname{rad} A / e_nAe_{n-1}A$$

is a projective $\bar{A}$-module. In fact, in view of our choice of $\mathbf{e}$, it is a projective A-module. It turns our that the canonical homomorphism $e_n \operatorname{rad} A \to P$ splits and we obtain

$$e_n \operatorname{rad} A \simeq P \oplus e_nAe_{n-1}A$$

with both direct summands projective. The proof of the theorem is completed. □

Let us point out that the assumption of Theorem A.2.9 is equivalent to the fact that every (saturated) chain of the Boolean lattice of all idempotent ideals of A is a heredity chain.

A.3 Basic Properties

We have already seen a close relationship between a quasi-hereditary algebra A with respect to $\mathbf{e} = (e_1, e_2, \ldots, e_n)$ and the individual centralizer algebras $C_i = \varepsilon_iA\varepsilon_i$. Indeed, there is a pair of functors

$$\Phi^{(i)} : \text{mod-}A \longrightarrow \text{mod-}C_i$$

and

$$\Psi^{(i)} : \text{mod-}C_i \longrightarrow \text{mod-}A$$

defined by $\Phi^{(i)}X = X\varepsilon_i$ and $\Psi^{(i)}Y = Y \otimes_{C_i} \varepsilon_iA$. Denote by mod-$A^{(i)}$ the full subcategory of all $X \in \mathcal{F}(\Delta_A)$ for which $X = X^{(i)}(= \tau_{\varepsilon_iA}X)$.

Proposition A.3.1. *Let A be a quasi-hereditary algebra with respect to* $\mathbf{e}$. *Then the restrictions of the functors $\Phi^{(i)}$ and $\Psi^{(i)}$ define an equivalence of* mod-$A^{(i)}$ *and* $\mathcal{F}(\Delta_{C_i}) \subseteq$ mod-C_i.

The statement follows immediately from the following proposition.

Proposition A.3.2. *Let A be a quasi-hereditary algebra with respect to* $\mathbf{e}$. *Let $X \in \mathcal{F}(\Delta)$. Then the multiplication map $\mu_{X,i} : X\varepsilon_i \underset{C_i}{\otimes} \varepsilon_i A \to X^{(i)}$ is bijective for all $1 \leq i \leq n$. In particular, all multiplication maps $A\varepsilon_i \underset{C_i}{\otimes} \varepsilon_i A \to I_i = A\varepsilon_i A$ are bijective.*

Proof. For $i = n$, $\mu_{X,i}$ is trivially bijective for $X = e_n A$. For $X \in \mathcal{F}(\Delta)$, since $X^{(n)} = X e_n A$ is projective, $X^{(n)} \simeq \oplus\, e_n A$ and everything follows.

Proceed by induction and write

$$\bar{A} = A/A\varepsilon_{i+1}A\,, \quad \bar{X} = X/X\varepsilon_{i+1}A = X/X^{(i+1)}\,.$$

Consider the short exact sequences of right and left C_i-modules

$$0 \longrightarrow X^{(i+1)}\varepsilon_i \longrightarrow X\varepsilon_i \longrightarrow \bar{X}\varepsilon_i \longrightarrow 0$$

and

$$0 \longrightarrow \varepsilon_i A\varepsilon_{i+1}A \longrightarrow \varepsilon_i A \longrightarrow \varepsilon_i\bar{A} \longrightarrow 0\,.$$

Tensoring the first one by $\varepsilon_i A$ and the second one by $\bar{X}\varepsilon_i$ and by $X^{(i+1)}\varepsilon_i$, we get

$$0 \longrightarrow X^{(i+1)}\varepsilon_i \underset{C_i}{\otimes} \varepsilon_i A \longrightarrow X\varepsilon_i \underset{C_i}{\otimes} \varepsilon_i A \longrightarrow \bar{X}\varepsilon_i \underset{C_i}{\otimes} \varepsilon_i A \longrightarrow 0 \qquad (A.3.1)$$

and

$$0 = \bar{X}\varepsilon_i \underset{C_i}{\otimes} \varepsilon_i A\varepsilon_{i+1}A \longrightarrow \bar{X}\varepsilon_i \underset{C_i}{\otimes} \varepsilon_i A \overset{\sim}{\longrightarrow} \bar{X}\varepsilon_i \underset{C_i}{\otimes} \varepsilon_i\bar{A} \longrightarrow 0 \qquad (A.3.2)$$

and

$$0 \longrightarrow X^{(i+1)}\varepsilon_i \underset{C_i}{\otimes} \varepsilon_i A\varepsilon_{i+1}A \overset{\sim}{\longrightarrow} X^{(i+1)}\varepsilon_i \underset{C_i}{\otimes} \varepsilon_i A \longrightarrow X^{(i+1)}\varepsilon_i \underset{C_i}{\otimes} \varepsilon_i\bar{A} = 0\,. \qquad (A.3.3)$$

Hence, from (A.3.2),

$$\bar{X}\varepsilon_i \underset{C_i}{\otimes} \varepsilon_i A \simeq \bar{X}\varepsilon_i \underset{C_i}{\otimes} \varepsilon_i\bar{A}\,,$$

which may be identified with $\bar{X}\bar{\varepsilon}_i \underset{\bar{C}_i}{\otimes} \bar{\varepsilon}_i\bar{A}$, where $\bar{C}_i = \bar{\varepsilon}_i\bar{A}\bar{\varepsilon}_i$. There is a canonical surjective map

$$X\varepsilon_{i+1} \underset{C_{i+1}}{\otimes} \varepsilon_{i+1}A \to X\varepsilon_{i+1}A\varepsilon_i \underset{C_i}{\otimes} \varepsilon_i A\varepsilon_{i+1}A \simeq X^{(i+1)}\varepsilon_i \otimes C_i\varepsilon_i A\,.$$

The last isomorphism comes from (A.3.3), since $X\varepsilon_{i+1}A = X^{(i+1)}$. Thus, we get from (A.3.1) the first row of the following commutative diagram with exact rows connected by the multiplication maps:

$$\begin{array}{ccccccccc}
0 & \to & X\varepsilon_{i+1} \underset{C_{i+1}}{\otimes} \varepsilon_{i+1}A & \to & X\varepsilon_i \underset{C_i}{\otimes} \varepsilon_i A & \to & \bar{X}\bar{\varepsilon}_i \underset{\bar{C}_i}{\otimes} \bar{\varepsilon}_i\bar{A} & \to & 0 \\
 & & \Big\downarrow{\scriptstyle \mu_{X,i+1}} & & \Big\downarrow{\scriptstyle \mu_{X,i}} & & \Big\downarrow{\scriptstyle \bar{\mu}_{\bar{X},i}} & & \\
0 & \to & X\varepsilon_{i+1}A = X^{(i+1)} & \to & X^{(i)} & \to & (\bar{X})^{(i)} \simeq X^{(i)}/X^{(i+1)} & \to & 0\,.
\end{array}$$

By induction, $\mu_{X,i}$ is bijective, as required. □

Let us remark that the condition for $i = n$, namely that the multiplication map $Ae_n \underset{C_n}{\otimes} e_nA \to I_n$ is bijective, together with the assumption that e_nA is Schurian (i. e. C_n is a division algebra) implies that I_n is projective (and thus a heredity ideal). This simple fact allows to formulate some characterizations of quasi-hereditary algebras in terms of bijectivity of multiplication maps (cf. [DR1]).

It is very important to realize that although the centralizer algebras $C_i = \varepsilon_i A \varepsilon_i$ of a quasi-hereditary algebra A with respect to $\mathbf{e}$ have such a close connection to A (and are, in particular, quasi-hereditary with respect to the induced order), there may be idempotents $e \in A$ such that eAe is arbitrary. This is the essence of the following theorem.

Theorem A.3.3. *Given an arbitrary K-algebra R, there is a quasi-hereditary K-algebra A and an idempotent $e \in A$ such that $R \simeq eAe$.*

Proof. We shall provide here only a sketch of the proof, referring the reader to [DR2].

Without loss of generality, assume that R is basic: $R_R = \bigoplus_{j=1}^{m} f_jR$. Consider all non-zero non-isomorphic (local) factor modules $M_{j,s} = f_jR/f_j(\operatorname{rad} R)^s$ $(1 \le j \le m,\ s \ge 1)$. Denote their number by n and order them as follows: $(j, s) \preceq (j', s')$ if and only if $s > s'$ or if $s = s'$ and $j \ge j'$. Then, indexing them in that order, consider their direct sum $M = \bigoplus_{i=1}^{n} M_i$. Thus, M_1 is the principal module f_jR of maximal Loewy length (with the largest j) and M_n is the simple R-module $M_{1,1} = f_1R/f_1 \operatorname{rad} R$. Put $A = E_R(M)$ and denote by e_i the canonical projections of M onto M_i. It is a routine (and tedious) calculation to show that $\mathbf{e} = (e_1, e_2, \dots, e_n)$ is a heredity sequence: A is a basic quasi-hereditary algebra with respect to $\mathbf{e}$. Using the notation $\varepsilon = e_{i_1} + e_{i_2} + \dots + e_{i_m}$, where e_{i_j} is the idempotent corresponding to the summand $M_{i_j} \simeq f_jR$, $1 \le j \le m$, we have obviously $\varepsilon A \varepsilon \simeq R$. □

Let us present an illustration of the previous theorem. Let $R_R = \begin{smallmatrix} 1 \\ 2 \end{smallmatrix} \oplus \begin{smallmatrix} 2 \\ 2 \\ 2 \end{smallmatrix}$;
then $M_A = \begin{smallmatrix} 2 \\ 2 \\ 2 \end{smallmatrix} \oplus \begin{smallmatrix} 2 \\ 2 \end{smallmatrix} \oplus \begin{smallmatrix} 1 \\ 2 \end{smallmatrix} \oplus 2 \oplus 1$ and $A_A = \begin{smallmatrix} & 1 & \\ & 2 & \\ 1 & & 4 \\ & 2 & \\ & 1 & \end{smallmatrix} \oplus \begin{smallmatrix} & 2 & \\ 1 & & 4 \\ & 2 & \\ & 1 & \end{smallmatrix} \oplus \begin{smallmatrix} 3 \\ 4 \\ 2 \\ 1 \end{smallmatrix} \oplus \begin{smallmatrix} 4 \\ 2 \\ 1 \end{smallmatrix} \oplus \begin{smallmatrix} 5 \\ 3 \end{smallmatrix}$,
$R \simeq (e_1 + e_3)A(e_1 + e_3)$. Observe that the quasi-hereditary algebra A is in no way minimal.

The endomorphism algebras constructed above were first considered by M. Auslander in his Queen Mary College Mathematics Notes (1971). There he shows that the global dimension of A is bounded by the Loewy length of R_R. In fact, every quasi-hereditary algebra is of finite global dimension.

Theorem A.3.4. *Let A be a quasi-hereditary algebra with respect to $\mathbf{e} = (e_1, e_2, \dots, e_n)$. Then*

$$proj.dim\, \Delta(i) \leq n - i\,,$$
$$proj.dim\, S(i) \leq n + i - 2\,,$$

and thus $gl.dim\, A \leq 2(n-1)$. *This bound is optimal.*

Proof. There is a short exact sequence

$$0 \longrightarrow V(i) \longrightarrow P(i) \longrightarrow \Delta(i) \longrightarrow 0\,,$$

where $V(i) = \big(V(i)\big)^{(i+1)} = V(i)\varepsilon_{i+1}A$; hence

$$proj.dim\, \Delta(i) \leq 1 + \max_{j>i}\{proj.dim\, \Delta(j)\}\,.$$

Since $proj.dim\, \Delta(n) = 0$, the first inequality follows by induction.

Again, there is a short exact sequence

$$0 \longrightarrow U(i) \longrightarrow \Delta(i) \longrightarrow S(i) \longrightarrow 0\,,$$

where $\big(U(i)\big)^{(i)} = U(i)\varepsilon_i A = 0$. Thus,

$$proj.dim\, S(i) \leq 1 + \max_{j<i}\{proj.dim\, S(j), proj.dim\, \Delta(i)\}\,.$$

For $i = 1$, $S(i) \simeq \Delta(i)$ and thus $proj.dim\, S(1) \leq n + 1 - 2$. By induction, for $i > 1$,

$$proj.dim\, S(i) \leq n + (i-1) - 2 + 1 = n + i - 2\,.$$

In order to show that the bound on the global dimension is the best possible, consider the path algebra of the graph

$$\overset{1}{\bullet} \rightleftarrows \overset{2}{\bullet} \rightleftarrows \overset{3}{\bullet} \rightleftarrows \cdots \rightleftarrows \overset{n}{\bullet}$$

modulo $\langle \alpha_{i(i-1)}\alpha_{(i-1)i}$ for $2 \leq i \leq n$, $\alpha_{i(i+1)}\alpha_{(i+1)(i+2)}$ for $1 \leq i \leq n-2$, $\alpha_{(i+2)(i+1)}\alpha_{(i+1)i}$ for $1 \leq i \leq n-2\rangle$, where α_{ij} denotes the arrow from i to j. Then

$$A_A = \begin{smallmatrix} 1 \\ 2 \\ 1 \end{smallmatrix} \oplus \begin{smallmatrix} & 2 & \\ 1 & & 3 \\ & 2 & \end{smallmatrix} \oplus \ldots \oplus \begin{smallmatrix} & n-1 & \\ n-2 & & n \\ & n-1 & \end{smallmatrix} \oplus \begin{smallmatrix} n \\ n-1 \end{smallmatrix}$$

and $gl.dim\, A = 2(n-1)$. The algebra A is the canonic shallow algebra of the next section.

The bound for the global dimension of a quasi-hereditary algebra stated in Theorem A.3.4 can also be easily obtained by applying the following lemma.

Lemma A.3.5. *Let e be a primitive idempotent of an algebra A such that AeA is a heredity ideal, i. e. that eA is Schurian and AeA is a projective A-module. Write $B = A/AeA$. Let X be a B-module, also considered canonically as an A-module. Then*

$$proj.dim\, X_B \leq proj.dim\, X_A \leq proj.dim\, X_B + 1\,,$$

and hence

$$gl.dim\, B \leq gl.dim\, A \leq gl.dim\, B + 2\,.$$

Proof. Let $\{e_1, e_2, \ldots, e_n = e\}$ be a complete set of primitive orthogonal idempotents of A. For every $1 \leq i \leq n-1$ we have the following exact sequence:

$$0 \to e_iAeA \to e_iA \simeq P(i) \to e_iA/e_iAeA \to 0\,,$$

and $e_iA/e_iAeA \simeq \bar{P}(i)$, the ith principal module of B; moreover, $e_iAeA = P(i) \cap AeA = \oplus P(n)$ is projective. Thus we can see easily by induction that, for a B-module X, B-projective resolutions

$$0 = \bar{P}_{d+1} \longrightarrow \bar{P}_d \longrightarrow \ldots \longrightarrow \bar{P}_t \longrightarrow \ldots \longrightarrow \bar{P}_1 \longrightarrow \bar{P}_0 \longrightarrow X_B \longrightarrow 0$$

correspond to A-projective resolutions

$$0 \longrightarrow P_{d+1} \longrightarrow P_d \longrightarrow \ldots \longrightarrow P_t \longrightarrow \ldots \longrightarrow P_1 \longrightarrow P_0 \longrightarrow X_A \longrightarrow 0$$

in the following manner. If $\bar{P}_t = \overset{n-1}{\underset{i=1}{\oplus}} s_{ti}\bar{P}(i)$, then $P_t = \overset{n-1}{\underset{i=1}{\oplus}} s_{ti}P(i) \oplus s_{tn}P(n)$ for $0 \leq t \leq d+1$ (clearly, $s_{0n} = 0$ and $s_{(d+1)n}$ may be positive while all $s_{(d+1)i} = 0$ for $1 \leq i \leq n-1$). The first inequalities follow.

Now, in the exact sequence of A-modules $0 \to U(n) \to P(n) \to S(n) \to 0$, the module $U(n)$ is annihilated by AeA and thus is a B-module. Therefore

$$\begin{aligned} proj.dim\, S(n) &\leq proj.dim\, U(n)_A + 1 \leq \\ &\leq proj.dim\, U(n)_B + 1 + 1 \leq \\ &\leq gl.dim\, B + 2\,. \end{aligned}$$

Since, for $1 \leq i \leq n-1$, $proj.dim\, S(i) \leq gl.dim\, B + 1$, the last inequality holds as well. □

There is an important consequence of Lemma A.3.5. We are going to formulate it now.

Lemma A.3.6. *Let e be a primitive idempotent of an algebra A such that AeA is a heredity ideal. Write $B = AeA$. Then for any two B-modules X, Y,*

$$\mathrm{Ext}^t_A(X,Y) \simeq \mathrm{Ext}^t_B(X,Y) \quad \textit{for all } t \geq 0\,.$$

In particular, $\mathrm{Ext}^t_A(B,B) = 0$ *for all* $t \geq 1$.

Theorem A.3.7. *Let A be a quasi-hereditary algebra with respect to $\mathbf{e} = (e_1, e_2, \ldots, e_n)$ and let $B_i = A/A\varepsilon_{i+1}A$ for all $1 \leq i \leq n$. For any two B_i-modules X, Y,*

$$\mathrm{Ext}^t_A(X,Y) \simeq \mathrm{Ext}^t_{B_i}(X,Y) \quad \textit{for all } t \geq 0\,.$$

In particular, $\mathrm{Ext}^t_A(B_i,B_i) = 0$ *for all* $t \geq 1$.

Proof of Lemma A.3.6. Trivially, $\mathrm{Hom}_A(X,Y) \simeq \mathrm{Hom}_B(X,Y)$. Moreover, every extension Z_A of X_A by Y_A (as A-modules) is annihilated by $AeA = (AeA)^2$: $Z(AeA)^2 \subseteq YAeA = 0$. Thus, it turns out that $\mathrm{Ext}^1_A(X,Y) \simeq \mathrm{Ext}^1_B(X,Y)$.

Now, applying successively the functors $\mathrm{Hom}_B(-,Y_B)$ and $\mathrm{Hom}_A(-,Y_A)$ to the exact sequence of B-modules

$$0 \longrightarrow U \longrightarrow P \longrightarrow X \longrightarrow 0$$

with P_B projective, we get, for all $t \geq 0$:

$$\mathrm{Ext}^t_B(P,Y) \longrightarrow \mathrm{Ext}^t_B(U,Y) \longrightarrow \mathrm{Ext}^{t+1}_B(X,Y) \longrightarrow \mathrm{Ext}^{t+1}_B(P,Y) \qquad (A.3.4)$$

and

$$\mathrm{Ext}^t_A(P,Y) \longrightarrow \mathrm{Ext}^t_A(U,Y) \longrightarrow \mathrm{Ext}^{t+1}_A(X,Y) \longrightarrow \mathrm{Ext}^{t+1}_A(P,Y). \qquad (A.3.5)$$

Since P_B is projective, we obtain from (A.3.4)

$$\mathrm{Ext}^t_B(U,Y) \simeq \mathrm{Ext}^{t+1}_B(X,Y).$$

Furthermore, $proj.dim\, P_A \leq 1$ yields $\mathrm{Ext}^{t+1}_A(P,Y) = 0$ for $t \geq 1$. Since also $\mathrm{Ext}^1_A(P,Y) = \mathrm{Ext}^1_B(P,Y) = 0$, (A.3.5) implies

$$\mathrm{Ext}^t_A(U,Y) \simeq \mathrm{Ext}^{t+1}_A(X,Y).$$

The statement of the theorem follows by induction. □

Of course, not all algebras of finite global dimension are quasi-hereditary.

Example. Consider the path algebra of the graph with two vertices, 1 and 2, with k arrows, $\alpha_1, \alpha_2, \dots \alpha_k$ from 1 to 2 and ℓ arrows, $\beta_1, \beta_2, \dots, \beta_\ell$, $\ell = k-1$ or $\ell = k$, from 2 to 1. Let $d = k + \ell$ and F_d the path algebra modulo the ideal

$$I_d = \langle \alpha_i \beta_j \text{ for } i > j \text{ and } \beta_i \alpha_j \text{ for } i \geq j \rangle.$$

Denoting by f_m, $m \geq 1$, the mth Fibonacci number (i. e. $f_1 = f_2 = 1$, $f_m = f_{m-2} + f_{m-1}$ for $m \geq 3$), one can calculate easily that $\dim_K F_d = f_{d+3}$ and $gl.dim\, F_d = d$. For $d \geq 3$, F_d is not a quasi-hereditary algebra.

Theorem A.3.8. *Let A be a K-algebra of global dimension ≤ 2. Then A is quasi-hereditary with respect to a suitable order of the complete set of primitive orthognal idempotents.*

Proof. Assume that A is basic and write $A_A = \bigoplus_{i=1}^{n} e_i A$, where

$$\text{Loewy length } e_i A \geq \text{Loewy length } e_{i+1} A \quad \text{for all } 1 \leq i \leq n-1.$$

Now, since the kernels of homomorphisms between projective modules are projective, there are no non-zero homomorphisms from $e_n A$ to $e_n \,\mathrm{rad} A$, and thus $e_n A$ is Schurian.

We want to show that, for each $1 \leq i \leq n-1$, $e_i A e_n A$ is projective. Consider the exact sequence

$$0 \longrightarrow X \longrightarrow P \longrightarrow e_i A e_n A \longrightarrow 0$$

with the projective cover $P \simeq \bigoplus e_n A$. Note that X is the kernel of a homomorphism between projective modules P and $e_i A$ and that $X \subseteq \operatorname{rad} P$. But then X is a projective module whose indecomposable direct summands are of Loewy length smaller than the Loewy length of $e_n A$. Therefore $X = 0$ and $Ae_n A$ is projective.

Since, in view of Lemma A.3.5, $gl.dim\, A/Ae_n A \leq gl.dim\, A = 2$, we conclude by induction that $A/Ae_n A$ is quasi-hereditary and the theorem follows. □

We have already seen that, for an A-module $X \in \mathcal{F}(\Delta)$, the multiplicities $[X : \Delta(i)]$, being the numbers of the indecomposable direct summands (isomorphic to $\Delta(i)$) in a decomposition of $X^{(i)}/X^{(i+1)}$, are well defined. We have also defined the dimension $\underline{\dim}\, X$ of X as the integral vector $(x_1, x_2, \ldots, x_n)$, where x_i is the number $[X : S(i)]$ of factors isomorphic to $S(i)$ in a composition series of X.

Clearly, $\{\underline{\dim}\, \Delta(i) \mid 1 \leq i \leq n\}$ forms an integral basis of $\mathbb{Z}^n$ and thus

$$\underline{\dim}\, X = \sum_{i=1}^{n} \rho_i \underline{\dim}\, \Delta(i)$$

with integers ρ_i. In particular, if $X \in \mathcal{F}(\Delta)$ then ρ_i are non-negative integers equal to $[X : \Delta(i)]$:

$$\underline{\dim}\, X = \sum_{i=1}^{n} [X : \Delta(i)]\, \underline{\dim}\, \Delta(i)\,.$$

As before, write $D_i = E_A(\Delta(i)) = E_A(S(i))$ and $d_i = \dim_K D_i$ for $1 \leq i \leq n$. The following lemma is an immediate consequence of Lemma A.1.4.

Proposition A.3.9. *The functors* $\operatorname{Hom}(-, \nabla(j))$ *are exact on* $\mathcal{F}(\Delta)$ *and* $\operatorname{Hom}(X, \nabla(j)) \simeq \operatorname{Hom}(X^{(j)}/X^{(j+1)}, \nabla(j))$. *Hence*

$$\dim_K \operatorname{Hom}(X, \nabla(j)) = d_j [X : \Delta(j)]\,.$$

Taking $X = P(i)$ we get the Bernstein-Gelfand-Gelfand reciprocity law.

Corollary A.3.10 *For every* $1 \leq i, j \leq n$,

$$d_j [P(i) : \Delta(j)] = d_i [\nabla(j) : S(i)]\,.$$

The reciprocity law can be reformulated for split algebras (in particular, for algebras over an algebraically closed field) in terms of factorization of the

Cartan-matrix $C(A)$ into unipotent triangular matrices. Recall that $C(A)$ is, by definition, the $n\times n$ integral matrix whose ith row equals $\underline{\dim}\, P(i)$, $1\le i\le n$. Indeed, since all $d_i=1$ and $\left(\underline{\dim}\, P(i)\right)_k=\sum_{j=1}^{n}\left[P(i):\Delta(j)\right]\left[\Delta(j):S(k)\right]$, we have

$$C(A)=\nabla(A)^{tr}\cdot\Delta(A)\,,$$

where $\nabla(A)$ and $\Delta(A)$ are the $n\times n$ matrices whose rows equal $\underline{\dim}\,\nabla(i)$ and $\underline{\dim}\,\Delta(i)$, respectively.

Let us point out that Corollary A.3.10 can be rewritten as

$$d_j\left[P(i):\Delta(j)\right]=d_i\left[\Delta^{\circ}(j):S^{\circ}(i)\right].$$

A.4 Canonical Constructions

There are two recursive constructions of quasi-hereditary algebras described in the literature:
(i) the construction via "not so trivial extensions" of [PS] and
(ii) the construction based on extensions of centralizers [DR1].

Here we just briefly describe the inductive steps and illustrate both (in some sense opposite) procedures on an example.

(i) Given a quasi-hereditary K-algebra B, a division K-algebra D, bimodules ${}_DM_B$, ${}_BN_D$ and an extension $\tilde{B}$ of B by $N\otimes_D M$, the K-algebra ("not so trivial extension")

$$B'=\begin{pmatrix}\tilde{B} & N\\ M & D\end{pmatrix}$$

with trivial multiplication $M\otimes N\to D$ can easily seen to be again quasi-hereditary (with respect to an extended order of idempotents): For $e=\left(\begin{smallmatrix}0&0\\0&1\end{smallmatrix}\right)$, $eB'\simeq(M\ \ D)$ is Schurian, $B'eB'\simeq\left(\begin{smallmatrix}N\otimes M & N\\ M & D\end{smallmatrix}\right)$ is projective and $B\simeq B'/B'eB'$. Clearly, having a quasi-hereditary K-algebra A with respect to $\mathbf{e}=(e_1,e_2,\dots,e_n)$, then denoting $B_i=A/A\varepsilon_{i+1}A$, $0\le i\le n$, each B_i can be obtained by the above construction from $B=B_{i-1}$, $D=\varepsilon_{i-1}A\varepsilon_{i-1}/\varepsilon_{i-1}A\varepsilon_iA\varepsilon_{i-1}$ and the respective bimodules. We have $B_0=0, B_1=A/A\varepsilon_2A,\dots,B_{n-1}=A/A\varepsilon_nA$, $B_n=A$; each consecutive step simply extends the principal modules in accordance with the Δ-filtration of the regular module A_A.

(ii) Given a quasi-hereditary K-algebra C, a division K-algebra D, bimodules ${}_DE_C$, ${}_CF_D$ such that $E_C\in\mathcal{F}(\Delta_C)$ and ${}_CF\in\mathcal{F}(\Delta_C^{\circ})$, and a multiplication map $\mu: F\otimes_D E\to \operatorname{rad} C$, denote by $\tilde{D}=D\ltimes(E\otimes_C F)$ the split extension of $E\otimes_C F$ by D and consider

$$C'=\begin{pmatrix}\tilde{D} & E\\ F & C\end{pmatrix}.$$

Here the multiplication in $\tilde{D}$ is given by $(d, e\otimes f)(d', e'\otimes f') = \big(dd', de'\otimes f' + e\otimes fd' + e\otimes \mu(f\otimes e')f'\big)$. The K-algebra C' is again quasi-hereditary (with respect to an extended order of idempotents): $\Delta_{C'}(1) \simeq D$ and $\Delta_{C'}(i) = \Delta_C(i)\otimes_C(F\ C)_{C'}$ for $\Delta_C(i) \in \Delta_C$.

Now, having a quasi-hereditary K-algebra A with respect to $\mathbf{e} = (e_1, e_2, \ldots, e_n)$ such that $A/\operatorname{rad} A$ is separable (e.g. over a perfect field K), then denoting $C_i = \varepsilon_i A \varepsilon_i$, $1 \le i \le n+1$, each C_i can be obtained by the above construction from $C = C_{i+1}$, $D = \varepsilon_i A\varepsilon_i/\varepsilon_i A\varepsilon_{i+1}A\varepsilon_i \simeq e_i A e_i/e_i A\varepsilon_{i+1}Ae_i$, $E = e_i A\varepsilon_{i+1}$, $F = \varepsilon_{i+1}Ae_i$ and the respective multiplication map $\mu = \mu_i$. We have $C_{n+1} = 0, C_n = \varepsilon_n A\varepsilon_n, \ldots, C_2 = \varepsilon_2 A\varepsilon_2, C_1 = A$.

Let us illustrate the described procedures on the following simple example: A is the path K-algebra of

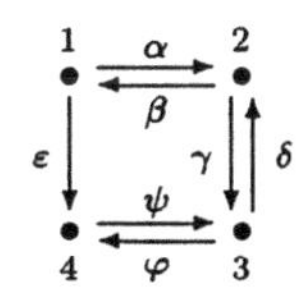

modulo $\langle \alpha\gamma, \beta\varepsilon, \beta\alpha-\gamma\delta, \gamma\varphi, \delta\gamma, \psi\varphi, \psi\delta\beta\rangle$. Thus the composition series of the regular representation can be described by

$$A_A = \begin{array}{cccc} & 1 & & \\ & 2 & 4 & \\ 1 & & 3 & \\ & & 2 & \end{array} \oplus \begin{array}{ccc} & 2 & \\ 1 & & 3 \\ & 2 & \\ & 1 & \end{array} \oplus \begin{array}{cccc} & 3 & & \\ & 2 & 4 & \\ 1 & & 3 & \\ & & 2 & \end{array} \oplus \begin{array}{c} 4 \\ 3 \\ 2 \end{array}.$$

The algebras B_1, B_2, B_3 and B_4 are successively described by the filtration $A/A\varepsilon_2 A$, $A\varepsilon_2 A/A\varepsilon_3 A$, $A\varepsilon_3 A/A\varepsilon_4 A$ and $A\varepsilon_4 A$:

$$(e_1),\ \begin{pmatrix} \alpha\beta & \alpha \\ \beta & e_2 \end{pmatrix},\ \begin{pmatrix} \gamma\delta\beta & \gamma\delta & \gamma \\ \delta\beta & \delta & e_3 \end{pmatrix},\ \begin{pmatrix} \varepsilon\psi\delta & \varepsilon\psi & \varepsilon \\ \varphi\psi\delta & \varphi\psi & \varphi \\ \psi\delta & \psi & e_4 \end{pmatrix}.$$

The algebras $C_1 = A$, C_2, C_3 and C_4 are marked as the right lower blocks:

$$\left(\begin{array}{c|c|c|c} \substack{e_1 \\ \alpha\beta} & \substack{\varepsilon\psi\delta \\ \alpha} & \varepsilon\psi & \varepsilon \\ \hline \substack{\beta \\ \beta\alpha\beta} & \substack{e_2 \\ \beta\alpha = \gamma\delta} & & \gamma \\ \hline \delta\beta & \substack{\delta \\ \varphi\psi\delta} & \substack{e_3 \\ \varphi\psi} & \varphi \\ \hline & \psi\delta & \psi & e_4 \end{array}\right)$$

Now, the K-species of the algebras in the first construction simply extend gradually to the K-species of the quasi-hereditary algebra A:

If $\mathcal{S}(B) = (\Lambda = D_1 \times D_2 \times \ldots \times D_r;\ {}_\Lambda W_\Lambda)$ is the K-species of B, then $\mathcal{S}(B') = \big(\Lambda' = D_1 \times D_2 \times \ldots \times D_r \times D;\ {}_{\Lambda'}(W \oplus M/\operatorname{rad} M \oplus N/\operatorname{rad} N)_{\Lambda'}\big)$

is the K-species of B'. Thus, the K-species of the algebras B_i are simply the restrictions of the K-species of A.

The situation in the case of the second construction is, in general, more complex. Already in the very simple situation of the (hereditary) path K-algebra A of the graph $\overset{2}{\bullet}\longrightarrow\overset{1}{\bullet}\longrightarrow\overset{3}{\bullet}$, the graph of the corresponding (hereditary) algebra C_2 is $\overset{2}{\bullet}\longrightarrow\overset{3}{\bullet}$, and thus not just a restriction of the original graph. It is therefore natural to consider quasi-hereditary algebras A with respect to $\mathbf{e} = (e_1, e_2, \dots, e_n)$ such that this does not happen. Clearly, this will not happen if the images of the multiplication maps μ_i will be in $\operatorname{rad}^2 C_i$. Let us formalize the condition in the following concept of being lean.

Definition A.4.1. A basic K-algebra A is called *lean* with respect to $\mathbf{e} = (e_1, e_2, \dots, e_n)$ if, for every $1 \le i \le n$, the K-species $\mathcal{S}(C_{i+1})$ is a restriction of the K-species $\mathcal{S}(C_i)$, i.e. if

$$\mathcal{S}(C_i) = (D_i, D_{i+1}, \dots, D_n\,;\ {}_rW_s\,,\ i \le r, s \le n)$$

then

$$\mathcal{S}(C_{i+1}) = (D_{i+1}, \dots, D_n\,;\ {}_rW_s\,,\ i+1 \le r, s \le n)\,.$$

Here, as before, $C_i/\operatorname{rad} C_i = D_i \times D_{i+1} \times \dots D_n$ and $\operatorname{rad} C_i/\operatorname{rad}^2 C_i = \bigoplus_{i\le r,s\le n} {}_rW_s$; thus $D_i = e_iAe_i/e_i \operatorname{rad}A\, e_i$ and ${}_rW_s = e_r \operatorname{rad}A\, e_s/e_r \operatorname{rad}^2 A\, e_s$.

We get immediately the following characterization of being lean.

Lemma A.4.2. *Let A be a K-algebra and $\mathbf{e} = (e_1, e_2, \dots, e_n)$. Then A is lean with respect to $\mathbf{e}$ if and only if*

$$e_i \operatorname{rad}^2 A\, e_j = e_i \operatorname{rad}A\, \varepsilon_m \operatorname{rad}A\, e_j \quad \textit{for all} \quad 1 \le i,j \le n \ \textit{and} \ m = \min\{i,j\}\,.$$

Lean quasi-hereditary algebras can be characterized in terms of top filtrations (see Theorem A.4.10 below); here is a definition.

Definition A.4.3. A monomorphism $\alpha : X \to Y$ is said to be a *top embedding* of X into Y if the induced homomorphism $\bar\alpha : \operatorname{top} X = X/\operatorname{rad} X \to \operatorname{top} Y = Y/\operatorname{rad} Y$ is monic; or equivalently, if $\alpha(\operatorname{rad} X) = \alpha(X) \cap \operatorname{rad} Y$.

There is, of course, also a dual notion of a *socle epimorphism.* We shall write simply $X \overset{t}{\subseteq} Y$ if the embedding is a top embedding.

Definition A.4.4. A filtration

$$0 = X_{s+1} \subseteq X_s \subseteq \dots \subseteq X_j \subseteq \dots \subseteq X_2 \subseteq X_1 = X \qquad (A.4.1)$$

is said to be a *top filtration* of X if $X_j \overset{t}{\subseteq} X$ for all $2 \le j \le s$.

We have the following obvious lemma.

Lemma A.4.5. *Let* $X \subseteq Y \subseteq Z$. *Then:*

1) $X \overset{t}{\subseteq} Z$ *implies* $X \overset{t}{\subseteq} Y$ *and*
2) $X \overset{t}{\subseteq} Y$ *and* $Y \overset{t}{\subseteq} Z$ *implies* $X \overset{t}{\subseteq} Z$.

Thus, (A.4.1) is a top filtration of X if and only if $X_j \overset{t}{\subseteq} X_{j-1}$ for all $2 \leq j \leq s$. In fact, (A.4.1) is a top filtration of X if and only if $X_j/X_{j+1} \overset{t}{\subseteq} X/X_{j+1}$ for all $2 \leq j \leq s$. This is an immediate consequence of the following lemma.

Lemma A.4.6. *Let* $X \subseteq Y \subseteq Z$. *Let* $X \overset{t}{\subseteq} Z$. *Then* $Y/X \overset{t}{\subseteq} Z/X$ *if and only if* $Y \overset{t}{\subseteq} Z$.

Proof. First, $X \overset{t}{\subseteq} Z$ implies that

$$\operatorname{rad}(Z/X) \simeq (X + \operatorname{rad} Z)/X \simeq \operatorname{rad} Z/(X \cap \operatorname{rad} Z) \simeq \operatorname{rad} Z/\operatorname{rad} X$$

and

$$\begin{aligned}\operatorname{top}(Z/X) &\simeq (Z/X)/((X + \operatorname{rad} Z)/X) \simeq (Z/\operatorname{rad} Z)/((X + \operatorname{rad} Z)/\operatorname{rad} Z) \\ &\simeq \operatorname{top} Z/\operatorname{top} X\,.\end{aligned}$$

Similarly for $X \overset{t}{\subseteq} Y$. Hence, denoting the given embeddings $X \subseteq Y$ and $Y \subseteq Z$ by α and β, respectively, we get a commutative diagram of exact sequences with induced embeddings $\bar{\alpha}$ and $\overline{\beta\alpha}$.

$$\begin{array}{ccccccccc}
 & & 0 & & 0 & & 0 & & \\
 & & \downarrow & & \downarrow & & \downarrow & & \\
0 & \longrightarrow & \operatorname{top} X & \xrightarrow{\bar{\alpha}} & \operatorname{top} Y & \longrightarrow & \operatorname{top}(Y/X) & \longrightarrow & 0 \\
 & & \| & & {\scriptstyle\bar{\beta}}\downarrow & & {\scriptstyle\bar{\bar{\beta}}}\downarrow & & \\
0 & \longrightarrow & \operatorname{top} X & \xrightarrow{\overline{\beta\alpha}} & \operatorname{top} Z & \longrightarrow & \operatorname{top}(Z/X) & \longrightarrow & 0 \\
 & & \downarrow & & \downarrow & & \downarrow & & \\
 & & 0 & \longrightarrow & \operatorname{top}(Z/Y) & = & top(Z/Y) & \longrightarrow & 0\,. \\
 & & & & \downarrow & & \downarrow & & \\
 & & & & 0 & & 0 & &
\end{array}$$

Clearly, $\bar{\beta}$ is a monomorphism if and only if $\bar{\bar{\beta}}$ is. □

The following two lemmas form an essential part of the proof of Theorem A.4.10 which will establish the relationship between lean algebras and algebras with top standard filtrations.

Lemma A.4.7. *Let A be a K-algebra and $\mathbf{e} = (e_1, e_2, \dots, e_n)$. Then, for all $1 \le i \le n$*

$$e_i \operatorname{rad}^2 A\, e_j = e_i \operatorname{rad} A\, \varepsilon_{i+1} \operatorname{rad} A\, e_j \quad \textit{for every} \quad j > i$$

if and only if

$$V(i) = \big(\operatorname{rad} P(i)\big)^{(i+1)} \overset{t}{\subseteq} \operatorname{rad} P(i)\,.$$

Proof. The proof follows from the following string of equivalent statements expressing the fact that $V(i) \overset{t}{\subseteq} \operatorname{rad} P(i)$:

$$e_i \operatorname{rad} A\, \varepsilon_{i+1} A \overset{t}{\subseteq} e_i \operatorname{rad} A$$
$$e_i \operatorname{rad} A\, \varepsilon_{i+1} A \cap e_i \operatorname{rad}^2 A = e_i \operatorname{rad} A\, \varepsilon_{i+1} \operatorname{rad} A$$
$$e_i \operatorname{rad} A\, \varepsilon_{i+1} A e_j \cap e_i \operatorname{rad}^2 A\, e_j = e_i \operatorname{rad} A\, \varepsilon_{i+1} \operatorname{rad} A\, e_j \quad \text{for all} \quad 1 \le j \le n\,.$$

However, the last equality is trivial for $j \le i$, since $\varepsilon_{i+1} A e_j = \varepsilon_{i+1} \operatorname{rad} A\, e_j$ and $\operatorname{rad} A\, \varepsilon_{i+1} \operatorname{rad} A \subseteq \operatorname{rad}^2 A$; moreover, for $j > i$, the left-hand side collapses to $e_i \operatorname{rad}^2 A\, e_j$ since $\operatorname{rad} A\, \varepsilon_{i+1} A e_j \supseteq \operatorname{rad} A e_j \supseteq \operatorname{rad}^2 A\, e_j$. □

Lemma A.4.8. *Let A be a K-algebra and $\mathbf{e} = (e_1, e_2, \dots, e_n)$. Then*

$$\big(\operatorname{rad} P^\circ(j)\big)^{(i)} / \big(\operatorname{rad} P^\circ(j)\big)^{(i+1)} \overset{t}{\subseteq} \operatorname{rad} P^\circ(j) / \big(\operatorname{rad} P^\circ(j)\big)^{(i+1)}$$

if and only if

$$e_i \operatorname{rad}^2 A\, e_j = e_i \operatorname{rad} A\, \varepsilon_i \operatorname{rad} A\, e_j\,.$$

Proof. As in the proof of the previous lemma, we write down equivalent statements, expressing the top embedding from the lemma:

$$A\varepsilon_i \operatorname{rad} A\, e_j / A\varepsilon_{i+1} \operatorname{rad} A\, e_j \overset{t}{\subseteq} \operatorname{rad} A\, e_j / A\varepsilon_{i+1} \operatorname{rad} A\, e_j\,,$$

$$A\varepsilon_i \operatorname{rad} A\, e_j \cap (\operatorname{rad}^2 A\, e_j + A\varepsilon_{i+1} \operatorname{rad} A\, e_j) = \\ = \operatorname{rad} A\, \varepsilon_i \operatorname{rad} A\, e_j + A\varepsilon_{i+1} \operatorname{rad} A\, e_j\,,$$

$$e_k A\varepsilon_i \operatorname{rad} A\, e_j \cap (e_k \operatorname{rad}^2 A\, e_j + e_k A\varepsilon_{i+1} \operatorname{rad} A\, e_j) = \\ = e_k \operatorname{rad} A\, \varepsilon_i \operatorname{rad} A\, e_j + e_k A\varepsilon_{i+1} \operatorname{rad} A\, e_j \quad \text{for all} \quad 1 \le k \le n\,.$$

For $k < i$, the last equality is trivial, since both sides equal $e_k \operatorname{rad} A\, \varepsilon_i \operatorname{rad} A\, e_j$. We can also verify easily that for $k > i$, both sides equal $e_k A\varepsilon_{i+1} \operatorname{rad} A\, e_j$; just observe that $e_k A\varepsilon_i \operatorname{rad} A \supseteq e_k A\varepsilon_{i+1} \operatorname{rad} A \supseteq e_k A e_k \operatorname{rad} A \supseteq e_k \operatorname{rad}^2 A$ and $e_k \operatorname{rad} A\, \varepsilon_i \operatorname{rad} A \subseteq e_k \operatorname{rad} A \subseteq e_k A\varepsilon_{i+1} \operatorname{rad} A$. Hence the only genuine condition remains for $k = i$: $e_i \operatorname{rad}^2 A\, e_j = e_i \operatorname{rad} A\, \varepsilon_i \operatorname{rad} A\, e_j$. □

Before formulating the main result, let us recall first the notation for the standard exact sequences

$$0 \longrightarrow V(i) \longrightarrow P(i) \longrightarrow \Delta(i) \longrightarrow 0$$

and

$$0 \longrightarrow U(i) \longrightarrow \Delta(i) \longrightarrow S(i) \longrightarrow 0, \ \ 1 \leq i \leq n.$$

Thus,

$$0 \longrightarrow V(i) = \left(\operatorname{rad} P(i)\right)^{(i+1)} \longrightarrow \operatorname{rad} P(i) \longrightarrow U(i) = \operatorname{rad} \Delta(i) \longrightarrow 0$$

is exact. Of course, there are similar sequences for the left modules $V^\circ(i)$, $P^\circ(i), \Delta^\circ(i), U^\circ(i)$ and $S^\circ(i)$.

Proposition A.4.9. *Let A be a quasi-hereditary algebra with respect to $\mathbf{e} = (e_1, e_2, \ldots, e_n)$. Then the following statements are equivalent:*

a) $e_i \operatorname{rad}^2 A\, e_j = e_i \operatorname{rad} A\, \varepsilon_i \operatorname{rad} A\, e_j$ *for all* $1 \leq i < j \leq n$.

b) $V(i) \overset{t}{\subseteq} \operatorname{rad} P(i)$ *for all* $1 \leq i \leq n$.

c) The trace filtration $\left\{ \left(U^\circ(j)\right)^{(i)} \mid 1 \leq i \leq j \right\}$ *of* $U^\circ(j) = \operatorname{rad} \Delta^\circ(j)$ *is a top filtration for all* $1 \leq j \leq n$.

Proof. Recall that $e_i \operatorname{rad} A\, \varepsilon_i \operatorname{rad} A = e_i \operatorname{rad} A\, \varepsilon_{i+1} \operatorname{rad} A$ for all $1 \leq i \leq n$. Then the equivalence of *a)* and *b)* follows from Lemma A.4.7 and the equivalence of *a)* and *c)* from Lemma A.4.8 and Lemma A.4.6, since clearly $\left(\operatorname{rad} \Delta^\circ(j)\right)^{(i)} / \left(\operatorname{rad} \Delta^\circ(j)\right)^{(i+1)} \simeq \left(\operatorname{rad} P^\circ(j)\right)^{(i)} / \left(\operatorname{rad} P^\circ(j)\right)^{(i+1)}$ for $i < j$. □

Let us point out that there is also a dual Proposition A.4.9°, whose formulation we leave to the reader.

Theorem A.4.10. *Let A be a quasi-hereditary algebra with respect to $\mathbf{e} = (e_1, e_2, \ldots, e_n)$. Then the following statements are equivalent:*

1) *A is lean (with respect to $\mathbf{e}$);*

2) *the trace filtration of $U(i)$ is a top filtration and $V(i) \overset{t}{\subseteq} \operatorname{rad} P(i)$ for all $1 \leq i \leq n$;*

2°) *the trace filtration of $U^\circ(i)$ is a top filtration and $V^\circ(i) \overset{t}{\subseteq} \operatorname{rad} P^\circ(i)$ for all $1 \leq i \leq n$;*

3) *$V(i) \overset{t}{\subseteq} \operatorname{rad} P(i)$ and $V^\circ(i) \overset{t}{\subseteq} \operatorname{rad} P^\circ(i)$ for all $1 \leq i \leq n$;*

4) *the trace filtrations of $\operatorname{rad} \Delta(i)$ and of $\operatorname{rad} \Delta^\circ(i)$ are top filtrations for all $1 \leq i \leq n$.*

Proof. The theorem is an immediate consequence of Proposition A.4.9 and its dual Proposition A.4.9°. In view of Lemma A.4.2, 1) is equivalent to *a)* and *a°)* (of Proposition A.4.9 and Proposition A.4.9°, respectively), 2) is equivalent to *b)* and *c°)*, 3) to *b)* and *b°)* and finally, 4) is equivalent to *c)* and *c°)*. □

Let us point out that 2) (and 2°)) can be reformulated as follows:

$$\operatorname{rad} P(i) \supseteq \left(\operatorname{rad} P(i)\right)^{(2)} \supseteq \left(\operatorname{rad} P(i)\right)^{(3)} \supseteq \dots$$
$$\supseteq \left(\operatorname{rad} P(i)\right)^{(i-1)} \supseteq \left(\operatorname{rad} P(i)\right)^{(i)} = \left(\operatorname{rad} P(i)\right)^{(i+1)} = V(i) \supseteq 0$$

is a top filtration.

We are going to define, and construct, some special classes of quasi-hereditary algebras.

Proposition A.4.11. *Let A be a quasi-hereditary algebra. Then the following statements are equivalent:*

1) *$V(i)$ is projective, i. e. proj.dim $\Delta(i) \leq 1$ for all $1 \leq i \leq n$;*
2) *proj.dim $T(A) \leq 1$, where $T(A)$ is the characteristic module of Theorem A.2.7;*
3) *$\mathcal{F}(\Delta^\circ)$ is closed under submodules;*
4) *$\operatorname{rad} \Delta^\circ(i) \in \mathcal{F}(\Delta^\circ)$ for all $1 \leq i \leq n$.*

Proof. Since $T = T(A) \in \mathcal{F}(\Delta)$, 1) implies 2). Moreover, 3) is equivalent to 4). To show the non-trivial implication of the latter equivalence, let M be a (left) A-module form $\mathcal{F}(\Delta^\circ)$ and N a maximal submodule of M; let $M/N \simeq S^\circ(i)$. Then, for their trace filtrations, $N^{(j)} = M^{(j)}$ for $j \geq i+1$, while $N^{(i)}/N^{(i+1)}$ is filtered by $\Delta^\circ(i)$'s and $\operatorname{rad} \Delta^\circ(i)$, and we have $N/N^{(i)} \simeq M/M^{(i)}$. Hence, $N \in \mathcal{F}(\Delta^\circ)$. By induction, every submodule of M belongs to $\mathcal{F}(\Delta^\circ)$.

Thus we need only to establish the implications 2) $\Rightarrow$ 3) and 3) $\Rightarrow$ 1). We are going to use the fact that 3) is equivalent to

3°) *$\mathcal{F}(\Delta)$ is closed under factor modules;*

furthermore, note that, in view of Theorem A.2.8, condition 2) yields

$$\mathcal{F}(\nabla) = \{Y \mid \operatorname{Ext}^1(T, Y) = 0\}.$$

To prove that 2) implies 3), take $Y \in \mathcal{F}(\nabla)$ and a short exact sequence $0 \to X \to Y \to Z \to 0$. From here, we get

$$0 = \operatorname{Ext}^1(T, Y) \longrightarrow \operatorname{Ext}^1(T, Z) \longrightarrow \operatorname{Ext}^2(T, X) = 0,$$

and thus $Z \in \mathcal{F}(\nabla)$.

Finally, assume 3). We want to show that $\operatorname{Ext}^2(\Delta(i), X) = 0$ for all modules X, $1 \leq i \leq n$. Consider the exact sequence $0 \to X \to Q \to Y \to 0$ with the injective hull Q of X. Since $Q \in \mathcal{F}(\nabla)$, also $Y \in \mathcal{F}(\nabla)$. Thus, the exact sequence yields

$$0 = \operatorname{Ext}^1\left(\Delta(i), Y\right) \longrightarrow \operatorname{Ext}^2\left(\Delta(i), X\right) \longrightarrow \operatorname{Ext}^2\left(\Delta(i), Q\right) = 0,$$

as required. □

In combination with the dual statement that all $V^\circ(i)$ are projective if and only if all $\operatorname{rad}\Delta(i) \in \mathcal{F}(\Delta)$, we get the following result.

Corollary A.4.12. *Let A be a quasi-hereditary algebra such that all $V(i)$ are projective right A-modules and all $V^\circ(i)$ are projective left A-modules ($1 \leq i \leq n$). Then*

$$gl.dim\, A \leq 2\,.$$

Indeed, the conditions are equivalent to $\operatorname{rad}\Delta(i) \in \mathcal{F}(\Delta)$ and $V(i)$ projective for all $1 \leq i \leq n$. Thus, since $proj.dim\, \Delta(i) \leq 1$, $proj.dim\, \operatorname{rad}\Delta(i) \leq 1$ for all i, and the corollary follows.

Taking into account Theorem A.4.10, we get also the following consequence.

Corollary A.4.13. *All* $\operatorname{rad}P(i)$, $1 \leq i \leq n$, *have top filtrations with factors $\Delta(j)$, $1 \leq j \leq i-1$ and $P(j)$, $i+1 \leq j \leq n$ if and only if all* $\operatorname{rad}P^\circ(i)$, $1 \leq i \leq n$, *have top filtrations with factors $\Delta^\circ(j)$, $1 \leq j \leq i-1$ and $P^\circ(j)$, $i+1 \leq j \leq n$.*

Definition A.4.14. A (quasi-hereditary) algebra satisfying the conditions of Corollary A.4.13 will be called *replete.*

Of course, all hereditary algebras are replete with respect to a suitable order of idempotents (given by the lengths of principal modules). Let us point out that replete algebras are lean.

Theorem A.4.15. *Let A be a quasi-hereditary algebra. Then the modules $V(i)$ have top filtrations with factors $\Delta(j)$, $i+1 \leq j \leq n$ for all $1 \leq i \leq n-1$ if and only if* $\operatorname{rad}\Delta^\circ(i)$ *are semisimple for all* $1 \leq i \leq n$.

Proof. This follows immediately from the reciprocity law formulated in Corollary A.3.10. Indeed, if $\mathcal{S}(A) = (D_i\,;\ {}_iW_j\,,\ 1 \leq i,j \leq n)$ is the K-species of A, write $\dim_{D_j}{}_iW_j = u_{ij}$ and $\dim_{D_i}{}_iW_j = v_{ij}$.

The standard filtration of a semisimple $\operatorname{rad}\Delta^\circ(j)$ is a top filtration with the factors equal to v_{ij} copies of $S^\circ(i)$, $1 \leq i \leq j-1$, for all $1 \leq j \leq n$. In view of the reciprocity law

$$d_i\big[\Delta^\circ(j) : S^\circ(i)\big] = d_j\big[P(i) : \Delta(j)\big]\,,$$

we have

$$\big[P(i) : \Delta(j)\big] = (d_i/d_j)v_{ij} = u_{ij} \quad \text{for all } i < j\,.$$

Hence, in view of Proposition A.4.9, the condition on $\operatorname{rad}\Delta^\circ(j)$ for $1 \leq j \leq n$ is equivalent to the fact that $V(i)$ has a top filtration with the factors equal to u_{ij} copies of $\Delta(j)$ for all $1 \leq i < j \leq n$. □

Corollary A.4.16. *All* $\operatorname{rad} P(i)$, $1 \le i \le n$, *have top filtrations with factors* $S(j)$, $1 \le j \le i-1$ *and* $\Delta(j)$, $i+1 \le j \le n$ *if and only if all* $\operatorname{rad} P^\circ(i)$, $1 \le i \le n$ *have top filtrations with factors* $S^\circ(j)$, $1 \le j \le i-1$ *and* $\Delta^\circ(j)$, $i+1 \le j \le n$.

Definition A.4.17. A (quasi-hereditary) algebra satisfying the conditions of Corollary A.4.16 is called *shallow.*

Thus, a quasi-hereditary algebra is shallow if and only if all $\operatorname{rad} \Delta(i)$ and all $\operatorname{rad} \Delta^\circ(i)$ are semisimple. In particular, shallow algebras are lean.

Definition A.4.18. A quasi-hereditary algebra is called *deep* if every $\operatorname{rad} \Delta(i)$ is a projective (right) B_{i-1}-module and every $\operatorname{rad} \Delta^\circ(i)$ is a projective (left) B_{i-1}-module, for all $2 \le i \le n$. (Here B_{i-1} denotes, as before, the quotient algebra $A/A\varepsilon_i A$.)

Both replete algebras and deep algebras have global dimension ≤ 2. Deep algebras are however, in general, not lean. There is a class of lean algebras which seems to be of importance for applications (see Sect. A.6); let us give the definition.

Definition A.4.19. A (quasi-hereditary) algebra is said to be *right medial* if all $\operatorname{rad} P(i)$, $1 \le i \le n$, have top filtrations with the factors $\Delta(j)$, $1 \le j \le n$, $j \ne i$. An algebra A is said to be *left medial* if the opposite algebra A^o is right medial, i.e. if all $\operatorname{rad} P(i)$, $1 \le i \le n$, have top filtrations with the simple factors $S(j)$, $1 \le j \le i-1$ and the projective factors $P(j)$, $i+1 \le j \le n$.

Now we are going to present canonical constructions of the quasi-hereditary algebras defined above, over a given ordered species. Let $\mathcal{S} = (D_1, D_2, \dots, D_n; {}_iW_j,\ 1 \le i,j \le n)$ be an ordered species with ${}_iW_i = 0$ for all $1 \le i \le n$. Let $T(\mathcal{S})$ be the tensor algebra over $\mathcal{S}$:

$$T(\mathcal{S}) = \Lambda \oplus W \oplus W^{\otimes 2} \oplus W^{\otimes 3} \oplus \dots,$$

where $\Lambda = D_1 \times D_2 \times \dots \times D_n$, $W = \bigoplus_{i,j} {}_iW_j$ is a Λ-Λ-bimodule with Λ operating via the projections, all tensor products are over Λ and the multiplication is induced by $W^{\otimes r} \otimes_\Lambda W^{\otimes s} \simeq W^{\otimes r+s}$. Of course, $T(\mathcal{S})$ is, in general, infinite dimensional.

Define the following ideals in $T(\mathcal{S})$:

$$\begin{aligned}
I_S &= \langle {}_iW_j \otimes {}_jW_k \mid j < \max\{i,k\} \rangle \\
I_{M_r} &= \langle {}_iW_j \otimes {}_jW_k \mid j < k \rangle \\
I_{M_l} &= \langle {}_iW_j \otimes {}_jW_k \mid i > j \rangle \\
I_R &= \langle {}_iW_j \otimes {}_jW_k \mid j < \min\{i,k\} \rangle \quad \text{and} \\
I_D &= \langle {}_{i_0}W_{i_1} \otimes {}_{i_1}W_{i_2} \dots \otimes {}_{i_{s-1}}W_{i_s} \mid i_0 = i_s \text{ and } i_r < i_0 \text{ for } 1 \le r \le s-1 \rangle .
\end{aligned}$$

Put

$$H(\mathcal{S}) = T(\mathcal{S})/I_H \text{ for } H = S, M_r, M_\ell, R, \text{ and } D.$$

Theorem A.4.20. *The algebras $S(\mathcal{S})$, $M_r(\mathcal{S})$, $M_\ell(\mathcal{S})$, $R(\mathcal{S})$ and $D(\mathcal{S})$ are quasi-hereditary algebras with the ordered species $\mathcal{S}$. The algebra $S(\mathcal{S})$ is shallow, $M_r(\mathcal{S})$ right medial, $M_\ell(\mathcal{S})$ left medial, $R(\mathcal{S})$ replete and $D(\mathcal{S})$ deep.*

In fact,

$$S(\mathcal{S}) \simeq \Lambda \oplus W \oplus (\bigoplus_{\substack{t>i \\ t>j}} {}_iW_t \otimes {}_tW_j),$$

and $M_r(\mathcal{S})$, $M_\ell(\mathcal{S})$ and $R(\mathcal{S})$ are isomorphic to

$$\Lambda \oplus W \oplus (\oplus\, {}_{i_0}W_{i_1} \otimes \ldots \otimes {}_{i_t}W_{i_{t+1}} \otimes \ldots \otimes {}_{i_{m-1}}W_{i_m}),$$

where the summation runs through all sequences $(i_0, i_1, \ldots, i_t, \ldots, i_{m-1}, i_m)$ subject to

$$i_1 > i_2 > \ldots > i_m, \quad m \geq 2,$$
$$i_0 < i_1 < \ldots < i_{m-1}, \quad m \geq 2 \text{ and}$$
$$i_0 < i_1 < \ldots < i_t > \ldots > i_{m-1} > i_m, \quad 0 \leq t \leq m,\ m \geq 2,$$

respectively.

Theorem A.4.21 *Let A be a basic quasi-hereditary K-algebra with the ordered species $\mathcal{S} = \mathcal{S}(A)$. Then*

$$\dim_K S(\mathcal{S}) \leq \dim_K A \leq \dim_K D(\mathcal{S}).$$

Moreover, $\dim_K S(\mathcal{S}) = \dim_K A$ *if and only if A is shallow and* $\dim_K D(\mathcal{S}) = \dim_K A$ *if and only if A is deep.*

If A is lean, then

$$\dim_K A \leq \dim_K R(\mathcal{S})$$

and $\dim_K R(\mathcal{S}) = \dim_K A$ *if and only if A is replete.*

For the proof of the statements concerning the shallow and deep algebras, we refer to [DR4]. The proof of the remaining statements is similar and is left to the reader. Let us point out that two shallow, or medial, or replete, or deep algebras over the same ordered species do not have to be isomorphic.

Typically, the algebra A whose regular representation is $A_A = \begin{smallmatrix}1\\2\\1\end{smallmatrix} \oplus \begin{smallmatrix}&2&\\1&&3\\&2&\end{smallmatrix} \oplus \begin{smallmatrix}3\\2\end{smallmatrix}$ is shallow, but $A \not\simeq S(\mathcal{S}(A))$ (cf. the example after Theorem A.3.4).

Let us insert the following observation.

Proposition A.4.22. *An algebra is replete if and only if it is lean and*

$$\operatorname{Ext}^2(\Delta(i), S(j)) = 0 = \operatorname{Ext}^2(S(j), \nabla(i)) \quad \textit{for all } 1 \leq i, j \leq n. \qquad (A.4.2)$$

Proof. This follows from the following fact and its dual. Consider, as before, the exact sequence

$$0 \longrightarrow V(i) \longrightarrow P(i) \longrightarrow \Delta(i) \longrightarrow 0\,,$$

which yields for every $1 \leq j \leq n$

$$0 = \mathrm{Ext}^1\big(P(i), S(j)\big) \longrightarrow \mathrm{Ext}^1\big(V(i), S(j)\big) \xrightarrow{\sim} \mathrm{Ext}^2\big(\Delta(i), S(j)\big) \longrightarrow$$
$$\longrightarrow \mathrm{Ext}^2\big(P(i), S(j)\big) = 0\,.$$

Thus $V(i)$ is projective if and only if $\mathrm{Ext}^2\big(\Delta(i), S(j)\big) = 0$ for all $1 \leq j \leq n$.
□

In general, there are many algebras over the same species which satisfy (A.4.2); for instance, it also holds for deep algebras. It may be worth pointing out that among lean algebras over a given species the right medial (left medial) algebras are exactly those which have the least K-dimension and satisfy $Ext^2\big(S(j), \nabla(i)\big) = 0$ (or $\mathrm{Ext}^2\big(\Delta(i), S(j)\big) = 0$, respectively) for all $1 \leq i, j \leq n$.

In order to get an idea of the size of the algebras constructed above, let us give their K-dimensions in the case of the "complete" ordered species

$$\mathcal{S}_n = (D_1 = D_2 = \ldots = D_n = K\,;\ {}_iW_j = K \text{ for all} 1 \leq i, j \leq n,\ i \neq j)\,.$$

These are easy to compute:

$$\begin{aligned}
s_n &= \dim_K S(\mathcal{S}_n) = \frac{1}{6} n(n+1)(2n+1); \\
m_n &= \dim_K M_r(\mathcal{S}_n) = \dim_K M_\ell(\mathcal{S}) = (n-1)2^n + 1; \\
r_n &= \dim_K R(\mathcal{S}_n) = \frac{1}{3}(2^{2n} - 1); \text{ and} \\
d_n &= \dim_K D(\mathcal{S}_n) = d_n \text{ satisfies the recursion: } d_{n+1} = d_n + (d_n + 1)^2.
\end{aligned}$$

Thus, $s_2 = m_2 = r_2 = d_2 = 5$, however already for $n = 10$, $d_{10} \approx 2.7 \times 10^{208}$ (!), while $s_{10} = 385$, $m_{10} = 9217$ and even r_{10} is "only" 349525.

A.5 Characterization of the Category $\mathcal{F}(\Delta)$

In Sect. A.3, we have seen the importance of the full subcategory $\mathcal{F}(\Delta)$ of mod-A in the theory of quasi-hereditary algebras. The module categories of quasi-hereditary algebras have been abstractly described by Cline, Parshall and Scott in terms of the highest weight categories with a finite number of weights [PS]. In the same spirit, we are going to give a characterization of the categories $\mathcal{F}(\Delta)$, called "standardization" in [DR5].

Let $\mathcal{C}$ be an abelian K-category and

$$\Delta = \{\Delta(i) \mid 1 \leq i \leq n\}$$

a finite ordered set of (non-isomorphic) objects of $\mathcal{C}$.

Definition A.5.1. The ordered set Δ is called a *standard sequence* if

1) $\dim_K \mathrm{Hom}(\Delta(i), \Delta(j)) < \infty$ and
$\dim_K \mathrm{Ext}^1(\Delta(i), \Delta(j)) < \infty$ for all $1 \le i, j \le n$;
2) $\mathrm{rad}\,(\Delta(i), \Delta(j)) = 0$ and
$\mathrm{Ext}^1(\Delta(i), \Delta(j)) = 0$ for $i \ge j$.

Here $\mathrm{rad}\,(\Delta(i), \Delta(j))$ equals $\mathrm{Hom}(\Delta(i), \Delta(j))$ for $i \neq j$ and $\mathrm{rad}\,\mathrm{End}\Delta(i)$ for $i = j$.)

Note that 2) implies that all $\Delta(i)$ are Schurian. Denote by $\mathcal{F}(\Delta)$ the full subcategory of $\mathcal{C}$ consisting of all objects which have filtrations with factors from Δ.

Theorem A.5.2. *Let Δ be a standard sequence in an abelian K-category $\mathcal{C}$. Then there exists a unique basic quasi-hereditary algebra A (with an order of idempotents given by the standard sequence) such that the subcategories $\mathcal{F}(\Delta)$ of $\mathcal{C}$ and $\mathcal{F}(\Delta_A)$ of* mod-A *are equivalent.*

Proof. The proof has several steps. First, we are going to construct for every $1 \le i \le n$ an indecomposable Ext-projective object $P(i)$ of $\mathcal{F}(\Delta)$ with an exact sequence

$$0 \longrightarrow V(i) \longrightarrow P(i) \longrightarrow \Delta(i) \longrightarrow 0\,, \quad V(i) \in \mathcal{F}(\Delta)\,.$$

In general, we show that for every $X \in \mathcal{F}(\Delta)$, there is a finite direct sum $P(X)$ of suitable $P(i)$'s, $1 \le i \le n$, such that the exact sequence

$$0 \longrightarrow X' \xrightarrow{\mu_X} P(X) \xrightarrow{\pi_X} X \longrightarrow 0 \tag{A.5.1}$$

satisfies $X' \in \mathcal{F}(\Delta)$.

Finally, taking $P = \bigoplus_{i=1}^{n} P(i)$ and putting

$$A = \mathrm{End} P\,,$$

we shall establish that A is the desired algebra with

$$P_A(i) = \mathrm{Hom}(P, P(i)) \quad \text{and} \quad \Delta_A(i) = \mathrm{Hom}(P, \Delta(i))\,.$$

The proof of the equivalence $\mathcal{F}(\Delta) \simeq \mathcal{F}(\Delta_A)$ will use the existence of the exact sequences (A.5.1). The fact that A is a quasi-hereditary algebra with the sequence $\Delta_A = \{\Delta_A(i) \mid 1 \le i \le n\}$ of standard modules then follows immediately: $P_A(i)$ has a Δ_A-filtration with the top factor $\Delta_A(i)$ and the remaining factors $\Delta_A(j)$, $j > i$. Moreover, since $\mathrm{Hom}(P_A(j), \Delta_A(i)) = 0$ for $j > i$, $\Delta_A(i)$ is the maximal factor module of $P_A(i)$ whose composition factors are $S(j)$ for $j \le i$.

Let us construct the objects $P(i)$, $1 \le i \le n$. Proceed inductively: For $i \le k \le n$, construct $P_k(i)$ such that $\operatorname{Ext}^1(P_k(i), \Delta(j)) = 0$ for all $1 \le j \le k$, and such that there is an exact sequence

$$0 \longrightarrow V_k(i) \longrightarrow P_k(i) \longrightarrow \Delta(i) \longrightarrow 0,$$

where $V_k(i)$ is filtered by $\Delta(j)$, $i < j \le k$. The condition 2) on the standard sequence gives $P_i(i) = \Delta(i)$. Assume that $V_{k-1}(i) \to P_{k-1}(i)$ are already constructed. Denote

$$d_{ki} = \dim_{D_k} \operatorname{Ext}^1(P_{k-1}(i), \Delta(k)),$$

where $D_k = \operatorname{End}\Delta(k)$, and consider the "universal extension"

$$0 \longrightarrow d_{ki}\Delta(k) \longrightarrow P_k(i) \longrightarrow P_{k-1}(i) \longrightarrow 0.$$

Thus $\operatorname{Ext}^1(P_k(i), \Delta(j)) = 0$ (since the homomorphism $\operatorname{Hom}(d_{ki}\Delta(k), \Delta(k)) \to \operatorname{Ext}^1(P_{k-1}(i), \Delta(k))$ is surjective) for all $1 \le j \le k$, and $P_k(i)$ is indecomposable (since $\operatorname{Hom}(\Delta(k), P_{k-1}(i)) = 0$). Furthermore, the corresponding $V_k(i)$ is easily seen to be an extension of $V_{k-1}(i)$ by $d_{ki}\Delta(k)$ and thus the inductive step is completed. Put $P(i) = P_n(i)$ and $P = \bigoplus_{i=1}^{n} P(i)$.

Now, let X be an arbitrary object of $\mathcal{F}(\Delta)$. We claim that there is $P(X)$ in $\operatorname{add} P$ such that (A.5.1) holds. For $X = \Delta(i)$ this is clearly true: take $P(\Delta(i)) = P(i)$. In general, X is an extension of $Z \in \mathcal{F}(\Delta)$ by $Y \in \mathcal{F}(\Delta)$, and by induction, using the fact that $\operatorname{Ext}^1(P(Z), Y) = 0$, we get the following commutative diagram of exact sequences:

$$\begin{array}{ccccccccc}
 & & 0 & & 0 & & 0 & & \\
 & & \downarrow & & \downarrow & & \downarrow & & \\
0 & \longrightarrow & Y' & \longrightarrow & X' & \longrightarrow & Z' & \longrightarrow & 0 \\
 & & \downarrow & & \downarrow & & \downarrow & & \\
0 & \longrightarrow & P(Y) & \longrightarrow & P(Y) \oplus P(Z) & \longrightarrow & P(Z) & \longrightarrow & 0 \\
 & & \downarrow & & \downarrow & & \downarrow & & \\
0 & \longrightarrow & Y & \longrightarrow & X & \longrightarrow & Z & \longrightarrow & 0; \\
 & & \downarrow & & \downarrow & & \downarrow & & \\
 & & 0 & & 0 & & 0 & &
\end{array}$$

here X' is an extension of Z' by Y', and therefore in $\mathcal{F}(\Delta)$. In what follows, we will keep the notation of (A.5.1) and of the related exact sequence

$$0 \longrightarrow X'' \longrightarrow P(X') \xrightarrow{\pi_{X'}} P(X) \xrightarrow{\pi_X} X \longrightarrow 0. \qquad (A.5.2)$$

Let $A = \operatorname{End}(P)$ and $h_P = \operatorname{Hom}(P, -) : \mathcal{C} \to \text{mod-}A$. In view of the condition 1), A is a finite dimensional algebra and all $h_P(X)$ with $X \in \mathcal{F}(\Delta)$ are finite dimensional A-modules. Let $P_A(i) = h_P(P(i))$ and $\Delta_A(i) = h_P(\Delta(i))$, $1 \le i \le n$. Since $\operatorname{Ext}^1(P, X) = 0$ for $X \in \mathcal{F}(\Delta)$, h_P is exact on exact sequences

$0 \to X \to Y \to Z \to 0$ with $X \in \mathcal{F}(\Delta)$ in $\mathcal{C}$. In consequence, h_P maps $\mathcal{F}(\Delta)$ into $\mathcal{F}(\Delta_A) \subseteq$ mod-A.

We are going to show that the restriction of h_P to $\mathcal{F}(\Delta)$ is an equivalence of $\mathcal{F}(\Delta)$ and $\mathcal{F}(\Delta_A)$. First, we claim that on $\mathcal{F}(\Delta)$, h_P is faithful. Let $X, Y \in \mathcal{F}(\Delta)$ and $\varphi : X \to Y$ with $h_P(\varphi) = 0$. Using the sequence (A.5.2), φ can be lifted to φ_0 and φ_1 to get a commutative diagram

$$\begin{array}{ccccccc} P(X') & \longrightarrow & P(X) & \xrightarrow{\pi_X} & X & \longrightarrow & 0 \\ {\scriptstyle\varphi_1}\downarrow & & {\scriptstyle\varphi_0}\downarrow & & \downarrow{\scriptstyle\varphi} & & \\ P(Y') & \xrightarrow{\pi_{Y'}} & P(Y) & \xrightarrow{\pi_Y} & Y & \longrightarrow & 0. \end{array}$$

Applying h_P, we get projective presentations for $h_P(X)$ and $h_P(Y)$

$$\begin{array}{ccccccc} h_P(P(X')) & \longrightarrow & h_P(P(X)) & \xrightarrow{h_P(\pi_X)} & h_P(X) & \longrightarrow & 0 \\ {\scriptstyle h_P(\varphi_1)}\downarrow & & {\scriptstyle h_P(\varphi_0)}\downarrow & & \downarrow{\scriptstyle h_P(\varphi)=0} & & \\ h_P(P(Y')) & \xrightarrow{h_P(\pi_{Y'})} & h_P(P(Y)) & \xrightarrow{h_P(\pi_Y)} & h_P(Y) & \longrightarrow & 0. \end{array}$$

Since $h_P(\pi_Y)h_P(\varphi_0) = h_P(\varphi)h_P(\pi_X) = 0$, there is a homomorphism $g : h_P(P(X)) \to h_P(P(Y'))$ such that $h_P(\pi_{Y'})g = h_P(\varphi_0)$. Furthermore, since the restriction of h_P to add P is obviously faithful and full, $g = h_P(\psi)$ with $\psi : P(X) \to P(Y')$ satisfying $\varphi_0 = \pi_{Y'}\psi$. But then $\varphi\pi_X = \pi_Y\pi_{Y'}\psi = 0$ and thus $\varphi = 0$.

To complete the proof of the theorem, we can proceed in a similar manner, making use of (A.5.1) and (A.5.2), to show that the restriction of h_P to $\mathcal{F}(\Delta)$ is full and dense. The details of the proof are left to the reader. □

In conclusion, let us point out that any subsequence of a standard sequence is again standard. Given a quasi-hereditary algebra A, its sequence of standard modules $\Delta = (\Delta(i) \mid 1 \le i \le n)$ is obviously standard. Thus, any subsequence of Δ leads to a quasi-hereditary algebra derived from A. For instance, if we choose $(\Delta(i) \mid 1 \le i \le r) \subseteq \Delta$, we obtain $B_r = A/A\varepsilon_{r+1}A$. If we take $(\Delta(i) \mid r \le i \le n) \subseteq \Delta$, we get $C_r = \varepsilon_r A \varepsilon_r$. These two special cases relate to the recursive constructions (i) and (ii) of Sect. A.4.

A.6 Final Remarks

In this last section we want to make several brief comments, concerning some particular classes of quasi-hereditary algebras which are closely related to current developments in a number of applications. Quasi-hereditary algebras have now become a central concept of the Kazhdan-Lusztig theory as developed by Cline, Parshall and Scott [CPS2], as well as an important tool in the work on the Berstein-Gelfand-Gelfand category $\mathcal{O}$; this category is the sum of blocks which are equivalent to module categories over quasi-hereditary algebras. Here the Yoneda Ext$^\bullet$-algebras seems to play a fundamental role. Recently Beilinson, Ginsburg and Soergel [BGS] have established an isomorphism between the principal block algebra for $\mathcal{O}$ and its own $\mathrm{Ext}^\bullet$-algebra. Recent studies of Dyer [D] should also be mentioned in this connection.

We are going to illustrate some of the relevant notions. Let us recall the definition of an $\mathrm{Ext}^\bullet$-algebra. Given a finite dimensional algebra A,

$$\mathrm{Ext}^\bullet\text{-}A = \bigoplus_{t\geq 0} \mathrm{Ext}^t_A(\bar{A}, \bar{A}),$$

where $\bar{A} = A/\mathrm{rad}\,A$ and multiplication is induced by the Yoneda product of exact sequences. Thus $\mathrm{Ext}^\bullet\text{-}A$ is finite dimensional if and only if A is of finite global dimension. If $P_\bullet = (P_t \mid t \geq 0)$ is a minimal projective resolution of a module M, then $\mathrm{Ext}^t_A(M, S(j)) \simeq \mathrm{Hom}_A(P_t, S(j)) \simeq \mathrm{Hom}_A(\mathrm{top}\,P_t, S(j))$. Clearly, we always have for the K-species $\mathcal{S}(\mathrm{Ext}^\bullet\text{-}A) \supseteq (\mathcal{S}(A))^*$, where * denotes the *dual species* to $\mathcal{S}(A)$. The latter is defined as follows: Given a K-species $\mathcal{S} = (\Lambda, W)$, then $\mathcal{S}^* = (\Lambda, W^* = \mathrm{Hom}_K(W, K))$. If in particular $\mathcal{S}$ is an ordered K-species, then $\mathcal{S}^*$ is an ordered K-species with the reverse order. An important role is played in the theory by the so-called quadratic algebras: Given a K-species $\mathcal{S} = (\Lambda, W)$, a K-algebra $A = T(\mathcal{S})/\langle\Omega\rangle$, where $T(\mathcal{S})$ is the tensor algebra and $\Omega \subseteq W^{\otimes 2}$, is said to be *quadratic*. Set

$$\Omega^\perp = \{f \in (W^{\otimes 2})^* \simeq W^* \otimes_\Lambda W^* \mid f(\Omega) = 0\}$$

and define

$$A^\perp = T(\mathcal{S}^*)/\langle\Omega^\perp\rangle .$$

A quadratic algebra A is said to be *formal* if $A^\perp \simeq \mathrm{Ext}^\bullet\text{-}A$. It is characterized by the fact that the inlcusion $\mathcal{S}(\mathrm{Ext}^\bullet\text{-}A) \supseteq (\mathcal{S}(A))^*$ turns into equality.

Of course, formal (quadratic) algebras (of finite global dimension) do not have to be quasi-hereditary; as an illustration, consider the "Fibonacci" algebras F_d of the Example in Sect. A.3. On the other hand, the following example shows that a quasi-hereditary quadratic algebra is not, in general, formal: consider the path algebra of the graph

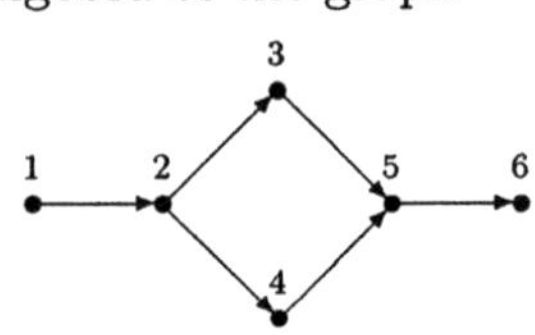

modulo the ideal $\langle \alpha_{12}\alpha_{23}, \alpha_{45}\alpha_{56}, \alpha_{23}\alpha_{35} - \alpha_{24}\alpha_{45}\rangle$; thus

$$A_A = \begin{smallmatrix}1\\2\\4\end{smallmatrix} \oplus \begin{smallmatrix} & 2 & \\ 3 & & 4 \\ & 5 & \end{smallmatrix} \oplus \begin{smallmatrix}3\\5\\6\end{smallmatrix} \oplus \begin{smallmatrix}4\\5\end{smallmatrix} \oplus \begin{smallmatrix}5\\6\end{smallmatrix} \oplus 6$$

and $A^\perp \not\simeq \mathrm{Ext}^\bullet\text{-}A$. Furthermore, the $\mathrm{Ext}^\bullet$-algebra of a formal quasi-hereditary algebra does not have to be quasi-hereditary. Take the path algebra of the graph $\overset{2}{\bullet} \rightleftarrows \overset{1}{\bullet} \rightleftarrows \overset{3}{\bullet}$ modulo $\langle \alpha_{21}\alpha_{12}, \alpha_{21}\alpha_{13}, \alpha_{31}\alpha_{13}\rangle$. Then

$$A_A = \begin{smallmatrix} & & 1 & & \\ & 2 & & 3 & \\ 1 & & & & 1 \\ & & & 2 & \\ & & & 1 & \end{smallmatrix} \oplus \begin{smallmatrix}2\\1\end{smallmatrix} \oplus \begin{smallmatrix}3\\1\\2\\1\end{smallmatrix}$$

and

$$(\mathrm{Ext}^{\bullet}\text{-}A)_{\mathrm{Ext}^{\bullet}\text{-}A} = \begin{smallmatrix} & 1 & \\ 2 & & 3 \end{smallmatrix} \oplus \begin{smallmatrix} 2 \\ 1 \\ 2 \end{smallmatrix} \oplus \begin{smallmatrix} & 3 & \\ & 1 & \\ 2 & & 3 \end{smallmatrix} \, ;$$

thus A is a quasi-hereditary, quadratic formal algebra ($\mathrm{Ext}^{\bullet}\text{-}A \simeq A^{\perp}$). In fact, $gl.dim\, A = 2$. Observe, however, that A is not lean (!). Indeed, $\mathrm{rad}\, P(3)$ lacks top filtration by $\Delta(1)$ and $\Delta(2)$ (cf. Theorem A.4.10).

On the other hand, we can verify easily the following statements.

Lemma A.6.1. *Let A be a quasi-hereditary algebra such that every* $\mathrm{rad}\, P(i)$, *$1 \leq i \leq n$, is a direct sum of some $S(j)$'s, $\Delta(j)$'s for $1 \leq j \leq n$ and of some $P(j)$'s for $i+1 \leq j \leq n$. Then A is quadratic and formal.*

Here a need for restricting the range of $P(j)$'s can be seen easily: The path algebra of the graph $\overset{2}{\bullet} \longrightarrow \overset{3}{\bullet} \longrightarrow \overset{1}{\bullet} \longrightarrow \overset{4}{\bullet}$, modulo the ideal generated by the path $\alpha_{23}\alpha_{31}\alpha_{14}$ is a non-quadratic quasi-hereditary algebra whose radical (as a right module) is a direct sum of standard and projective modules.

Proposition A.6.2. *Given an ordered species $\mathcal{S}$, $S(\mathcal{S})$, $M_r(\mathcal{S})$, $M_\ell(\mathcal{S})$ and $R(\mathcal{S})$ are formal. In fact,*

$$\begin{array}{lll} \mathrm{Ext}^{\bullet}\text{-}S(\mathcal{S}) \simeq R(\mathcal{S}^*)\,, & & \mathrm{Ext}^{\bullet}\text{-}R(\mathcal{S}) \simeq S(\mathcal{S}^*)\,, \\ \mathrm{Ext}^{\bullet}\text{-}M_r(\mathcal{S}) \simeq M_\ell(\mathcal{S}^*) & \textit{and} & \mathrm{Ext}^{\bullet}\text{-}M_\ell(\mathcal{S}) \simeq M_r(\mathcal{S}^*)\,. \end{array}$$

Note that all the above $\mathrm{Ext}^{\bullet}$-algebras are quasi-hereditary algebras with respect to the opposite order of the order of $\mathcal{S}$.

Proposition A.6.3. *Let $\mathcal{S} = (D_1, D_2, \ldots, D_n\, ;\; {}_iW_j\,,\; 1 \leq i,j \leq n)$ be an ordered species such that $D_i \simeq D_{n-i+1}$ and ${}_iW_j \simeq {}_{n-i+1}W_{n-j+1}$ for all i,j; thus $\mathcal{S} \simeq \mathcal{S}^*$ with the opposite order. Then*

$$\mathrm{Ext}^{\bullet}\text{-}M_r(\mathcal{S}) \simeq M_\ell(\mathcal{S}^*) \simeq M_r(\mathcal{S})^{\circ}$$

and

$$\mathrm{Ext}^{\bullet}\text{-}M_\ell(\mathcal{S}) \simeq M_r(\mathcal{S}^*) \simeq M_\ell(\mathcal{S})^{\circ}\,.$$

References to the Appendix

[ADL] Ágoston, I., Dlab, V., Lukács, E., Lean quasi-hereditary algebras, to appear in *Can. Math. Soc. Proceedings* series **13** (1993).

[BGS] Beilinson, A., Ginsburg, V., Soergel, W., Koszul duality patterns in representation theory, preprint.

[CPS1] Cline, E., Parshall, B.J., Scott, L.L., Finite dimensional algebras and highest weight categories, *J. Reine Angew. Math.* **391** (1988), 85–99.

[CPS2] Cline, E., Parshall, B.J., Scott, L.L., The homological dual of a highest weight category, preprint.

[D] Dyer, M.J., Kazhdan-Lusztig-Stanley polynomials and quadratic algebras I, preprint.

[DR1] Dlab, V., Ringel, C.M., A construction for quasi-hereditary algebras, *Compositio Math.* **70** (1989), 155–175.

[DR2] Dlab, V., Ringel, C.M., Every semiprimary ring is the endomorphism ring of a projective module over a quasi-hereditary ring, *Proc. Amer. Math. Soc.* **107** (1989), 1–5.

[DR3] Dlab, V., Ringel, C.M., Auslander algebras as quasi-hereditary algebras, *J. London Math. Soc.* **39** (1989), 457–466.

[DR4] Dlab, V., Ringel, C.M., The dimension of a quasi-hereditary algebra, *Banach Center Publ.*, Vol **26**, Warsaw 1990, 263–271.

[DR5] Dlab, V., Ringel, C.M., The module theoretical approach to quasi-hereditary algebras, *London Math. Soc.*, Lecture Note Series **168**, Cambridge Univ. Press 1992, 200–224.

[PS] Parshall, B.J., Scott, L.L., Derived categories, quasi-hereditary algebras, and algebraic groups, *Proc. Ottawa-Moosonee Workshop*, Carleton-Ottawa Math. Lecture Note Series **3** (1988), 1–105.

[R] Ringel, C.M., The category of modules with good filtrations over a quasi-hereditary algebra has almost split sequences, *Math. Z.* **208** (1991), 209–223.

Index

I

J

K

L

M

N

T

U

V

W

Y

MIX
Papier aus verantwortungsvollen Quellen
Paper from responsible sources
FSC® C105338

If you have any concerns about our products,
you can contact us on
ProductSafety@springernature.com

In case Publisher is established outside the EU,
the EU authorized representative is:
Springer Nature Customer Service Center GmbH
Europaplatz 3, 69115 Heidelberg, Germany

Printed by Libri Plureos GmbH
in Hamburg, Germany